T0143336

Understanding Real Analysis

Second Edition

Paul Zorn

St. Olaf College
Northfield, Minnesota

CRC Press
Taylor & Francis Group
Boca Raton London New York

CRC Press is an imprint of the
Taylor & Francis Group, an **informa** business

A CHAPMAN & HALL BOOK

CRC Press
Taylor & Francis Group
6000 Broken Sound Parkway NW, Suite 300
Boca Raton, FL 33487-2742

First issued in paperback 2022

© 2018 by Taylor & Francis Group, LLC
CRC Press is an imprint of Taylor & Francis Group, an Informa business

No claim to original U.S. Government works

Version Date: 20171025

ISBN 13: 978-1-03-247645-2 (pbk)
ISBN 13: 978-1-138-03301-6 (hbk)

DOI: 10.1201/b21844

This book contains information obtained from authentic and highly regarded sources. Reasonable efforts have been made to publish reliable data and information, but the author and publisher cannot assume responsibility for the validity of all materials or the consequences of their use. The authors and publishers have attempted to trace the copyright holders of all material reproduced in this publication and apologize to copyright holders if permission to publish in this form has not been obtained. If any copyright material has not been acknowledged please write and let us know so we may rectify in any future reprint.

Except as permitted under U.S. Copyright Law, no part of this book may be reprinted, reproduced, transmitted, or utilized in any form by any electronic, mechanical, or other means, now known or hereafter invented, including photocopying, microfilming, and recording, or in any information storage or retrieval system, without written permission from the publishers.

For permission to photocopy or use material electronically from this work, please access www.copyright.com (http://www.copyright.com/) or contact the Copyright Clearance Center, Inc. (CCC), 222 Rosewood Drive, Danvers, MA 01923, 978-750-8400. CCC is a not-for-profit organization that provides licenses and registration for a variety of users. For organizations that have been granted a photocopy license by the CCC, a separate system of payment has been arranged.

Trademark Notice: Product or corporate names may be trademarks or registered trademarks, and are used only for identification and explanation without intent to infringe.

Publisher's Note
The publisher has gone to great lengths to ensure the quality of this reprint but points out that some imperfections in the original copies may be apparent.

**Visit the Taylor & Francis Web site at
http://www.taylorandfrancis.com**

**and the CRC Press Web site at
http://www.crcpress.com**

TEXTBOOKS in MATHEMATICS

Series Editors: Al Boggess and Ken Rosen

Contents

Preface

About This Edition: What's Old and What's New?

This second edition retains all the main mathematical and pedagogical attributes and strategies of the first edition, as outlined below. For example, this edition retains—and perhaps increases—the first edition's strong focus on helping students acquire and use mathematical language. Many students at the level envisioned for this text find the linguistic challenge of parsing complicated definitions and theorems just as difficult as the mathematics itself. This edition, therefore, contains additional problems and exercises that ask students to unpack, instantiate, or even "perturb" the statements of definitions and theorems. (See, e.g., Problem 7, page 112.)

This book, like any second edition, incorporates repair of first edition typos (and possibly the introduction of some brand-new ones). The narrative has also been "smoothed" or (in the author's view) otherwise improved in places identified either by the author or (thanks!) by other users.

Following are some more significant new features and content.

New sections on topology and compactness. Two entirely new sections, Sections 3.5 and 3.6, introduce basic ideas of topology and compactness. Some general definitions and principles are discussed, but the setting is almost always the real line, treated (when helpful) as a metric space. Compactness is defined in terms of open covers (which pose their own linguistic challenges!) but there is strong emphasis on closed and bounded sets in $\mathbb{R}$, via the Heine–Borel theorem. This material is very occasionally alluded to, but not really depended on, in later sections. In this sense these sections are essentially self-contained, and could be used for independent study or enrichment projects.

New material on series of functions and "Taylor stuff." Treatment of function series in general and of Taylor series in particular has been beefed up, both in Section 4.4 and in the new Section 4.5, on Taylor series. Section 4.5 joins Section 2.7 (on upper and lower limits) in the mode of "guided discovery". Both sections are designed for students to encounter as projects or other forms of enrichment. Their content, though mathematically important, is not explicitly required in the sequel.

New problems and exercises. This edition includes new problems and exercises in most of the sections. Many new problems focus on helping students understand and "unpack" (in the sense discussed above) definitions and theorems. Students are often asked to prove special cases, explore concrete instances of general results, and the like. Many problem sets have also been more carefully ordered to distinguish between odd- and even-numbered exercises; most of the former have hints or solutions in the back.

Thanks

This book, in both of its editions, owes its existence to a large (uncounted but surely countable) set of teachers, academic colleagues, St. Olaf College students, publishing company professionals, friends, advisors, critics, "competitors," family, and others. It is a pleasure to acknowledge some of them by name—and to claim sole credit for errors that survived their best efforts to help.

Among local colleagues I thank Bruce Hanson, Paul Humke, Loren Larson, the late Arnold Ostebee, Matt Richey, the late Lynn Steen, Ted Vessey, and many others with whom I have discussed matters mathematical and pedagogical for many years. On matters of taste and usage—both linguistic and mathematical—my friend Barry Cipra is a reliable resource. His advice, when I have taken it, has always proved correct. These gifted teachers, expositors, and mathematicians have helped me think about the deep connections and subtle relationships among teaching, telling, and doing mathematics. They have also demonstrated, by argument and example, that serious engagement with post-calculus mathematics is possible, and valuable, for a broad range of students, not just a small elite. This "big tent" approach to our discipline informs and underpins St. Olaf's very successful undergraduate program, and I have kept it in mind when writing this book.

Academic colleagues elsewhere have taught me a lot, too. These include my own mathematics teachers at Washington University and the University of Washington; authors and referees from my work with *Mathematics Magazine* and other publications of the Mathematical Association of America; and authors of *other* real analysis textbooks from which I have learned, taught, or read. Real analysis can be viewed usefully from many angles; doing so offers depth and perspective.

The late Klaus Peters and Charlotte Henderson worked with remarkable care, diligence, and intelligence to make the first edition of this book better and its production easier. I'm all the more grateful to them because, as a sometime editor myself, I know how difficult and exacting that work can be.

For encouragement and equally diligent and meticulous work on the second edition, I thank Karen Simon, Robert Ross, Sunil Nair, and Shashi Kumar, all of Taylor & Francis.

Last, but hardly least important, are the support and forbearance of my wife, Janet, in everything that book writing entails for anyone unlucky enough to observe and be around the process. Thanks.

The remainder of this preface is from the first edition.

What This Book Is, and Isn't

This book aims to do what its title suggests: help students—especially those new to real analysis—really understand what it is all about.

Who could oppose such a goal? A better question is how, concretely, this textbook tries to achieve it. Following are some more specific goals, strategies, and features.

Building on calculus basics. Students in the beginning real analysis course that this book supports may have little or no course experience beyond single-variable calculus (a year, say) and perhaps some exposure to linear algebra, differential equations, or multivariate calculus. What *can* be expected from such students is a general, though probably informal, sense of the big ideas of calculus—function, limit, derivative, and integral—and some curiosity about the rigorous theory that lies behind the techniques. This book's main strategy is to exploit students' prior experience with and curiosity about calculus in working toward deeper understanding.

Focus on mathematical proof and structure. Beginning real analysis students may or may not already have taken a "proofs" or "bridge" or "transition" course. This book makes no hard-wired assumption either way. The long and relatively leisurely Chapter 1 includes enough of this material (e.g., a fairly complete introduction to sets and cardinality) to help less "mature" students make their way efficiently into real analysis. A class of students who do have this background can simply move more quickly through these preliminaries; there is enough material here to support a full semester course in either case.

Real analysis is, arguably, a challenging area in which to learn proof techniques, as I hope many students will do. Definitions and theorems, for instance, can be quite technical, with multiple layers of quantification. On the flip side (I'm convinced!), real analysis has the advantage of being about ideas most students will have recently encountered—and many students really are eager to accept a challenge so closely connected to their recent experience.

Focus on mathematical language. Outsiders and newcomers find mathematical language surprisingly subtle, sophisticated, and sensitive to word order, usage, and punctuation. (Run the definition of uniform continuity by a colleague in the humanities sometime.) Most mathematics students need guidance and explicit attention to linguistic conventions in the discipline. This book aims to offer such help, especially in earlier sections, before the mathematical and linguistic subtlety ramp up—together.

Redoing some elementary calculus topics "right." A beginning analysis course can help students deepen and consolidate their understanding of calculus ideas. For example, few students seem to make deep sense of infinite series on that first dreaded pass, usually in Calculus 2. Students who study sequence convergence carefully, as in the course envisioned here, are in a far better position to make sense of abstract questions of series convergence and divergence and to go deeper into such topics as estimation of series "tails."

Focus on the basics. This book focuses on (what I take to be) basic elements of the theory; neither compactness nor the Lebesgue integral is covered, for instance. This is intentional: I've willingly traded "coverage" for "simplicity." (Instructors who want to go deeper can readily do so; see below for suggestions.) Covering fewer topics, moreover, leaves room for more narrative discussion and concrete examples, of which there are many. To put it another way, I try to look closely at relatively simple things.

Many examples, many solutions. The book contains many worked-out examples and quite a few detailed solutions to exercises, especially in earlier sections. I believe that many students learn theory largely "inductively," from examples, and that detailed solutions to a substantial number of problems can usefully illustrate, for newcomers, the language and conventions of mathematical discourse.

Reader-friendly style. I've aimed for a friendly-but-serious narrative style. The marginal notes, for instance, are intended mainly as brief nudges, pointers to "dangerous curves," or just friendly asides. It takes two to achieve friendliness, of course, so the verdict is ultimately up to the reader.

Increasing sophistication. Mathematical depth and technical sophistication gradually increase over the chapters. Chapter 1, for instance, is quite leisurely, while Chapter 5, on the Riemann integral, is relatively technical (like the integral itself) and business-like.

Using This Book

Following are some suggestions for using this book.

A one-semester course? This book is designed for use in one typical college semester, but there is more material here than I would cover in that time, except perhaps with veterans of a proofs course. With a less experienced class I might omit the coverage of infinite series, for instance, to concentrate more heavily on integrals. Or one could omit integrals entirely, and go into more depth on series, sequences, and derivatives.

Supplementary topics. Real analysis offers many possibilities for group or individual projects on supplementary topics. These can be especially useful with unusually well-prepared or highly talented students, who can otherwise be bored in a class of less highly selected students. Here are some ideas; all are readily researchable in books or online by motivated students:

- The Cantor–Banach–Bernstein theorem: if $f : A \to B$ and $g : B \to A$ are *injective*, then there is a *bijection* $h : A \to B$.

- Exploring the limit superior and the limit inferior (a "guided discovery" on this topic appears in the book).

- Exploring Taylor polynomials and remainder theorems (Taylor's theorem is mentioned, but only briefly).

- Exploring compactness of subsets of $\mathbb{R}$, the Heine–Borel theorem, and connections with boundedness of functions and convergence of sequences.

- Exploring deeper properties of the Riemann integral. Suppose, for instance, that a function f is integrable on $[0, 1]$ and $f(x) > 0$ for all x. Showing rigorously that $\int_0^1 f > 0$ is harder than it might seem, and it raises good questions about monotonicity, points of continuity, and integrability.

CHAPTER 1

Preliminaries: Numbers, Sets, Proofs, and Bounds

1.1 Numbers 101: The Very Basics

Real analysis is all about *real*-valued functions of a *real* variable, so an obvious early step is to understand basic properties of the set of real numbers. Many of these properties are related to or inherited from slightly simpler sets, such as the integers and the rational numbers. We reserve some special symbols for these important sets:*

$$\mathbb{N} = \{1, 2, 3, \dots\} : \quad \text{the } \textit{natural} \text{ numbers;}$$
$$\mathbb{Z} - \{\dots, \ 2, \ 1, 0, 1, 2, \dots\} . \quad \text{the } \textit{integers;}$$
$$\mathbb{Q} : \quad \text{the } \textit{rational} \text{ numbers;}$$
$$\mathbb{R} : \quad \text{the } \textit{real} \text{ numbers.}$$

Using these symbols we can discuss these sets clearly and efficiently, with sentences like the following:

$$\mathbb{N} \subseteq \mathbb{Z} \subseteq \mathbb{Q} \subseteq \mathbb{R}; \qquad \frac{42}{43} \in \mathbb{Q}; \qquad \sqrt{152399025} \in \mathbb{Z}; \qquad \sqrt{93} \in \mathbb{R} \setminus \mathbb{Q}.$$

What each of these symbol strings says should be clear. Whether each claim is true or false is less obvious—but that's a matter of mathematics, not of symbolic clarity. As it turns out, all four claims are true; proving the last assertion takes some effort.

The last one says that $\sqrt{93}$ is real but not rational. Notice the funny "setminus" sign.

We'll see a proof soon; for now the proof is beside the point.

Set notation is useful and efficient, but only when used correctly. Something is amiss, for instance, with each of the following expressions:

$$\text{(i) } \mathbb{R} \in \mathbb{Q}; \qquad \text{(ii) } 3.1 \supset \mathbb{N}; \qquad \text{(iii) } \sqrt{\mathbb{N}} \subseteq \mathbb{R}.$$

---

*The special font can help avoid confusion with other sets. The choices $\mathbb{N}$ and $\mathbb{R}$ are made for obvious reasons; $\mathbb{Q}$ and $\mathbb{Z}$ remind us of *quotient* and *Zahlen* (German for "numbers"), respectively. The symbol $\mathbb{C}$ can denote the set of complex numbers.

A charitable reader might try to make sense of (i)–(iii), but their problems are ultimately fatal—none really makes sense. (A nicer way to say this is to refer to errors of *syntax*.) The problem with (i), for instance, is not the (true!) fact that real numbers may be irrational, but rather that $\mathbb{R}$ is a *set* of numbers, and hence not even in the running to be an *element* of the set of rationals. Expression (ii) is meaningless for similar reasons, and (iii) is even worse: the expression $\sqrt{\mathbb{N}}$ has no clear meaning, so the question of containment in $\mathbb{R}$ is moot.

EXAMPLE 1.   Does each of the following symbolic claims make sense? If a claim makes sense, is it true or false?

(i) $\mathbb{Q} \in \mathbb{R}$;   (ii) $10^6 \pi \in \mathbb{Q}$;   (iii) if $a \in \mathbb{Q}$ and $b \in \mathbb{Q}$ then $a + b \in \mathbb{Q}$.

SOLUTION.   Statement (i) makes no sense—$\mathbb{Q}$ is the wrong type of object to be an element of $\mathbb{R}$. Claim (ii) asserts—meaningfully but falsely—that $10^6 \pi$ is rational. Claim (iii) also makes good sense; it says that the sum of two rational numbers is another rational number. This, too, is true, because the sum of two "fractions" is another "fraction." $\diamondsuit$

> If $10^6 \pi$ were a quotient of integers, π would be rational, too, which it isn't.

Integers, Rationals, and Reals

What exactly *is* a real number?

The question may seem absurd. Nobody reading this book will balk at expressions like

$$\sin\left(\frac{\pi}{4}\right) = \frac{\sqrt{2}}{2} \quad \text{and} \quad \int_0^1 e^x \, dx = e - 1,$$

both of which are rich stews of integers, fractions, and irrational numbers. We will not deny or ignore your earlier experience—all of it will be useful—but aim to sharpen intuition and make assumptions explicit.

As a look ahead consider, for instance, the expression

$$\lim_{x \to \pi} \frac{f(x) - f(\pi)}{x - \pi} = f'(\pi),$$

> It's $f'(\pi)$, as we know from beginning calculus.

which describes a certain derivative. Making clear sense of such an expression, as we'll need to do later in studying derivatives, depends on various properties of real numbers. Here are some early hints:

- *Real arithmetic:* The fractional expression

$$\frac{f(x) - f(\pi)}{x - \pi}$$

is built up using subtraction and division of the real numbers x, π, $f(x)$, and $f(\pi)$, at least some of which are irrational. Depending on its definition, the function f may require additional arithmetic operations on reals.

- *Approximation and distance:* The idea of limit entails one quantity "approaching" another, in a subtle sense that we'll soon work hard to make precise. But it is clear already that we won't get far without some notion of "distance" between numbers, and we'll need to know that any real number, say π, has plenty of "near" neighbors along which an "approach" to π can be made.

- *No gaps:* For any limit to exist there needs to *be* a value for the quantity in question to approach; if the real numbers had "gaps" there could be trouble. This modest-seeming requirement, called *completeness*, turns out to be subtler than it might first seem. We'll revisit it soon.

Integers. We'll build up our description of the real numbers by starting with simpler sets: first $\mathbb{Z}$ and then $\mathbb{Q}$. What's "simpler" depends, of course, on the situation. To a number theorist, for instance, there is nothing simple about the integers. In real analysis, however, we'll need only basic, familiar properties of the integers, and we'll usually take these as "known." We'll accept without fuss, for instance, that equations like

$$a + b = b + a, \qquad ab = ba, \qquad (a + b)c = ac + bc, \quad \text{and}$$
$$(3a - 5b)(2a - 7b) = 6a^2 - 31ab + 35b^2$$

make sense, and hold true, for any integers a, b, and c. We also "know" that every integer a has an additive inverse, $-a$, but that a multiplicative inverse, $1/a$, need not be an integer.

What happens if $a = 0$?

The well-ordering property. Here's another basic fact we'll just assume: *Every nonempty set of positive integers has a least element.* This fact is known formally as the *well-ordering property* of the positive integers. The name may be mysterious but the property itself should be familiar and believable from experience. Among all even positive multiples of 7, for instance, 14 is smallest. We'll take the well-ordering property as an axiom, not as a theorem to be proved.

It comes from set theory—not our subject here.

In more formal discussion the property is sometimes proved to follow from even simpler axioms.

Rational numbers. Basic algebraic properties of $\mathbb{Q}$ are equally familiar. All four equations above, for example, hold just as well for *rational* numbers a, b, and c as for integers. Because nonzero rational numbers have (rational) reciprocals, equations and expressions that involve *division* can now make sense. Here are three examples:

$$\frac{p+q}{r} = \frac{p}{r} + \frac{q}{r}; \qquad \frac{pq}{r} = p\,\frac{q}{r}; \qquad \frac{p+q}{pq} = \frac{1}{q} + \frac{1}{p}.$$

All are true, of course, whenever p, q, and r are rational numbers and all denominators are nonzero.

In later work we'll freely use such familiar facts, usually without comment. Right now it is a good exercise to *derive* some properties of $\mathbb{Q}$, *assuming* basic properties of $\mathbb{Z}$.

For us a rational number r is just a *ratio* of integers:

$$r = \frac{a}{b}, \quad \text{where } a, b \in \mathbb{Z} \text{ and } b \neq 0.$$

Of course, any given rational number has many possible forms:

$$\frac{2}{3} = \frac{14}{21} = \frac{-222}{-333} = \frac{88}{132} = \cdots;$$

the first form, in which numerator and denominator have no common factors, is called *reduced*.

EXAMPLE 2.   Prove—assuming basic properties of *integer* arithmetic—that if p and q are any rational numbers, then $p + q \in \mathbb{Q}$ and $p + q = q + p$.

SOLUTION.   The proof is direct: We start with the hypothesis and derive the conclusions. Because p and q are rational, we can write

$$p = \frac{a}{b} \quad \text{and} \quad q = \frac{c}{d},$$

where $a, b, c, d \in \mathbb{Z}$ and $b \neq 0 \neq d$. Now we add fractions:

$$p + q = \frac{a}{b} + \frac{c}{d} = \frac{ad + cb}{bd},$$

which exhibits $p + q$ as the desired ratio of integers. (Both numerator and denominator are integers, and the denominator is nonzero, thanks to basic integer arithmetic, which we take as known.)

Also,

$$q + p = \frac{c}{d} + \frac{a}{b} = \frac{cb + ad}{db} = \frac{ad + cb}{bd},$$

where the last equality follows from commutativity of *integer* operations. The last expression is just what we calculated earlier for $p + q$, so we're done.   ◊

EXAMPLE 3.   Prove that if $p \in \mathbb{Q}$ and $p \neq 0$, then there is a rational number q with $pq = 1$. (In other words, q deserves the name $1/p$.)

SOLUTION.   We'll give an explicit formula for q.

The hypothesis means that there are *nonzero* integers a and b with $p = a/b$. Now $q = b/a$ does the job, since

$$pq = \frac{ab}{ab} = 1. \qquad \diamond$$

$\mathbb{Q}$ is a field.   The following theorem says, in mathematical parlance, that $\mathbb{Q}$ is a *field*.

Theorem 1.1.   *The set $\mathbb{Q}$ has the following properties:*

Nothing in the theorem should surprise you; the point is to collect and formalize key properties.

- Two operations: *Addition and multiplication are operations on $\mathbb{Q}$: if p and q are any rational numbers, then so are $p + q$ and pq.*

- Commutativity: *Addition and multiplication are commutative operations: if p and q are any rational numbers, then $p + q = q + p$ and $pq = qp$.*

- Associative operations: *Addition and multiplication are associative operations: if p, q, and r are any rational numbers, then $(p+q)+r = p+(q+r)$ and $(pq)r = p(qr)$.*

- Identities: *The rational number zero is an additive identity: $0 + p = p$ holds for all $p \in \mathbb{Q}$. The rational number one is a multiplicative identity: $1p = p$ holds for all $p \in \mathbb{Q}$.*

- Inverses: *For every rational number p, the rational number $-p$ is an additive inverse: $p + (-p) = 0$. For every nonzero rational number p, the rational number $1/p$ is a multiplicative inverse: $p \cdot 1/p = 1$.*

- Distributivity: *Multiplication distributes across addition: For any rationals p, q, and r, $p(q + r) = pq + pr$.*

All parts of the theorem can be proved in the spirit of the preceding examples, assuming similar properties of $\mathbb{Z}$. But ...

EXAMPLE 4. $\mathbb{Z}$ is *not* a field. Why not?

SOLUTION.   To be a field, a set must satisfy each of the long list of properties in Theorem 1.1. Proving all this can be daunting. But there's a happy flip side: *Disproving* that $\mathbb{Z}$ is a field is easy. It is enough to find just one property that fails for $\mathbb{Z}$, and for this even a single counterexample suffices. We could just observe, for instance, that the integer 42 has no multiplicative inverse *among the integers*, and leave it at that. $\diamond$

ℚ is not ℝ. The set ℚ is clearly infinite—it contains all the integers, for one thing. A bit less obvious, perhaps, is the fact that *between any two different rationals lie infinitely many more rationals.* Between 3.14 and 3.15, for instance, lie

$$3.141, \ 3.142, \ \ldots, \ 3.149, \ 3.1411, \ 3.1412, \ \ldots, \ 3.14237568, \ 3.14237569, \ \ldots$$

and so on. With so many rational numbers around, one might wonder whether there are any *irrationals*—numbers that *cannot* be written as ratios of integers.

Nowadays everybody "knows" the answer: $\sqrt{2}$, $\sqrt{3}$, $\sqrt[3]{2}$, π, e, and many other favorite numbers are all irrational. But none of these facts is completely trivial. Indeed, the Pythagoreans (around 500 BCE) saw *all* numbers as rational. By some accounts, they drowned the philosopher Hippasus as a heretic after he demonstrated that a square of side 1 has irrational diagonal. We'll take that risk and still give a proof.

Theorem 1.2. $\sqrt{2}$ *is irrational.*

> Like other proofs we've given, this one exploits basic properties of integers.

Proof: The proof is by contradiction: Assuming $\sqrt{2}$ is rational, we'll derive an absurdity.

Assume, then, that $\sqrt{2}$ is rational. We can write $\sqrt{2} = \frac{a}{b}$, for some positive integers a and b, *not both even.* (If both were even we could cancel a common factor of 2.) Next we do some algebra:

$$\sqrt{2} = \frac{a}{b} \quad \Longrightarrow \quad 2 = \frac{a^2}{b^2} \quad \Longrightarrow \quad 2b^2 = a^2.$$

Now $2b^2$ is certainly even, and so a^2 must be even, too. But then a itself must be even; otherwise, a^2 would be odd. Thus we can write $a = 2c$ for some integer c, and we have

$$2b^2 = a^2 = 4c^2 \quad \Longrightarrow \quad b^2 = 2c^2.$$

This implies that b (like a) is even, which contradicts our earlier assumption, and completes the proof. □

> See the exercises for details.

The following more general theorem can be proved by similar means. It shows, too, that the set of irrational numbers, or $\mathbb{R} \setminus \mathbb{Q}$, has infinitely many members.

Theorem 1.3. *If the positive integer n is not a perfect square, then $\sqrt{n}$ is irrational. In symbols, $\sqrt{n} \in \mathbb{R} \setminus \mathbb{Q}$.*

Infinitely many irrationals for the price of one. It took some work to find even one irrational number. This done, however, it is surprisingly easy to find many, many more irrationals.

Theorem 1.4. *If x is irrational and $r \neq 0$ is rational, then*

$$xr, \quad x+r, \quad x-r, \quad 1/x, \quad \frac{x}{r}, \quad and \quad \frac{r}{x}$$

are all irrational.

Proof: There is less to this than meets the eye. Everything boils down to the field properties of $\mathbb{Q}$, listed in Theorem 1.1. For example, suppose (aiming for a contradiction) that xr is rational. We know from Theorem 1.1 that $1/r$ is also rational, and so the product

$$xr \times \frac{1}{r} = x$$

is rational, too. This contradicts our hypothesis, and so xr must be irrational. The remaining parts are similar, and are left as exercises.   □

Real numbers. So what *is* a real number?

For us, the set $\mathbb{R}$ is the union of all the rational and all the irrational numbers. Admittedly, this is not a logically airtight definition. It ignores, for example, such difficult questions as how to be sure we've included *all* the needed irrationals. But our informal view is a good start, and we'll deepen it in the next section as we study properties and alternative ways of viewing the real numbers.

> The reals can be defined rigorously, starting from the rationals; this is sometimes done in more advanced courses.

Exercises

1. As in Example 1, page 2, decide whether each of the following claims makes sense. If a claim makes sense, is it true or false?

   (a) If $r \in \mathbb{Q}$ then $5r \in \mathbb{Q}$.

   (b) $\sqrt{8} \subset \mathbb{R} \setminus \mathbb{Q}$.

   (c) $\{\sqrt{8}\} \subset \mathbb{R} \setminus \mathbb{Q}$.

   (d) If $a \in \mathbb{Q}$ and $b \in \mathbb{R} \setminus \mathbb{Q}$ then $ab \in \mathbb{R} \setminus \mathbb{Q}$.

   (e) If $a \in \mathbb{Q}$ and $b \in \mathbb{N}$ then $a/b \in \mathbb{Q}$.

   (f) If $a \in \mathbb{Q}$ then there exists $n \in \mathbb{N}$ such that $na^2 > 100$.

2. Decide whether each of the following claims makes sense. If not, why not? If so, is the claim true or false? (No proofs needed.)

   (a) If $a \in \mathbb{R} \setminus \mathbb{Q}$, then $a^2 \in \mathbb{R} \setminus \mathbb{Q}$.

   (b) $\mathbb{Q}^2 > 0$.

   (c) If $a \in \mathbb{R} \setminus \mathbb{Q}$, then there exists $n \in \mathbb{N}$ such that $na^2 > 100$.

   (d) If $a \in \mathbb{Q}$ then $a + \sqrt{2} \notin \mathbb{Q}$.

(e) If $a \in \mathbb{Q}$ and $a \neq 0$, then $\sqrt{2} \cdot a \notin \mathbb{Q}$.

3. For which integers a is $1/a$ an integer? For which integers a is $1/a$ rational? For which integers a is $1/a = a$? No proofs necessary, but be sure to consider all possibilities.

4. We said in this section that $\mathbb{Z}$ is not a field. Which of the several requirements in Theorem 1.1 does $\mathbb{Z}$ fail? Which does it pass? Give an example to illustrate each *failed* requirement.

5. The set of irrationals is not a field because (among many other reasons) the irrationals contain no multiplicative identity. Give one reason—as briefly as possible—why each of the following sets is *not* a field.

    (a) The set S_1 of nonnegative rationals.

    (b) The set S_2 consisting of all the irrationals and all the integers.

    (c) The set S_3 of all numbers of the form $a + b\sqrt{2}$, where a and b are integers.

    (d) The set S_4 of real numbers of the form πr, where r is rational.

6. Let p and q be any two rational numbers. In each part following, decide whether the given expression *must* be a rational number, regardless of the values of p and q. If so, explain why, referring to theorems in this section. If not, give a counterexample.

    (a) $\dfrac{p+q}{2}$ (the average of p and q)

    (b) $\dfrac{p+q}{p^2+q^2}$

    (c) $\sqrt{p^2+q^2}$

    (d) $\sqrt{p^2+2pq+q^2}$

7. Let both x and y denote irrational numbers, and consider the quantities (a) xy; (b) $x + y$; (c) $x - y$; (d) x/y. Give examples (i.e., specific values of x and y) to show that each quantity can be either rational or irrational.

8. Suppose $x \in \mathbb{R} \setminus \mathbb{Q}$, $r \in \mathbb{Q}$, and $r \neq 0$. What (if anything) can be said about rationality or irrationality of (a) xr; (b) $x + r$; (c) x/r; (d) $\sqrt{x}$; (e) $\sqrt{r}$; (f) $\sqrt{xr}$

9. Imitate the proof of Theorem 1.2 to show that $\sqrt{3}$ is irrational. (Hint: If a^2 is divisible by 3, then a is divisible by 3, too. This is easy to prove, but just assume it here.)

10. Use Theorem 1.4 in this problem.

   (a) Show that if x^2 is irrational, then x is irrational, too.

   (b) We know that $\sqrt{2}$ and $\sqrt{3}$ are irrational. Show that $\sqrt{2} + \sqrt{3}$ is irrational, too.

   (c) All of $\sqrt{2}$, $\sqrt{3}$, and $\sqrt{5}$ are irrational. Show that $\sqrt{2} + \sqrt{3} + \sqrt{5}$ is irrational, too. (This is harder.)

11. Add details (explain anything that isn't obvious) to the following hints to give another proof that $\sqrt{2}$ is irrational. (The proof's very clever idea is attributed to Fermat.)

   If $\sqrt{2} = a/b$ holds for any positive integers a and b, then there must be a *smallest* possible b, in the sense that no smaller denominator works. Let such a and b be given; note that $2b^2 = a^2$. Now consider the new fraction

   $$\frac{2b - a}{a - b} = \frac{a'}{b'}.$$

   Explain why (i) both a' and b' are positive integers; (ii) $b' < b$; (iii) $(a'/b')^2 = 2$, and so deduce that $\sqrt{2} = a/b$ is impossible.

12. Consider the set $\mathbb{Z}_2 = \{0, 1\}$.

   (a) Show that $\mathbb{Z}_2$ *is not* a field with the usual operations of multiplication and addition.

   (b) Show that $\mathbb{Z}_2$ *is* a field if we use the two operations (i) ordinary multiplication; (ii) addition "mod 2": $0 + 0 = 0$ and $0 + 1 = 1 + 0 = 1$, but $1 + 1 = 0$.

13. The set $M_{2 \times 2}$ of 2×2 matrices with real number entries permits matrix addition and matrix multiplication, so we can ask about the properties mentioned in Theorem 1.1. No proofs needed, but give examples to illustrate any properties that fail.

   (a) Are addition and multiplication in $M_{2 \times 2}$ commutative?

   (b) Which elements in $M_{2 \times 2}$ have additive and/or multiplicative inverses?

   (c) Does addition and multiplication in $M_{2 \times 2}$ satisfy the distributive property?

14. Let F be the set of real numbers of the form $a + b\sqrt{2}$, with a and b rational numbers. Show that F is a field. (Hint: The hardest part concerns multiplicative inverses.)

15. We know from elementary calculus that $\ln n$ tends to infinity ("blows up") slowly as n tends to infinity. It follows that $\ln \ln \ln n$ also tends to infinity, but slower still. Use the well-ordering property to explain why there must be a *least* positive integer n_0 such that $\ln \ln \ln n_0 > 2$. (Notes: Finding n_0 exactly would be tricky, since n_0 is very, very large. For fun, try to estimate how many digits n_0 has.)

16. The well-ordering property (see page 3) says something special about the set $\mathbb{N}$:

    If $S \subset \mathbb{N}$ and $S \neq \emptyset$, then S has a least element.

    Give informal answers, not formal proofs, in each part following.

    (a) Does $\mathbb{Q}$ have the well-ordering property? (In other words, does every nonempty subset of $\mathbb{Q}$ have a least element?) Why or why not?

    (b) Does the set $R = \{1, 10, 100, 1000, \dots\}$ have the well-ordering property? Why or why not?

    (c) Does the set $T = \{-3, -2, -1, \dots, 41, 42\}$ have the well-ordering property? Why or why not?

    (d) Replace the word "least" with "greatest" in the well-ordering property above. Does the result hold for $\mathbb{N}$? For T? For $\mathbb{Z} \setminus \mathbb{N}$?

1.2   Sets 101: Getting Started

This section is a brief introduction to (or review of) rudiments of set theory that play some role in real analysis. An important purpose is to establish vocabulary and notations that will help us streamline later discussions.

A *set* is any collection of objects, called *elements* of the set. Here are some samples:

$$A = \{\text{January, February, March}, \dots, \text{December}\},$$
$$B = \{\text{January, March, May, July, August, October, December}\},$$
$$C = \{1, \{2,3\}, x^2 + 1, \text{November}, \text{elephant}\},$$
$$D = \{2, 3\},$$
$$I = [2, 3] = \{x \mid 2 \leq x \leq 3\},$$
$$\mathbb{N} = \{1, 2, 3, \dots\},$$
$$\emptyset = \text{the empty set.}$$

Observe:

- *Names:* It is common, but not required, to use uppercase letters to denote sets. For a few very special sets, like $\mathbb{N}$ and $\emptyset$ above, we use special characters.

- *Looks can deceive:* The sets $D = \{2,3\}$ and $I = [2,3]$ bear some typographical resemblance but are otherwise very different—D has only two elements while I, like every nonempty interval, has infinitely many.

- *Different descriptions:* Sets can be described in various ways. For B, C, and D, for instance, we listed each set's elements explicitly. For A, I, and $\mathbb{N}$ we described a pattern or condition that elements must satisfy. What really matters is clarity.

- *Weird sets:* The strange set C illustrates that a set's elements may be quite different from each other. In particular, a set may contain other sets as elements. We'll usually stick to relatively tame sets in studying real analysis.

Notations and operations. Standard symbols let us describe set properties and operations clearly and concisely. Just a few go a long way:

- *Subsets and containment:* The expression $B \subset A$ says (truthfully for the sets above) that B is a *subset* of A; the C-like symbol suggests *containment*. If $B \subsetneq A$, then B is a *proper subset* of A.

  In a similar spirit, all of these expressions:

  $$A \supset B; \qquad B \subseteq A; \qquad B \subsetneq A; \qquad I \supseteq D; \qquad \emptyset \subseteq D$$

  describe various containment relations.

  Are they all true of the sets above?

- *Membership:* The expression $a \in A$ means that a is an *element* of A. Here are some (true) examples for the sets above:

  Read $\in$ as "in."

  $$\text{March} \in A; \qquad 2 \in D; \qquad \text{April} \notin B; \qquad \{2,3\} \in C; \qquad \{2,3\} \notin I.$$

  The last two claims might be surprising. Notice that the *set* $\{2,3\}$ is indeed an *element* of the peculiar set C, but not of I, which contains only numbers.

  On the other hand, $\{2,3\} \subset I$.

- *New sets from old:* Any two sets S and T can be combined in various ways to form new sets:

$$
\begin{aligned}
\textit{union}: \quad & S \cup T = \{x \mid x \in S \text{ or } x \in T\}; \\
\textit{intersection}: \quad & S \cap T = \{x \mid x \in S \text{ and } x \in T\}; \\
\textit{set difference}: \quad & S \setminus T = \{x \mid x \in S \text{ and } x \notin T\}; \\
\textit{Cartesian product}: \quad & S \times T = \{(x,y) \mid x \in S \text{ and } y \in T\}.
\end{aligned}
$$

Here are some (true) examples from the sets above; note that all quantities in question are *sets*:

$$A \cap C = \{\text{November}\}; \qquad A \cup B = A; \qquad A \cap B = B;$$
$$A \setminus B = \{\text{February, April, June, September, November}\}; \qquad B \setminus A = \emptyset;$$
$$A \times \mathbb{N} = \{(m, n) \mid m \text{ is a month and } n \text{ is a positive integer}\}.$$

- *Handling complements:* The sets $\mathbb{Q}$ of rationals and P of irrationals are *complementary* in $\mathbb{R}$ in an obvious way:

$$P \cap \mathbb{Q} = \emptyset \quad \text{and} \quad P \cup \mathbb{Q} = \mathbb{R}.$$

Notice, too, that $P = \mathbb{R} \setminus \mathbb{Q}$ and $Q = \mathbb{R} \setminus P$.

For *any* set $A \subseteq \mathbb{R}$, we call $\mathbb{R} \setminus A$ the *complement of A in $\mathbb{R}$*. In the same spirit, A is the complement of $\mathbb{R} \setminus A$, and it's easy to check that $\mathbb{R} \setminus (\mathbb{R} \setminus A) = A$.

De Morgan's laws connect complements with unions and intersections. For any two subsets A_1 and A_2 of $\mathbb{R}$, De Morgan's laws assert the following identities:

$$\mathbb{R} \setminus (A_1 \cup A_2) = (\mathbb{R} \setminus A_1) \cap \mathbb{R} \setminus A_2$$
$$\mathbb{R} \setminus (A_1 \cap A_2) = (\mathbb{R} \setminus A_1) \cup \mathbb{R} \setminus A_2$$

Similar rules hold for unions and intersections of more than two subsets of $\mathbb{R}$. We'll use these rules in coming chapters; a proof is in the exercises.

EXAMPLE 1. What do De Morgan's laws say if $A_1 = \mathbb{Q}$ and A_2 is the set of negative numbers?

SOLUTION. Here $A_1 \cap A_2$ is the set of negative rationals, so $\mathbb{R} \setminus (A_1 \cap A_2)$ is the set of numbers that are *not* negative rationals. Similarly, $\mathbb{R} \setminus A_1$ is the set of irrationals, and $\mathbb{R} \setminus A_2$ is the set of nonnegative numbers. The relevant De Morgan law asserts that

$$\mathbb{R} \setminus (A_1 \cap A_2) = (\mathbb{R} \setminus A_1) \cup \mathbb{R} \setminus A_2).$$

In our situation, this means, plausibly enough, that a number is *not* a negative rational if and only if it is either irrational or nonnegative or both.

The set $A_1 \cup A_2$ is the set of numbers that are either rational or negative or both. De Morgan's other law,

$$\mathbb{R} \setminus (A_1 \cup A_2) = (\mathbb{R} \setminus A_1) \cap \mathbb{R} \setminus A_2),$$

also makes good sense: numbers not in $A_1 \cup A_2$ are neither rational nor negative. $\Diamond$

Named for an influential Victorian-era English mathematician—and textbook writer.

Analogous rules hold for infinitely many A_i, too.

The notation seems formal, but the claim makes common sense.

Decoding set language. Statements about sets may involve complicated or tricky symbolic combinations. Making sense of them takes careful reading.

EXAMPLE 2.   Let $\mathbb{Z}$, $\mathbb{Q}$, and $\mathbb{R}$ denote (as usual) the sets of integers, rationals, and reals, respectively, and consider the sets

$$S = \left\{ x \in \mathbb{R} \mid x^2 - x - 1 = 0 \right\}; \qquad T = \left\{ x \in \mathbb{R} \mid x^2 - x - 1 > 5 \right\}.$$

The sets S and T illustrate *set-builder notation*: they are "built" from a larger set using a selection rule or membership criterion. (The vertical bar $\mid$ means something like "such that.")

What does each of the following assertions mean? Which are true?

(i) $S \in \mathbb{Z}$;   (ii) $S \subset \mathbb{Q}$;   (iii) $S \subset \mathbb{R} \setminus \mathbb{Q}$;   (iv) $4 \in T$.

SOLUTION.   Statement (i) says that S—a *set* of numbers—is itself an integer; this is clearly false. Statements (ii) and (iii) make better sense; (ii) says that all roots of $x^2 - x - 1$ are rational, while (iii) says that these same roots are irrational. Who's right? Well, by the quadratic formula, the two roots are $(1 \pm \sqrt{5})/2$, both of which are real but irrational, so (iii) is true and (ii) is false. Statement (iv) boils down to the claim that $4^2 - 4 - 1 > 5$, which is clearly true.

We might, by the way, have saved some work by first rewriting S and T in simpler or different forms. As we've seen,

$$S = \left\{ \frac{1 + \sqrt{5}}{2}, \frac{1 - \sqrt{5}}{2} \right\}.$$

The quadratic formula also shows that $x^2 - x - 1 = 5$ has the two roots $x = -2$ and $x = 3$, and so

$$T = \left\{ x \in \mathbb{R} \mid x < -2 \text{ or } x > 3 \right\}. \qquad \Diamond$$

Check for yourself.

Intervals. Intervals in the real line are familiar but useful sets in studying real analysis. Calculus veterans have seen countless examples; we collect a few as reminders of the possible variety and of some useful descriptive language.

interval	description	inequality form
$[-2, 3]$	closed, bounded	$\{ x \in \mathbb{R} \mid -2 \leq x \leq 3 \}$
$(-2, 3)$	open, bounded	$\{ x \in \mathbb{R} \mid -2 < x < 3 \}$
$[-2, \infty)$	closed, unbounded	$\{ x \in \mathbb{R} \mid -2 \leq x \}$
$(-\infty, 3)$	open, unbounded	$\{ x \in \mathbb{R} \mid x < 3 \}$
$(-2, 3]$	half-open, bounded	$\{ x \in \mathbb{R} \mid -2 < x \leq 3 \}.$

In particular, *closed intervals* contain their endpoints, if any, while *open intervals* do not. All intervals, open or closed, bounded or unbounded, share two defining properties:

Definition 1.5. A set $I \subset \mathbb{R}$ is an *interval* if (i) I contains at least two points; (ii) if a and b are in I and $a < x < b$, then $x \in I$, too.

EXAMPLE 3. Let I and J be any two intervals. What possible forms can the intersection $I \cap J$ take?

SOLUTION. The intervals I and J might miss each other entirely; then $I \cap J = \emptyset$. Or I and J might intersect in a single point, as do $I = [1, 3]$ and $J = [3, 5]$. More interesting is the fact that only one other possibility exists: If $I \cap J$ contains at least two points, then $I \cap J$ is an *interval*.

The hard way to prove this is to handle many special cases, depending on whether each of I and J is open, closed, bounded, unbounded, etc. The easy way is to use Definition 1.5, in which only (ii) is a live question.

Suppose, then, that both a and b are in $I \cap J$, and $a < x < b$. Since I is an interval, Definition 1.5 guarantees that $x \in I$. For the same reason, we must have $x \in J$, too. Thus $x \in I \cap J$, and we're done.   ◊

Open sets: a look ahead. Open *intervals* are sets like $(1, 3)$ and $(-\infty, 42)$ that don't contain any endpoints. More generally, *any* set $U \subset \mathbb{R}$ is called (topologically) *open* if U is the union of any collection (empty, finite, or infinite) of open *intervals*. A set $A \subseteq \mathbb{R}$ is called *closed* if $A = \mathbb{R} \setminus U$, where U is open. In other words, closed sets are complements of open sets.

We get serious about basics of topology later. This is just a taste.

Example 4 shows that this new terminology plays well with what's already familiar.

EXAMPLE 4. Among the sets

$$(1, 3), \quad \mathbb{R}, \quad [1, 3], \quad \{42\}, \quad \{42\} \cup [1, 3],$$

which are open or closed? What about $\emptyset$?

SOLUTION. The sets $(1, 3)$ and $\mathbb{R} = (-\infty, \infty)$ are open intervals, and therefore also open in the topological sense. For the other sets, we study *complements*:

$$[1, 3] = \mathbb{R} \setminus ((-\infty, 1) \cup (3, \infty)) \quad \text{and} \quad \{42\} = \mathbb{R} \setminus ((-\infty, 42) \cup (42, \infty))$$

It would be troubling if $[1, 3]$ were not closed.

Both of these complements are unions of open intervals and are therefore open sets; hence both $[1, 3]$ and $\{42\}$ are closed sets. The complement of $\{42\} \cup [1, 3]$ is the union of *three* open intervals, so $\{42\} \cup [1, 3]$ is also closed.

Here comes a twist: As the union of an *empty* collection of open intervals, the empty set $\emptyset$ is open. As the complement of the open set $\mathbb{R}$, $\emptyset$ is also closed. As the complement of $\emptyset$, $\mathbb{R}$ is also closed. Thus, $\mathbb{R}$ and $\emptyset$ turn out to be *both* open and closed. $\diamond$

Exercises

1. Consider several sets discussed in this section:

$$A = \{\text{January, February, March}, \ldots, \text{December}\};$$
$$B = \{\text{January, March, May, July, August, October, December}\};$$
$$C = \{1, \{2, 3\}, x^2 + 1, \text{November, elephant}\};$$
$$D = \{2, 3\}; \quad I = [2, 3] = \{x \mid 2 \le x \le 3\}.$$

   (a) Clearly, $B \subset A$. Are there any other *subset* relations among A, B, C, D, and I? Is any of these sets an *element* of another?

   (b) Find a rule or pattern to define B in terms of A. (We could use such a rule to write B in set-builder notation: $B = \{m \in A \mid \ldots \text{something} \ldots\}$.

   (c) Describe the set $A \times D$. How many elements are there?

   (d) Find (list the elements) each of the following sets: $A \setminus B$; $B \setminus A$; $A \cap C$; $B \cap A$; $D \cap I$; $D \cup I$.

2. Consider the sets $S = \{x \in \mathbb{R} \mid x^2 + x = 0\}$ and $T = \{x \in \mathbb{R} \mid x^2 + x < 5\}$. (We used similar sets in Example 2, page 13.)

   (a) Rewrite S and T in simpler forms. (One is a finite set and the other an interval.)

   (b) Decide whether each of the following statements is true or false, and explain: $S \subset \mathbb{N}$; $S \subset T$; $T \cap \mathbb{Q} \ne \emptyset$; $-2.8 \in \mathbb{Q} \setminus T$.

   (c) Give the simplest possible description of the set $U = \{x \in \mathbb{R} \mid x^2 + x < 0\}$.

3. For any $A \subseteq \mathbb{R}$, the sets A and $\mathbb{R} \setminus A$ are *complements* of each other. If $A = (1, 3)$, for instance, $\mathbb{R} \setminus A = (-\infty, 1] \cup [3, \infty)$. In each part, find the complement $\mathbb{R} \setminus A$; use interval notation if possible.

   (a) $A = [1, 3]$

   (b) $A = [1, \infty)$

   (c) $A = (1, 2) \cup (3, 4)$

(d) $A = (1, 2) \cap (3, 4)$

(e) $A = (-\infty, 0) \cup (0, \infty)$

4. (a) Show that the complement of a closed and bounded interval $[a, b]$ is the union of two unbounded open intervals.

   (b) Give examples to show that the complement of an open interval I can be empty, one closed interval, or the union of two closed intervals.

   (c) Write the complement of $\mathbb{Z}$ as a union of intervals.

5. We said in this section that for any set $A \subseteq \mathbb{R}$, $\mathbb{R} \setminus (\mathbb{R} \setminus A) = A$. Explain this in your own words.

6. This problem is about De Morgan's laws; see page 12. Let A and B be any subsets of $\mathbb{R}$, and consider the two claims

   (i) $\mathbb{R} \setminus (A \cup B) = (\mathbb{R} \setminus A) \cup (\mathbb{R} \setminus B)$;

   (ii) $\mathbb{R} \setminus (A \cup B) = (\mathbb{R} \setminus A) \cap (\mathbb{R} \setminus B)$.

   (a) One of (i) and (ii) is true and the other false. Identify the false claim and give specific sets A and B to show it is false.

   (b) What happens in the special case that $A = B$?

   (c) Prove the true statement above. (Hint: Show that every element of the left side is an element of the right side, and vice versa.)

7. This problem is about De Morgan's laws (page 12) when A_1 and A_2 are *intervals*. Let $A_1 = (1, 3)$ and $A_2 = (2, 5)$. Note that $\mathbb{R} \setminus A_1 = (-\infty, 1] \cup [3, \infty)$ and $\mathbb{R} \setminus A_2 = (-\infty, 2] \cup [5, \infty)$. Write $\mathbb{R} \setminus (A_1 \cup A_2)$ and $\mathbb{R} \setminus (A_1 \cap A_2)$ in interval notation, and check that De Morgan's laws hold as claimed.

8. This problem explores De Morgan's laws (page 12) when $A_1 = (0, 1)$ and $A_2 = (2, \infty)$. As in Problem 7, write all of the sets $\mathbb{R} \setminus A_1$, $\mathbb{R} \setminus A_2$, $\mathbb{R} \setminus (A_1 \cap A_2)$, and $\mathbb{R} \setminus (A_1 \cup A_2)$ using interval notation, and check that De Morgan's laws hold.

9. Suppose $S \subset T \subseteq \mathbb{R}$, and let S' and T' be the complements of S and T. Show that $T' \subset S'$.

10. Give specific examples of intervals I and J for which $I \cap J$ is (a) open; (b) closed; (c) half-open; (d) open and unbounded. Is it possible in each case to choose I and J so that neither $I \subset J$ nor $J \subset I$?

11. Example 3, page 14, is about *intersections* of intervals. This problem is about *unions* of intervals.

(a) Find any two intervals I and J such that $I \cup J$ is not an interval.

(b) Find *disjoint* intervals I and J such that $I \cup J$ is an interval.

(c) Consider open intervals $I = (a, b)$ and $J = (c, d)$ with $a < c$, $b < d$, and $0 \in I \cap J$. Show that $I \cup J$ is an open interval.

12. Suppose I and J are any intervals, and c any number such that $c \in I \cap J$. Use Definition 1.5 to show that $I \cup J$ is an interval.

13. Is it possible for both I and $\mathbb{R} \setminus I$ to be intervals? Can both I and $\mathbb{R} \setminus I$ be *bounded* intervals? In each case, give an example or explain why not.

14. Can any interval have exactly 123456789 points? Why or why not?

15. Can an interval I contain only rational numbers?

16. This problem links to Example 4, page 14; note the meanings there of "open" and "closed."

(a) Show that $(1, 2) \cup (3, \infty)$ is open. What related set is closed?

(b) Let $a \in \mathbb{R}$. Show that $\{a\}$ is closed.

(c) Let $a \in \mathbb{R}$. Show that $(-\infty, a)$ is open and $(-\infty, a]$ is closed.

(d) The interval $(0, 1)$ is obviously open. Show that it's not closed.

17. This problem links to Example 4, page 14; note the meanings there of "open" and "closed."

(a) Show that $\{1, 2, 3\}$ is a closed set.

(b) Show that $\mathbb{R} \setminus \mathbb{Z}$ is open (and so $\mathbb{Z}$ is closed).

(c) The set $\mathbb{Q}$ is neither closed nor open. Explain the "not open" part. (Hint: Can an interval contain only rational numbers?)

(d) Assuming (it's true!) that $\mathbb{Q}$ is neither closed nor open, explain why the set of *irrationals* is neither closed nor open.

(e) Show that the interval $(0, 1]$ (we've called it "half-open") is neither open nor closed.

18. From any set S we can create the new set $P(S)$, called the *power set* of S, consisting of all subsets of S. If $S = \{1, 2\}$, for instance, then $P(S) = \{\emptyset, \{1\}, \{2\}, \{1, 2\}\}$.

(a) Find $P(S)$ if $S = \{1, 2, 3\}$.

(b) Show that if $S \subset T$, then $P(S) \subset P(T)$.

(c) Let $N_{10} = \{1, 2, \ldots, 10\}$ and $N_{11} = \{1, 2, \ldots, 10, 11\}$. Explain why $P(N_{11})$ has twice as many members as $P(N_{10})$. (We'll prove this formally in a later section.)

19. Let $N_{10} = \{1, 2, \ldots, 10\}$.

(a) Let S be the set of all *three-member subsets* of N_{10}. How many elements does S have?

(b) Let T be the set of all *three-tuples* (a, b, c) with a, b, and c in N_{10}. How many elements does T have?

(c) Let S_{10} be the set of all *permutations* (i.e., orderings) of the elements of N_{10}. How many elements does S_{10} have?

(d) Do any two of the sets N_{10}, S, T, S_{10} have nonempty intersection?

20. Let S be a set with n elements. Let $\mathcal{S}$ be the set of all $(n-1)$-element subsets of S. How many elements does $\mathcal{S}$ have? Why?

21. Let $N_{100} = \{1, 2, \ldots, 100\}$. Let A_{42} and A_{58} be the sets of all 42- and 58-member subsets of N_{100}, respectively. Show that A_{42} and A_{58} have the same number of elements.

22. The xy-plane can be thought of as the Cartesian product $\mathbb{R} \times \mathbb{R}$ (hence the notation $\mathbb{R}^2$). Sketch each of the following subsets of $\mathbb{R} \times \mathbb{R}$.

(a) $\{1, 2, 3\} \times \mathbb{R}$

(b) $\mathbb{R} \times \{1, 2, 3\}$

(c) $\mathbb{Z} \times \mathbb{N}$

(d) $\{(x, y) \mid y = x^2\}$

(e) $\{(x, y) \mid x = \sin y\}$

(f) $\{(x, y) \mid x^2 + y^2 = -1\}$

23. A simple picture can be drawn in a $n \times m$ grid of squares (i.e., "checkerboard") by coloring each square either black or white. Think of the grid as the Cartesian product $G = H \times V$, where $H = \{1, 2, \ldots, n\}$ and $V = \{1, 2, \ldots, m\}$, with lower-left corner $(1, 1)$ and upper-right corner (n, m).

(a) Suppose the set of black squares is $\{(x, y) \mid x + y = m + n + 1\}$. Draw the picture.

(b) Describe the set of black squares in a "checkerboard" picture (with lower-left square black).

(c) What does the element $(2, 3, \text{black})$ in the set $G \times \{\text{black, white}\}$ represent in this context? What does the set $G \times \{\text{black, white}\}$ represent?

(d) A picture in the sense above can be thought of as an element of $P(G)$, the power set of G. Explain.

1.3   Sets 102: The Idea of a Function

In everyday mathematics we might discuss "the tangent function" or "the function $f(x) = x^2$." In a calculus course such language may be fine, but in other settings we'll need a more general description of functions—using the language of sets.

Definition 1.6. Let A and B be nonempty sets. A *function* $f : A \to B$ is a rule for assigning to each "input" element a in A one and only one "output" element, $f(a)$, in B. Here A is the *domain* of f and B is the *codomain*.

In other words. Alternative language is available. A function f is sometimes called a *mapping*, and we say that f *maps* a domain element a to a codomain element b, sometimes called the *image* of a. In symbols, we might write $f : a \mapsto b$, or just $a \mapsto b$ if the function name is understood.

EXAMPLE 1.   How is the "calculus description" $f(x) = x^2$ related to the formal definition?

SOLUTION.   The formula $f(x) = x^2$ makes the *rule* perfectly clear: For inputs -3, $\sqrt{2}$, and 1.2345, the corresponding outputs are $f(-3) = 9$, $f(\sqrt{2}) = 2$, and $f(1.2345) = 1.2345^2$. The *domain* and the *codomain*, on the other hand, are open to choice. In a calculus course we might, if pressed, use the *natural domain*—the set of all real numbers for which the rule makes sense. For $f(x) = x^2$, that's $\mathbb{R}$ itself. (For $g(x) = \tan(x)$, the natural domain omits some real numbers.) In a number theory course, we might use $\mathbb{N}$ as domain.

The codomain, too, is open to choice. For a given domain, we need only assure that the codomain contains all possible outputs. With domain $\mathbb{R}$, for instance, we could use as codomain for $f(x) = x^2$ any set that contains all the nonnegative reals. This might be $\mathbb{R}$ itself, the infinite interval $[0, \infty)$, or something stranger, like $(-42, \infty)$ or $\mathbb{C}$. With domain $\mathbb{N}$, we could reuse $\mathbb{N}$ as codomain, or choose any set of integers that contains all positive perfect squares.   ◊

The moral. The preceding example shows that a function is more than a formula. A function is a 3-part package: a domain A, a codomain B, and a rule (which may or may not be a symbolic formula) for assigning a unique output $b = f(a)$ in B to every input a in A. The notation $f : A \to B$ emphasizes this three-fold nature. In practice the domain or codomain or both are sometimes understood from context, or even ignored, but they're always waiting in the wings.

Range and codomain. The *range* (or *image set*) of a function $f : A \to B$ is the set of all outputs:

$$\text{range of } f = \{f(a) \mid a \in A\}.$$

Note that the range is always a *subset*—perhaps a proper subset—of the codomain. For $f : \mathbb{R} \to \mathbb{R}$ given by $f(x) = x^2$, for instance, the range is the interval $[0, \infty)$, which omits all the negative numbers.

Function examples. A good way to explore functions is through examples and non-examples, including some strange ones with strange names. (We'll usually stick with one-letter function names, although longer, more descriptive names are perfectly legal.)

EXAMPLE 2.   Consider the sets

$$A = \{\text{January, February, March}, \ldots, \text{December}\};$$
$$N_{12} = \{1, 2, 3, \ldots, 12\}; \qquad P = \{\text{Alice, Bob, Carol, Dave}\}.$$

Is each of the following a function?

- MONTHNUMBER : $A \to N_{12}$, where MONTHNUMBER(January) $= 1$, MONTHNUMBER(February) $= 2, \ldots,$ MONTHNUMBER(December) $= 12$.

- WORDLENGTH : $A \to N_{12}$, where WORDLENGTH(a) is the number of letters in a.

- BIRTHMONTH : $P \to A$, where BIRTHMONTH(p) is the month in which p was born.

- $j : P \to N_{12}$, where $j(p) =$ MONTHNUMBER (BIRTHMONTH(p)).

- SQUARE : $N_{12} \to N_{12}$, where SQUARE$(n) = n^2$.

SOLUTION.   All of MONTHNUMBER, WORDLENGTH, BIRTHMONTH, and j are functions. In each case, every member of the domain is paired with one and only one member of the codomain. Some of the rules are admittedly peculiar. To find BIRTHMONTH(Alice) and j(Alice), for instance, we would need more information. Still, the rules make sense because everyone has a unique birth month. By contrast, SQUARE has a problem: Squaring members of N_{12} produces results outside N_{12}. ◇

We could "fix" this problem by choosing a larger codomain, such as $\mathbb{N}$.

Seeing Functions

Graphs. Graphs are deservedly popular in elementary calculus. Properties like smoothness, steepness, rising vs. falling, concavity, and existence of asymptotes reflect and reveal a lot about functions.

A less familiar fact is that, like functions, graphs can also be described in the language of sets. The graph of a calculus-style function, say, $f(x) = x^2$, is a curve in the xy-plane, made up of points of the form $(x, f(x))$—in this case, (x, x^2). But the idea of a graph makes sense for *any* function:

Definition 1.7. Let $f : A \to B$ be a function. The *graph* of f is the set

$$G = \{(a, b) \mid a \in A \text{ and } b = f(a)\} .$$

Observe:

- *A graph is a set:* The graph of f is a certain *set of ordered pairs*, and thus a *subset* of the Cartesian product $A \times B$.

- *But not just any set:* For f to be a function with domain A, its graph G must contain *one and only one* point (a, b) for each $a \in A$. The graph of a function $f : \mathbb{R} \to \mathbb{R}$, for instance, must contain exactly one point of the form $(3, y)$.

  In pre-calculus lingo, this is the *vertical line test*.

- *Maybe a curve, maybe not:* Graphs of calculus-style functions are often nice curves. Indeed, a lot of beginning calculus is about connecting geometric properties of curves to analytic properties of functions. But some functions have graphs that are nothing like curves. The graph of j in Example 2, for instance, has points of the form (Carol, February)!

- *Other "graphs" out there:* Like other useful math words, "graph" has different meanings in different settings. The graph of an *equation*, for example, is the set of points (x, y) for which x and y satisfy the equation. The graph of $x^2 + y^2 = 1$ is a *circle*, and therefore not the graph of any function. In this book, "graph" will refer only to functions.

- *One idea, two views:* A function and its graph are so closely linked—either one completely determines the other—that functions are sometimes *defined*, rather than just visualized, as sets of ordered pairs. We'll use both of these viewpoints freely.

  From this perspective, functions are sets in their own right.

Other views. For some functions, geometric graphs make little sense, so we use other descriptive devices—tables, diagrams, etc. To describe the function BIRTHMONTH in Example 2, for instance, we could use a *table*:

The BIRTHMONTH function

input (person)	Alice	Bob	Carol	Dave
output (month)	May	December	December	August

We could also use a *diagram* to describe BIRTHMONTH, showing the domain, the codomain, and arrows connecting inputs p to their corresponding outputs $h(p)$. (In such a view, the *range* is the set of "arrowheads," where incoming arrows "land.")

Yet another, even less formal notation is sometimes useful. To describe the MONTHNUMBER function, we could just write

$$\text{January} \mapsto 1, \quad \text{February} \mapsto 2, \quad \text{March} \mapsto 3, \quad \dots, \quad \text{December} \mapsto 12$$

to convey the idea. (The $\mapsto$ symbol is usually read as "maps to.")

All Kinds of Functions

Functions come in many varieties. We will now define and illustrate, mainly using "toy" functions, some properties a given function may or may not have.

Definition 1.8 (One-to-one functions). A function $f : A \to B$ is *one-to-one* (or *injective*) if $f(a_1) \neq f(a_2)$ whenever a_1 and a_2 are in A and $a_1 \neq a_2$. In equivalent *words*: f maps different inputs to different outputs. In equivalent *symbols*:

$$f(a_1) = f(a_2) \quad \Longrightarrow \quad a_1 = a_2.$$

The next two examples will help illustrate the meaning of—and how to prove or disprove—injectivity.

EXAMPLE 3.   The BIRTHMONTH function from Example 2 is *not* one-to-one, because BIRTHMONTH(Bob) = BIRTHMONTH(Carol) = December, even though Bob $\neq$ Carol. By contrast, MONTHNUMBER, which maps each month to its number in the year, *is* one-to-one. The "inclusion" function $i : \mathbb{Q} \to \mathbb{R}$ with rule $i(x) = x$ is one-to-one, since outputs *are* inputs. The constant function $c : \mathbb{R} \to \mathbb{R}$ with rule $c(x) = 3$ is at the opposite extreme—*all* inputs are mapped to the same output.

The linear function $L : \mathbb{R} \to \mathbb{R}$ given by $L(x) = 3x + 5$ is also one-to-one. To prove this, let's check the equation in Definition 1.8, using some algebra:

$$L(x_1) = L(x_2) \quad \Longrightarrow \quad 3x_1 + 5 = 3x_2 + 5 \quad \Longrightarrow \quad 3x_1 = 3x_2$$
$$\Longrightarrow \quad x_1 = x_2,$$

as desired. $\Diamond$

Margin notes:

Draw your own diagram for BIRTHMONTH.

Convince yourself of the equivalence.

Finding even one pair of inputs that give the same output is enough.

Definition 1.9 (Onto functions). A function $f : A \to B$ is *onto* (or *surjective*) if, for every $b \in B$, there is some $a \in A$ with $f(a) = b$. In equivalent *words*: Every element of the codomain is also in the range. In equivalent *symbols*:

$$\{f(a) \mid a \in A\} = B.$$

Using "onto" as an adjective sounds ugly, but everybody does it.

EXAMPLE 4.   The function BIRTHMONTH from Example 2 is *not* onto, because no member of P was born in (say) March. The function MONTHNUMBER : $A \to N_{12}$ *is* onto, because the 12 months range in number from 1 to 12. The "inclusion" $i : \mathbb{Q} \to \mathbb{R}$ with rule $i(x) = x$ is not onto, because the codomain includes irrationals, but the range does not.

Finding even one such codomain member is enough.

Is the quadratic function $q : \mathbb{R} \to \mathbb{R}$ given by $q(x) = x^2 - 6x$ surjective? The short answer is no: A little calculus or algebra shows that $q(3) = -9$ is the minimum value, so the range of q is the interval $[-9, \infty)$, a smaller set than the codomain. We could *make* q surjective by using $[-9, \infty)$, not $\mathbb{R}$, as the codomain. $\Diamond$

Bijective functions. A function that is *both* injective and surjective is called *bijective,* or, equivalently, a *one-to-one correspondence.* Our MONTHNUMBER function is one bijection. Another is suggested by a few values:

$$1 \mapsto \text{a}, \quad 2 \mapsto \text{b}, \quad 3 \mapsto \text{c}, \quad \ldots, \quad 25 \mapsto \text{y}, \quad 26 \mapsto \text{z}.$$

What are the domain and codomain? What's a good name for this function?

Here are two bijections from calculus, this time in a matched pair.

$$f : \left(-\frac{\pi}{2}, \frac{\pi}{2}\right) \to \mathbb{R}, \quad \text{with rule} \quad f(x) = \tan(x)$$

$$g : \mathbb{R} \to \left(-\frac{\pi}{2}, \frac{\pi}{2}\right), \quad \text{with rule} \quad g(x) = \tan^{-1}(x).$$

You can readily convince yourself, perhaps with graphs, that f and g are indeed one-to-one and onto. But notice a little surprise: f and g are one-to-one correspondences between an interval of finite length and the entire real line. Such strange behavior is possible when *infinite* sets are involved; we will explore infinite sets further in a later section.

New Functions from Old

Elementary calculus courses are full of complicated functions, like

$$f(x) = \sin x + (1 + e^x) \sin (x + e^x),$$

concocted in various ways from simpler ingredients. Tools like the product rule, the chain rule, and u-substitution help us find derivatives and antiderivatives of such built-up functions as appropriate combinations of simpler derivatives and antiderivatives.

Composition. Among these familiar ways to combine functions, *composition* is not quite like the others, requiring some special care with domains and codomains. If $f : A \to B$ and $g : B \to C$ are functions, then we can *compose* f and g to form the new function

$$g \circ f : A \to C, \quad \text{with rule} \quad g \circ f(a) = g(f(a)).$$

Recall, especially, that *order matters*: The notation $g \circ f$ means that g *follows* f. The two compositions $g \circ f$ and $f \circ g$ are seldom equal—even if both make good sense.[†]

This makes sense with respect to nested parentheses—but it goes against the usual grain of reading from left to right.

EXAMPLE 5.   Which compositions make sense for the functions

$$\text{MONTHNUMBER} : A \to N_{12},$$
$$\text{WORDLENGTH} : A \to N_{12}, \quad \text{and}$$
$$\text{BIRTHMONTH} : P \to A$$

from Example 2?

SOLUTION.   The compositions

$$\text{WORDLENGTH} \circ \text{BIRTHMONTH} : P \to N_{12}$$

and

$$\text{MONTHNUMBER} \circ \text{BIRTHMONTH} : P \to N_{12}$$

make sense (the former is the function we called j in Example 2). For instance,

$$\text{WORDLENGTH} \circ \text{BIRTHMONTH}(\text{Carol}) = \text{WORDLENGTH}(\text{December}) = 8$$

and

$$\text{MONTHNUMBER} \circ \text{BIRTHMONTH}(\text{Carol}) = \text{MONTHNUMBER}(\text{December}) = 12.$$

Another possibility, $\text{SQUARE} \circ \text{MONTHNUMBER} : A \to N_{12}$, fails right out of the box, because $\text{SQUARE} : N_{12} \to N_{12}$ isn't a function at all. But there is an easy fix: If we use the almost-identical function

We explained why in Example 2.

$$\text{SQUARE} : \mathbb{N} \to \mathbb{N}, \quad \text{with rule} \quad \text{SQUARE}(n) = n^2,$$

[†] A minor technical point: Above, to avoid extra notation, we used the same symbol, B, both for the codomain of f and for the domain of g. In fact, these sets need not be identical. What really matters is that the composition rule $g \circ f(a) = g(f(a))$ make good sense. This occurs as long as the *range* of f is contained in the domain of g. As we have seen, a function's codomain is often open to some choice, so this subtlety seldom causes difficulty.

then the composition SQUARE ∘ MONTHNUMBER $: A \to N$ makes good sense, and (say)

$$\text{SQUARE} \circ \text{MONTHNUMBER}(\text{March}) = \text{SQUARE}(5) = 25.$$

Many other compositions, like

$$\text{BIRTHMONTH} \circ \text{BIRTHMONTH} \quad \text{and} \quad \text{BIRTHMONTH} \circ \text{WORDLENGTH},$$

make no sense.   ◇

Properties of composite functions. If both f and g are, say, injective, must $g \circ f$ be injective too? We answer several such questions in the following theorem, leaving some proofs to the exercises.

Theorem 1.10. *Let $f : A \to B$ and $g : B \to C$ be functions.*

(i) If both f and g are one-to-one, then so is $g \circ f$.

(ii) If both f and g are onto, then so is $g \circ f$.

(iii) If both f and g are bijective, then so is $g \circ f$.

(iv) If $g \circ f$ is one-to-one, then so is f.

(v) If $g \circ f$ is onto, then so is g.

Notes on proofs. We'll prove (ii) and (iv), leaving (i) and (v) to the exercises. Statement (iii) just combines (i) and (ii), so there is nothing new to prove.

To prove (ii) we need to show that for any $c \in C$ there is some $a \in A$ with $g \circ f(a) = c$. We can do this directly. For given $c \in C$, we know (because g is onto) there exists $b \in B$ with $g(b) = c$. Because f is onto there exists $a \in A$ with $f(a) = b$, and this a does the job: $g \circ f(a) = g(f(a)) = g(b) = c$, as desired.

To prove (iv) we'll show that if f is *not* one-to-one, then $g \circ f$ cannot be one-to-one either. (In math-speak, this is an *indirect proof;* more details on such things in later sections.) Suppose, then, that $a_1 \neq a_2$, but $f(a_1) = f(a_2)$. Then we'd have $g(f(a_1)) = g(f(a_2))$, which is just another way of saying that $g \circ f(a_1) = g \circ f(a_2)$. Thus $g \circ f$ is not one-to-one, and the proof is done.

Note, finally, that g need not be one-to-one just because $g \circ f$ is. The calculus formula $y = (e^x + 1)^2$ illustrates this. (Let $f(x) = e^x + 1$ and $g(x) = x^2$; further details are left to exercises.)

Inverse Functions

Another important definition:

Definition 1.11 (Inverse functions). Let $f : A \to B$ and $g : B \to A$ be functions. We say f and g are *inverse functions* if both

$$f(g(b)) = b \quad \text{for all } b \in B \quad \text{and} \quad g(f(a)) = a \quad \text{for all } a \in A.$$

In this case we can write $f = g^{-1}$ or $g = f^{-1}$.

Recall, first, that "inverse" has several meanings in mathematics. For instance, the numbers 3 and -3 are *additive inverses* because $3 + (-3) = 0$. Similarly, 3 and $1/3$ are *multiplicative inverses* because $3 \times 1/3 = 1$, the *multiplicative identity*. In a similar spirit, two functions are inverses if *composing* them (rather than adding or multiplying) produces *identity functions*:

$$f \circ g(b) = b \quad \text{for all } b \in B; \qquad g \circ f(a) = a \quad \text{for all } a \in A$$

The function MONTHNUMBER $: A \to N_{12}$ of Example 2 has a natural *inverse function*—let's call it NUMBERMONTH $: N_{12} \to A$, with values as follows:

MONTHNUMBER(January) $= 1$;   NUMBERMONTH$(1) =$ January; $\ldots$
MONTHNUMBER(December) $= 12$;   NUMBERMONTH$(12) =$ December.

By contrast, WORDLENGTH $: A \to N_{12}$, which maps a month to its number of letters, has *no inverse* function LENGTHWORD $: N_{12} \to A$. One problem is that

$$\text{WORDLENGTH(June)} = 4 = \text{WORDLENGTH(July)},$$

so there's no clear choice for LENGTHWORD(4). Another problem is that *no* month has one letter, so there's no sensible value for LENGTHWORD(1).

These toy examples suggest an important general fact: a function $f : A \to B$ has an inverse $g : B \to A$ if—but only if—f is *both one-to-one and onto*. In this case, f and g simply "reverse" each other's work: if f maps a to b, then g maps b to a, and vice versa.

Relations

A function and its graph are essentially the same thing.

Let A be a nonempty set, and $f : A \to A$ any function on A. We can think of f as its graph $G_f = \{(a, f(a)) \mid a \in A\}$. Note that G_f is a *subset* of $A \times A$—but a subset with special properties that reflect the fact that f is a function. For instance, G_f cannot contain two different elements of the form (a_0, a_1) and (a_0, a_2).

Definition 1.12 (Relation on a set). A *relation* R on a set A is *any* subset of $A \times A$. If $(x, y) \in R$, we write $x \, R \, y$, and we say that x is *related to* y.

This may sound forbiddingly abstract, but familiar (if lightly disguised) examples are all around us.

- *Functions are relations:* The graph of any function $f : \mathbb{N} \to \mathbb{N}$ is a subset of $\mathbb{N} \times \mathbb{N}$, so f is also a relation, and we could write $x\,f\,y$ instead of $y = f(x)$.

- *Extreme examples:* The extreme (but uninteresting) cases are $R = N \times N$ and $R = \emptyset$. In the first case, *all* pairs of integers are related; in the latter, no integer has any relatives, even itself.

- *Equality:* The set

$$\text{EQUALS} = \{(n, n) \mid n \in \mathbb{N}\}$$

corresponds to the *equality relation:* each integer is related (i.e., equal) only to itself. In this case, of course, we usually write $n = n$, not $n\,\text{EQUALS}\,n$.

- *Order relations:* The set

$$\text{LESSTHAN} = \{(m, n) \mid m < n\}$$

is an *order relation:* each integer is related to every integer larger than itself. Again we have a handy symbol, and we write $3 < 4$ instead of the more cumbersome $3\,\text{LESSTHAN}\,4$. (Similar relations correspond to other ordering symbols: $\le$, $>$, and $\ge$.)

- *Divides:* The set

$$\text{DIVIDES} = \{(m, n) \mid m \text{ divides } n \text{ evenly}\}$$

defines the "divides" relation, in which we find $3\,\text{DIVIDES}\,123456$, $42\,\text{DIVIDES}\,42$, and $1\,\text{DIVIDES}\,n$ for all n. Number theorists use the handy symbol $m \mid n$ rather than $m\,\text{DIVIDES}\,n$.

Equivalence relations. The definition of a relation on A—*any* subset of $A \times A$—is very, very general. In practice, many useful relations have some special properties, with impressive names. An *equivalence relation R* on A is one that is

(i) *reflexive*: $a\,R\,a$ for all $a \in A$;

(ii) *symmetric*: if $a\,R\,b$, then $b\,R\,a$;

(iii) *transitive*: if $a\,R\,b$ and $b\,R\,c$, then $a\,R\,c$.

Equality (on any set) is the prototype and simplest—but not the only—example of an equivalence relation. Consider, for example, the SAMEBLOODTYPE relation on the set of humans, where x SAMEBLOODTYPE y means (of course) that x and y have the same blood type. It is easy to see that SAMEBLOODTYPE is indeed an equivalence relation, which sorts people into four "families" (called *equivalence classes*) based on their blood types: A, B, AB, or O. More examples are in the exercises and later in this book.

Exercises

1. Each part following gives the rule (implicit in the name) for a possible function. For each rule, find a reasonable domain A and codomain B to create a function. (Try to make A relatively large and then make B relatively small.) Is each function one-to-one? Onto?

    (a) MOTHER : $A \to B$
    (b) FIRSTBORNSON : $A \to B$
    (c) EYECOLOR : $A \to B$
    (d) BIRTHDAY : $A \to B$

2. As in the preceding problem, find reasonable domains A and codomains B for the given rules. Is each resulting function one-to-one? Onto?

    (a) BLOODTYPE : $A \to B$
    (b) NUMBEROFPAGES : $A \to B$
    (c) CUBEROOT : $A \to B$
    (d) FACTORIAL : $A \to B$

3. Find the natural domain for each of the following calculus-style functions.

    (a) $f(x) = \sin(x + e^x)$
    (b) $g(x) = \tan(x)$
    (c) $h(x) = \sqrt{x^2 - 1}$

4. Find the natural domain for each of the following calculus-style functions.

    (a) $f(x) = \sqrt{1 - e^x}$
    (b) $g(x) = \sqrt{x^2 + \pi x + 1}$ (give a decimal approximation)
    (c) $h(x) = \ln(e^x)$ (give a decimal approximation)

5. In each part, find an appropriate quadratic formula $q(x) = ax^2 + bx + c$ for the given function. (More than one answer may be correct.)

(a) $f(x) > 0$ for all $x \in \mathbb{R}$

(b) $f : \mathbb{R} \to [4, \infty)$ is onto.

(c) $f : (-5, \infty) \to \mathbb{R}$ is one-to-one. (Note the domain!)

6. In each part, find an appropriate quadratic formula $q(x) = ax^2 + bx + c$ for the given function. (It's allowed for one or more of a, b, c to be zero.)

(a) $f : \mathbb{R} \to \mathbb{R}$ has range $[42, \infty)$

(b) $f(0) = -1$, $f(1) = 1$, and $f(2) = 3$.

(c) $f(0) = -1$, $f(1) = 1$, and $f(2) = 5$.

(d) $f(x) = 3$ has one solution, $f(x) = 2$ has two solutions, and $f(x) = 4$ has no solutions.

7. Imagine a calculus-style function $f : [0, 1] \to [0, 1]$ and its graph.

(a) How can you tell from its graph whether f is one-to-one?

(b) How can you tell from its graph whether f is onto?

(c) Find formulas for three different calculus-style functions $f : [0, 1] \to [0, 1]$ that are one-to-one and onto, and such that $f(0) = 1$.

(d) Define $f : [0, 1] \to [0, 1]$ by setting $f(x) = x$ for $x \in \mathbb{Q}$ and $f(x) = x^2$ for $x \notin \mathbb{Q}$. Is f one-to-one? Onto?

8. In each case following, is the given function one-to-one? Is it onto? Sketch or describe its graph.

(a) $f : \mathbb{R} \to \mathbb{R}$, where $f(x) = x^2 + x$

(b) $g : [0, \infty) \to [0, \infty)$, where $g(x) = x^2 + x$

(c) $h : \mathbb{R} \to \mathbb{R}$, where $h(x) = x + e^x$

(d) $k :$ U.S. citizens $\to \mathbb{N}$, where $k(x) = $ last 4 digits of x's Social Security Number.

9. Consider the functions f and g with $f(x) = \ln x$ and $g(x) = e^x$. Find reasonable domains and codomains for f and g. Sketch their graphs. In what sense are f and g inverses?

10. Consider the function WORDLENGTH $: A \to N_{12}$, discussed in Example 2, page 20, and let G be its graph.

(a) Describe the set G. (Hint: G is a set of ordered pairs; for instance, (May, 3) $\in G$.)

(b) The WORDLENGTH function is *not* one-to-one. How does the graph G reveal this?

(c) The WORDLENGTH function is also not onto. How does the graph G reveal this?

11. A certain function f has graph $G = \{(a, 1), (e, 2), (i, 3), (o, 4), (u, 5)\}$.

   (a) What are the domain and the (smallest possible) codomain of f?

   (b) The function f is bijective. How can we tell this from the graph?

   (c) Because f is bijective there is an inverse function, f^{-1}. What is the graph of f^{-1}?

   (d) Can you think of more descriptive names than f and f^{-1} for these functions?

12. A certain function f has graph

$$G = \{ (\text{Montana}, 3), (\text{Maine}, 4), (\text{Maryland}, 10), (\text{Massachusetts}, 12),$$
$$(\text{Michigan}, 17), (\text{Minnesota}, 10), (\text{Mississippi}, 6), (\text{Missouri}, 11)\}.$$

   (a) What is the domain of f? What is the range? What is a reasonable codomain?

   (b) Is f injective? How do you know?

   (c) Is f surjective? (The answer depends on your choice of codomain.)

   (d) Can you think of a more descriptive name than f for this function?

13. Consider the *linear* function $L : \mathbb{R} \to \mathbb{R}$ given by $L(x) = 3x + 2$.

   (a) Explain why L is both one-to-one and onto.

   (b) Let $M : \mathbb{R} \to \mathbb{R}$ be given by $M(x) = \frac{x-2}{3}$. Show that $L \circ M(x) = M \circ L(x) = x$ for all $x \in \mathbb{R}$.

14. Consider the *linear* function $L : \mathbb{R} \to \mathbb{R}$ given by $L(x) = ax + b$, where a and b are any real numbers, with $a \neq 0$.

   (a) Explain why L is both one-to-one and onto.

   (b) Find a formula for an inverse function $M = L^{-1}$. Check your answer by showing that $L \circ M(x) = M \circ L(x) = x$ for all $x \in \mathbb{R}$.

15. (a) Consider the function $f : \mathbb{R} \to \mathbb{R}$ given by $f(x) = x^n$. For which positive integers n is f one-to-one and onto?

   (b) Consider the function $f : (0, \infty) \to (0, \infty)$ given by $f(x) = x^n$. For which positive integers n is f one-to-one? What if n is negative?

16. Let $f : A \to B$ and $g : B \to C$ be functions.

(a) Show that if both f and g are one-to-one, then so is $g \circ f$. (This is (i) of Theorem 1.10, page 25.)

(b) Show that if $g \circ f$ is onto, then so is g. (This is (v) of Theorem 1.10, page 25.)

(c) Consider the functions $f(x) = e^x + 1$ and $g(x) = x^2$, both with domain $\mathbb{R}$ and codomain $\mathbb{R}$. Show that $g \circ f$ is one-to-one, but g is not.

17. Consider $\ell : \mathbb{N} \to \mathbb{Z}$ given by $\ell(n) = \begin{cases} \frac{n}{2} & \text{if } n \text{ is even} \\ \frac{1-n}{2} & \text{if } n \text{ is odd} \end{cases}$. Is ℓ one-to-one? Onto? Does this seem strange given the "sizes" of $\mathbb{N}$ and $\mathbb{Z}$?

18. Consider the functions $f(x) = x^2$ and $g(x) = \sqrt{x}$, both with $[0, \infty)$ as domain and as codomain.

(a) Explain why f and g are inverse functions.

(b) Consider the function $h : \mathbb{R} \to [0, \infty)$ with rule $h(x) = x^2$ (formed from f by tinkering with the domain). Explain why h and g are *not* inverse functions.

19. The ordinary sine and arcsine functions are inverses as long as some care is taken with domains and codomains. Work out the details—that is, specify domains and codomains for these functions that make them inverses.

20. Explain why the SAMEBLOODTYPE relation, discussed at the end of this section, is reflexive, symmetric, and transitive.

21. Consider the relation MOD5 on $\mathbb{Z}$, defined by MOD5 $= \{(m, n) \mid 5 \text{ divides } m - n\}$.

(a) Show that MOD5 is an equivalence relation.

(b) Find the set, denoted by $[0]$, of all integers that are related (by Mod5, of course) to 0. Do the same for $[1]$, $[2]$, $[3]$, and $[4]$. (These sets are called the *equivalence classes* for MOD5.)

22. Let P be the set of (around 400,000) citizens of Minneapolis, Minnesota. One possible relation on P is BROTHER, defined by BROTHER $= \{(x, y) \mid y \text{ is } x\text{'s brother}\}$. Notice that BROTHER is transitive, but neither reflexive (nobody is his own brother) nor symmetric (think about sisters). Notice too that BROTHER is *not* a function, because some Minneapolitans have *no* Minneapolitan brothers, while others have more than one.

In each part following, decide whether the relation is (i) reflexive; (ii) symmetric; (iii) transitive; (iv) a function. Explain answers briefly.

(a) SAMELASTNAME

(b) ANDERSON $= \{(x, y) \mid x \text{ and } y \text{ are Andersons}\}$

(c) SIBLING

(d) MAYOR $= \{(x, y) \mid y \text{ is } x\text{'s mayor}\}$

(e) OLDESTNEIGHBOR

1.4　Proofs 101: Proofs and Proof-Writing

Without proofs, mathematics would not exist. Proof is as fundamental to working mathematicians as structured experiments are to chemists or biologists.

See also the coda at the end of this section.

This is not to say that mathematics is nothing *but* proof. Calculations, hunches, experiments, and searches for patterns are all indispensable in mathematics, as they are in the sciences. But mathematical certainty, obtained through rigorous proof, has no clear analogue in the other sciences, where theories are *always* open to revision or falsification through experiments. The Pythagorean theorem, by contrast, has been *known* to be true for at least 2500 years.

Pythagoras recognized his proof as a big deal. He celebrated, legend has it, by sacrificing 100 oxen.

A first course in real analysis is a natural setting for learning about, and for practicing, mathematical proof. Many concepts and objects of real analysis—sequences, functions, derivatives, etc.—are familiar, if perhaps only informally, from calculus courses. The main theorems are mathematically powerful and intuitively accessible. Finding and writing proofs takes care and practice, but the reward is proportional to the effort.

Mathematical Language and Literature

> The beginning of wisdom is to call things by their right names.

The old Chinese sage who supposedly said so may have had other things in mind, but the advice certainly applies to mathematics. Without clear and unambiguous language—"right names"—we can't know exactly what we are talking about, and therefore we can't produce really convincing proofs.

The moral is that care with language is just as important in mathematics as it is in, say, the literary arts. Granted, poets and mathematicians use language very differently: good poetry may rely on subtle allusions and shaded meanings, but the best mathematical proofs are always clear, direct, and straightforward, even when they convey difficult ideas. Both poems and proofs can be praised as elegant, but the judgment depends on different standards. The good news is that we mathematicians need not aspire to fancy artistry: clarity and directness of expression are less rarefied arts than practical skills, readily acquired and improved on the job.

Following are some samples, phrased as advice, of characteristics of mathematical writing.

Use standard symbols and notations. Using standard notations, of which we'll encounter many in this book, helps shorten, unclutter, and clarify mathematical discussion—but only if notations are used consistently, and with care. For instance, the notations $(1, 3)$, $[1, 3]$, and $\{1, 3\}$ all have precise but *different* meanings—one is an open interval, one is a closed interval, and one is a set with just two members. Straying from these conventions is asking for trouble. It is far from clear, for instance, what such notations as

$$[x \mid x^2 > 2] \quad \text{and} \quad \mathbb{Q} \mid (-\sqrt{2}, \sqrt{2})$$

really mean. By contrast, the expressions

$$\{x \mid x^2 > 2\} \quad \text{and} \quad \{x \in \mathbb{Q} \mid x^2 < 2\}$$

are clear and unambiguous.

> Or will be, once we have defined all the ingredients.

Define everything. Everyone expects theorems, proofs, and calculations in mathematical writing. Less expected, but equally important, are formal definitions. Words in everyday language have fluid meanings, with nuances that depend on context. By contrast, words in mathematics have fixed, precise, agreed-upon meanings. A typical theorem claims that some mathematical object has some mathematical property. Without clear definitions of the objects and properties under discussion, proof can't get started.

> There are some exceptions, but not many.

EXAMPLE 1.   Prove this claim: *The set of rational numbers is closed under addition.*

DISCUSSION.   The mathematical object in question is the set of rational numbers. The claimed property, closure under addition, means that the *sum* of any two rational numbers is another rational number.

It is easy enough to check that the claim holds for any *particular* pair of rational numbers. For the rationals $3/5$ and $2/7$, for instance, we use a common denominator to calculate the sum, $31/35$, and observe that it is rational, as claimed. A general proof, however, requires a general definition, and to this end we define a rational number as any *ratio* of integers a/b, where $b \neq 0$. While we're at it, let's also sharpen our description of the closure property: a set S is closed under addition if $r + s$ is a member of S whenever r and s are members of S. Now we can restate the claim for our convenience:

> *If x and y are rational numbers, then so is $x + y$.*

With clear definitions at hand we're ready for a concise proof.

PROOF.   Since x and y are rational, we can write $x = a/b$ and $y = c/d$, where a, b, c, and d are integers, and b and d are nonzero. Now we use some fraction arithmetic:

See the common denominator?

$$x + y = \frac{a}{b} + \frac{c}{d} = \frac{ad}{bd} + \frac{bc}{bd} = \frac{ad + bc}{bd},$$

a rational number, and the proof is done.   ◊

Fair game? Few proofs are completely from scratch. Even the simple proof above relies on some very basic properties of *integers*: sums and products of integers are always integers, and the product of *nonzero* integers is always nonzero. (In this book we will freely assume and use such basic properties of integer arithmetic.) Knowing just which assumptions are safe and which need proof can be tricky—especially when assumptions are "understood" rather than stated explicitly. Learning to sort out such matters is part of the craft of proof.

Write in complete "sentences." The quotes are there because mathematical sentences in mathematics may include not just ordinary words but also symbols, equations, inequalities, etc. For instance, it is fine to write

because $x > 3$ we know $x^2 > 9$ and $x^3 > 27$,

but it's bad form to write

$$x > 3, x^2 > 9, x^3 > 27.$$

One problem with the latter is the lack of "connective tissue": a reader can't tell whether you're asserting that something implies something else, or just listing your favorite inequalities. At any cost, be clear.

Use standard grammatical niceties. These devices—such as punctuation and capital letters—are used in ordinary English to help the reader see clearly what's being said or implied. Consider, for instance, a notice supposedly posted near an Australian beach:

Crocodiles don't swim here

Would you swim here or not? Is the sign intended for human or for reptile readers?

In practice, many mathematical sentences convey complex ideas, and so naturally have correspondingly complex structures. It is especially important, therefore, to write mathematics as clearly and unambiguously as possible, and to help the reader decipher your meaning.

Write sentences that "scan." Proofs and solutions must of course be correct, but they must also be intelligible to the intended reader. An excellent way to ensure the latter is to read each sentence back to yourself. (Doing this silently may reduce ridicule from neighbors.) A sentence with proper English grammar and syntax may be mathematically right or wrong; every mathematician has seen eloquent proofs that boil down to nonsense. But an ungrammatical sentence is almost surely wrong or, worse, meaningless to a reader.

Make sense. Ask carefully whether what you write makes sense by the strict standards of mathematical writing. For instance, the sentence

> For all $\mathbb{Q}$, we have $x^2 \geq 0$.

is nonsensical (do you see why?) and hence *neither* true nor false. Your first job is to write sentences that make sense. Your second job is to write sentences that are true.

Watch word order. Because mathematical language conveys complex thoughts, word order can be crucial. Consider these two statements:

> For every positive ϵ there exists an integer n such that $0 < 1/n < \epsilon$.

> There exists an integer n such that $0 < 1/n < \epsilon$ for every positive ϵ.

The first statement is true; it is a version of so-called *Archimedean principle* for real numbers. The second statement is meaningful, but false.

We will study the Archimedean principle in Section 1.6.

Just don't do "it." The harmless-looking pronoun "it" commits countless crimes in mathematics. Here, for example, is a confusing way to describe an important connection between a function f and its derivative f':

> It's maximum or minimum when it's flat, and that happens when it's a zero of its derivative.

That's way too many pronouns, and who knows what each refers to? Just say no:

> A function f may have a maximum or minimum value where the graph of f is flat. This can occur at a value of x for which $f'(x) = 0$.

Make it look easy. A musician planning a recital invests hours of practice and study, and hits plenty of false notes. The recital itself skips all of this practice and study, and most of the false notes. In the same way, a finished mathematical proof should be the polished result, rather than the basic process, of whatever informal thinking, experiments, and false starts may have happened along the way. It is sometimes helpful to hint at the investigative phase of proving a result, but it is important not to confuse such material with the proof itself. Good proofs are clear, concise, and couched in standard mathematical language.

Don't say too much—or too little. Respect, but don't overtax, your reader's intelligence and willingness to work. Ideas in your proof should be clear and accessible to your reader—someone with your own level of intelligence and knowledge.

Proof Language: A Lexicon

Mathematics, like every discipline, has its own body of technical language. Specialized vocabulary ("jargon" to its detractors) is sometimes impugned as willfully obscure, but exactly the opposite is true: well-chosen technical language is essential to clear and concise discussion of technical subjects. To get these benefits, of course, we'll need to know and agree on what technical words mean. The following little lexicon of important "proof words" begins that project. A general warning: Some technical words, such as "statement," have everyday meanings somewhat different from their mathematical uses.

We could, but won't, argue over whether this is for better or for worse.

Statement. A mathematical *statement* is a phrase or sentence that can reasonably be called either true or false. Here are some possibilities:

$$P : 3 > 5 \qquad Q : \text{It is cloudy.} \qquad R : \text{It is raining.} \qquad S : \text{Life is good.}$$

Statements P, Q, and R are all good examples. Statement P is obviously false, but mathematically meaningful. Statements Q and R are also fine because, at a given time and place, each is either clearly true or clearly false. Statement S is too vague to make our cut, and we'll leave the fun of debating it to colleagues across campus.

Complex statements can be built up from simpler ones. Using ingredients above, for instance, we can create blends like these:

$$\text{if } R \text{ then } Q : \text{If it is raining, then it is cloudy.}$$

and

$$Q \text{ and not } R : \text{It is cloudy but not raining.}$$

Variables. Some statements involve *variables:* symbols that stand in for unspecified inputs. For such statements we sometimes use names like $P(x)$ to emphasize the presence of variables. Here are some examples:

$$P(x) : x^2 < 3 \qquad Q(y) : y \text{ is Fred's brother.} \qquad R(x, y) : x^2 + y^2 \geq -3$$

The truth or falsity of such statements usually depends on the values of the variables involved. Here, for instance, $P(7)$ is false, $P(-1)$ is true, $Q(\text{Sue})$ is probably false, and $Q(\text{Ed})$ might be true. In this case $R(x, y)$ happens to be true for *all* real number inputs x and y.

Things might change if we allowed, say, imaginary number inputs.

Implication and equivalence. An implication is a statement of the form "if P then Q"; we often write $P \implies Q$. This means, of course, that Q is true whenever P is. With $P : x > 3$ and $Q : x^2 > 9$, for example,

$$P \implies Q \quad \text{means} \quad x > 3 \implies x^2 > 9$$

and

$$Q \implies P \quad \text{means} \quad x^2 > 9 \implies x > 3.$$

Here, clearly, the implication $P \implies Q$ is true: $x^2 > 9$ does indeed hold whenever $x > 3$. But the implication $Q \implies P$ is false; try $x = -42$. An important moral is that implication is, by default, a *one-way street*: $P \implies Q$ is no guarantee that $Q \implies P$.

If it happens that both $P \implies Q$ and $Q \implies P$, then P and Q always have the same truth values, and are therefore called *equivalent*. The statements $P(x) : 2x + 5 = 11$ and $Q(x) : x = 3$ are equivalent, for instance.

> Solving equations, as in linear algebra, is all about searching for equivalent, but simpler, equations.

And, or. Given statements P and Q, we can form new statements (P and Q) and (P or Q). (The parentheses aren't essential, but they help keep the right things together.) "And" is used mathematically much as it is in everyday speech: (P and Q) is true if, but only if, *both P and Q are true*. "Or" is a little different: We take (P or Q) to be true when *either or both* of P and Q is true. This convention, called the *inclusive or*, differs a bit from the *exclusive or*, sometimes written *xor*, that's common in everyday life: A child might be offered either candy *or* ice cream, but not both.

Negation. The negation of a statement P is a new statement, which we'll denote

$$\text{not } P,$$

whose truth value is *opposite* to that of P. Negating simple statements is easy. With

$$R : \text{It is raining,}$$

for instance, we have simply

$$\text{not } R : \text{It is not raining.}$$

Negating complicated statements can take some thought. Consider, for instance, *Goldbach's conjecture:*

$$G : \text{Every even integer greater than 2 is the sum of two primes.}$$

> This famous unsolved problem dates back to the 1740s, in correspondence between Christian Goldbach and Leonhard Euler.

How would we negate G? We could, admittedly, write something like

$$\text{It is not the case that } G,$$

but this is unhelpful. A better approach is to notice that G asserts that *every* even integer has a certain property. To negate such a claim is to say that *some* even integer does *not* have this property:

not G: Some even integer greater than 2 is not the sum of two primes.

In other words, to *disprove* G it is enough to find even *one* big even number that is *not* the sum of two primes.

In a similar spirit, *De Morgan's laws* tell how to negate statements that involve *and* and *or*:

$$\text{not } (P \text{ and } Q) \quad \text{is equivalent to} \quad (\text{not } P) \text{ or } (\text{not } Q);$$
$$\text{not } (P \text{ or } Q) \quad \text{is equivalent to} \quad (\text{not } P) \text{ and } (\text{not } Q).$$

Notice, especially, that negating *and* statements produces *or* statements, and vice versa.

The meaning of existence. Consider two typical, analysis-style statements about real numbers:

(i) For all $x > 3$, $x^2 + 9 > 6x$.

(ii) $\sin x = x/2$ for some $x \in [1, 2]$.

Statement (i) says that a certain property holds *for all* $x \in (3, \infty)$, while (ii) says that a number with a certain property *exists* in the set $[1, 2]$. The *for all* and *there exists* constructions are everywhere in mathematics, and have their own symbols: $\forall$ and $\exists$, respectively. Using these symbols we could rewrite

(i) $\forall x \in (3, \infty)$, we have $x^2 + 9 > 6x$;

(ii) $\exists x \in [1, 2]$ such that $\sin x = x/2$;

or even just

(i) $\forall x \in (3, \infty) \quad (x^2 + 9 > 6x)$;

(ii) $\exists x \in [1, 2] \quad (\sin x = x/2)$;

if brevity is really important.

The symbol $\forall$ is known as the *universal quantifier;* $\forall x P(x)$ means that the property $P(x)$ holds for *every* x in some "universe" of discussion. Similarly, $\exists$ is the *existential quantifier;* $\exists x\, P(x)$ means that $P(x)$ holds for *at least one* x.

The $\forall$ and $\exists$ symbols arise frequently in analysis, often together. For instance, the statement

$$\forall x \in \mathbb{R}\ \exists n \in \mathbb{N} \text{ such that } x < n$$

And achieve fame.

It's been tried . . . .

We alluded to them in slightly different form in an earlier section.

They happen to be true.

says—truthfully—that every real number x is smaller than some integer. The superficially similar-looking statement

$$\exists x \in \mathbb{R} \text{ such that } \forall n \in \mathbb{N} \ x < n$$

says something different—and false.   *What does it say to you?*

Converse and contrapositive. For some reason, C-words are disproportionately common among crucial components of concise corroborations. Can this be just curious coincidence?

An implication $P \implies Q$ has both a *converse* and a *contrapositive*. The   *Which we dare not confuse ....*
converse of $P \implies Q$ is the implication $Q \implies P$; the *contrapositive* is the
implication $(\text{not } Q) \implies (\text{not } P)$. If, say, $P : x > 3$ and $Q : x^2 > 9$, then we
have

$$P \implies Q : \quad x > 3 \implies x^2 > 9 \qquad \text{(the original)};$$
$$Q \implies P : \quad x^2 > 9 \implies x > 3 \qquad \text{(the converse)};$$
$$(\text{not } Q) \implies (\text{not } P) : \quad x^2 \leq 9 \implies x \leq 3 \qquad \text{(the contrapositive)}.$$

As noted earlier, an implication and its converse *need not* have the same truth value—and they *do not* in the present case. An implication and its contrapositive, on the other hand, always have the *same truth value*. This matters in mathematical practice, as we will see, because the contrapositive of a statement is sometimes easier to prove than the original.

Common sense bears this out; we explore it further in the exercises.

Counterexample. For which integer powers p is $2^p - 1$ prime? Back-of-the-envelope calculations show that this happens if p is 2, 3, 5, or 7; and so, running out of envelope, we might guess that $2^p - 1$ is prime for *all* prime powers p. Not so, alas: $2^{11} - 1 = 2047$, which is *not* prime. In proof talk, 11 is a *counterexample* to our hasty claim.

A counterexample is just what the name implies: a single *example* that *counters* an incorrect general claim. When, as above, the claim asserts that some property $P(x)$ holds for *all* x in a given set, a counterexample is simply *some* x for which $P(x)$ fails. Mathematicians love counterexamples, perhaps for the pleasure they can offer in toppling grand theories with humble facts.

Coda: new-age proofs. Proofs are no less important now than they've ever been. But modern viewpoints and, especially, modern technology have made new kinds of mathematics possible—and sometimes require new kinds of proof. For example, the four-color theorem (four colors suffice to color any planar map of "countries") was posed in 1852, but proved only in 1976, with aid from a computer to check hundreds of special cases. Computer-aided proofs are now common, but they remain controversial.

Exercises

1. Here are some statements or attempted statements about real numbers. Which are true? Which are false? Are any nonsensical? Are any negations of each other?

   P: $\exists n \in \mathbb{N}$ such that $2n^2 - n < 0$

   Q: $\forall n \in \mathbb{N}$ such that $2n^2 - n \geq 0$

   R: $\forall n \in \mathbb{N}, \quad 2n^2 - n \geq 0$

2. Here are some statements or attempted statements about real numbers. Which are true? Which are false? Are any nonsensical? Are any negations of each other?

   S: $\exists n \notin \mathbb{N}$ such that $2n^2 - n < 0$

   T: $\forall n \notin \mathbb{N}, \quad 2n^2 - n \leq 0$

   U: $n \in \mathbb{N} \implies 2n^2 - n \geq 0$

3. Following are several possibly true but poorly-stated claims from elementary calculus. Fix each statement by replacing all instances of "it" and "its" and "it's" with clearer words or phrases. (An example appears on page 35.)

   (a) If it's increasing for all x then its derivative is nonnegative.

   (b) At its maximum either its derivative is zero or it does not exist.

   (c) If its second derivative is positive for all x then it's concave up.

   (d) If a series converges then it goes to zero.

4. Each of the following sentences has one or more syntax errors. In each case, make a clear (and true) sentence with as little editing as possible.

   (a) If $x^2 \neq \mathbb{Q}$, then $x \neq \mathbb{Q}$ either.

   (b) $\sqrt{2}$ is not $\mathbb{Q}$ but $\sqrt{2}^2$ is $\mathbb{Q}$.

   (c) $\sqrt{\mathbb{N}} \in \mathbb{R}$.

   (d) $\sin(\mathbb{R}) \subseteq [-1, 1]$.

5. Decide whether each of the following is true or false. Explain answers as concisely as possible.

   (a) $\exists n \in \mathbb{N}$ such that $\forall x \in \mathbb{R}, x < n$

   (b) $\exists x \in \mathbb{R}$ such that $\forall n \in \mathbb{N}, x < n$

6. Decide whether each of the following is true or false. Explain answers as concisely as possible.

   (a) $\forall x > 0 \; \exists y > 0$ such that $xy = 1$

   (b) $\exists x > 0$ such that $\forall y > 0, \; \dfrac{2}{y^2} > x$

7. This problem is about three statements:

$$R : \text{It is raining.} \quad S : \text{It is sunny.} \quad C : \text{It is cloudy.}$$

In each part following, rewrite the given implication in ordinary language. Then do the same for the converse and the contrapositive. Label all statements as true or false.

   (a) $R \implies \text{not } S$

   (b) $R \implies C$

   (c) $R \implies (\text{not } S) \text{ and } C$

   (d) $C \implies \text{not } S$

8. This problem is about three statements:

$$P : x > 3; \quad Q : x < 4; \quad R : 9 < x^2 < 16.$$

In each part following, rewrite the given implication as simply as possible. Then do the same for the converse and the contrapositive. Label all statements as true or false; no proofs needed.

   (a) $P \implies R$

   (b) $Q \implies R$

   (c) $(P \text{ and } Q) \implies R$

9. *Negations and counterexamples.* Following are some statements. For each one, first write the negation. Then try to decide (no proofs needed) whether each is true or false. If possible and appropriate, give a counterexample.

   R: $\forall n \in \mathbb{N}, \sqrt{n^2 + 6} \in \mathbb{N}$

   S: $\exists n \in \mathbb{N}$ such that $\sqrt{n^2 + 6} \in \mathbb{N}$

   T: $\forall x \in [0, 1], \cos(x) > 0.6$

   U: Every even integer n with $5 < n < 31$ is the sum of two primes.

10. *Converses and contrapositives.* In each part, write (as simply as possible) both the converse and the contrapositive of the given implication. No proofs needed, but try to label each statement as true or false. In all parts, a, b, a_n, etc. all stand for real numbers.

   (a) If a and b are both rational, then $a + b$ is rational.

   (b) If a is irrational then $1/a$ is irrational, too.

   (c) If a and b are both irrational, then ab is irrational.

   (d) If a series $\sum a_n$ converges, then $\lim_{n \to \infty} a_n = 0$.

11. *Negations.* In each part, write (as simply as possible) the negation of the given statement.

   (a) At least one of a, b, and c is nonnegative.

   (b) $f(x) \leq 3$ for all $x \in [2, 7]$.

   (c) $\sin n$ is irrational for every positive integer n.

   (d) $\exists x \in \mathbb{R}$ such that $x^2 = -1$.

   (e) $\forall x \in \mathbb{R} \quad \exists y \in \mathbb{R}$ such that $xy = 1$.

12. *Negations.* For each statement following, write the *negation* as simply as possible. Then try to decide (some parts may be hard) whether each statement is true or false; no proofs needed.

   P: All dogs have fleas.

   Q: Several dogs have fleas.

   R: My dog, Spike, is flea-free.

   S: $3 < 5$.

   T: $\sqrt{n-1} + \sqrt{n+1}$ is irrational for every positive integer n.

   U: $\sqrt{n-1} + \sqrt{n+1}$ is rational for every positive integer n.

   V: $\cos(x) < 1.001$ for all $x > 0$.

   W: $\cos(x) < 0$ for all $x \in [2, 3]$.

   X: Every even integer $n > 2$ is the sum of two (not necessarily distinct) primes.

   Y: Every set of ten distinct numbers has a largest and a smallest element.

   Z: Every subset of the interval $[0, 1]$ contains a largest element.

13. *Counterexamples.* At least three of the statements P–Z in Problem 12 are false. Choose any three false statements. For each one, carefully state a *counterexample* that shows the statement to be false.

14. *Converses and contrapositives.* In each part following is an *implication*: a statement of the form "If P, then Q". In each part, first write the converse and the contrapositive of the given implication. Then label each of the three statements as true or false; no need to prove anything.

   (a) If $x > 3$ then $x^2 > 9$.

   (b) If $x > 3$ then $x^3 - 4x^2 + 3x > 0$.

   (c) If $a > 0$ and $b > 0$ then $|a + b| = |a| + |b|$.

15. What is the contrapositive of the contrapositive of an implication? Explain.

1.5   Types of Proof

Mathematical assertions come in enormous variety. Here is a tiny sample, including at least one that is false:

   I: If a and b are rational numbers, then so is ab.

   II: If a and b are irrational, then so is $a + b$.

   III: If a is irrational, then so is $1/a$.

   IV: $\sqrt{5}$ is irrational.

   V: There are infinitely many prime numbers.

   VI: Every even integer from 4 to 100 is the sum of two primes.

   VII: Every even integer greater than 2 is the sum of two primes.

   VIII: A finite set with n elements has 2^n subsets.

Proofs (and disproofs, when needed) come in similar variety; no simple taxonomy is possible. Still, a few standard types of argument are so common that they appear in every mathematician's toolkit. We name and illustrate several of them in this section.

Hypotheses and conclusions. A typical mathematical assertion (theorem, claim, proposition, lemma, corollary, etc.) starts with one or more *hypotheses* and derives a *conclusion*. This structure is often—but not always—explicit in the language itself. All of I–III above, for instance, have the general form of an implication: if [hypothesis], then [conclusion].

   For other assertions, including IV–VIII above, the hypotheses are implicit or understood rather than stated explicitly. If we prefer to state hypotheses more explicitly, we could rewrite Claim IV, for instance, as

IV': If $x^2 = 2$, then x is irrational.

Claim V is harder to restate in *if–then* form because its hypotheses, such as the definition of a prime number, would be tedious to list.

The form of a claim often suggests possible strategies for proving or refuting it. In the case of an *if P then Q* claim, for instance, we might attack either the claim itself or its contrapositive, *if not Q then not P*. For broad claims like VII and VIII, each of which covers infinitely many cases, we should expect to work harder for a proof—or maybe look for even one counterexample as a disproof.

Direct Proof

A *direct proof* addresses a claim *if P then Q* in the "obvious" way—it starts with P and derives Q. We illustrate with Claim I.

Remember, we're assuming integer basics.

EXAMPLE 1.   Prove directly (using basic properties of *integer* arithmetic) that if a and b are rational numbers, then ab is rational.

SOLUTION.   Because a and b are rational, we can write

$$a = \frac{x}{y} \quad \text{and} \quad b = \frac{z}{w},$$

where x, y, z, w are all integers, and $y \neq 0 \neq w$. Now we multiply:

$$ab = \frac{x}{y} \times \frac{z}{w} = \frac{xz}{yw},$$

which exhibits ab as the desired ratio of integers, and completes the proof.

Note, by the way, that with a few more words we could also establish that $ab = ba$. (We did something similar with *sums* in Example 2, page 4.)   ◊

Indirect Proof

An *indirect proof*, like a direct one, addresses a claim of the form *if P, then Q*. But there is a twist: instead of showing that P implies Q, we prove the *contrapositive* (but equivalent!) statement:

$$\text{if (not } Q\text{) then (not } P\text{)}.$$

An indirect proof may be the simplest choice when either P or Q is awkward, but (not P) or (not Q) is simpler. We illustrate with Claim III.

EXAMPLE 2.   Prove indirectly: if a is irrational, then $1/a$ is irrational too.

SOLUTION. A direct argument seems awkward here because the hypothesis says something negative about a. The contrapositive is simple and straightforward: if $1/a$ is rational, then a is rational.

This is easily shown. If $1/a$ is rational, then we can write $1/a = x/y$ for some (nonzero) integers x and y. But then we have $a = y/x$, which shows that a is rational, as desired. ◊

EXAMPLE 3. Discuss Claim II: If a and b are irrational, then so is $a + b$.

SOLUTION. The direct approach looks unpromising, as it did in Example 2. An indirect proof might seem the next resort. But there is a catch: Claim II is *false*, and *any* counterexample will demolish it. If, say, $a = \sqrt{2}$ and $b = -\sqrt{2}$, then both a and b are irrational, but $a + b = 0$, which is rational, so our counterexample is good.

Our counterexample depends, of course, on the fact that $\sqrt{2}$ is irrational, as we showed in Theorem 1.2, page 6. Observe, however, that *any* irrational number would do in place of $\sqrt{2}$. ◊

Proof by Contradiction

Proofs *by contradiction* are close kin to indirect proofs. In each case we first assume that the conclusion *fails*, and then try to deduce a contradiction, either of the hypothesis or of some other known fact. From the resulting absurdity we infer that the original conclusion must have been true all along. (The method is also known as *reductio ad absurdum*, Latin for "reduction to the absurd.")

EXAMPLE 4. Prove Claim IV: $\sqrt{5}$ is irrational.

SOLUTION. We could tweak the proof of Theorem 1.2, page 6, but for variety we'll take an approach based on *prime factorization*: Every positive integer n has a unique list of prime factors, some of which may be repeated. (For $n = 60$ the list is $\{2, 2, 3, 5\}$.) The key insight for our proof is about squaring: each prime factor of n appears *twice as often* among the prime factors of n^2. (For $n^2 = 60^2$, the prime factors are $\{2, 2, 2, 2, 3, 3, 5, 5\}$.) In particular, every square integer has an *even* number of prime factors. So much said, we're ready for a crisp proof.

PROOF. Assume, toward contradiction, that $\sqrt{5}$ is rational. Then we can write $\sqrt{5} = a/b$ for some positive integers a and b, and so

It is traditional, and helpful, to label contradiction proofs up front.

$$\sqrt{5} = \frac{a}{b} \implies 5 = \frac{a^2}{b^2} \implies 5b^2 = a^2.$$

The last equation provides our contradiction. The right side is a square, and therefore has an even number of prime factors. But the left side has an *odd* number of prime factors—an even number coming from b^2 and one more from the 5. This absurdity completes the proof. ◊

EXAMPLE 5. Prove Claim V: There are infinitely many prime numbers.

SOLUTION. The following beautiful proof is attributed to Euclid, a Greek mathematician who worked in Alexandria, Egypt, around 300 BCE. Euclid's idea can be phrased as a proof by contradiction.

N is one more than a multiple of each p_i.

PROOF. Assume, toward contradiction, that there are only finitely many primes, say $p_1, p_2, p_3, \ldots, p_n$. Now consider the number $N = p_1 p_2 p_3 \cdots p_n + 1$. By its construction, N is *not* divisible by any of the primes $p_1, p_2, p_3, \ldots, p_n$. Hence either N is itself prime or N has at least one prime factor *not* among $p_1, p_2, p_3, \ldots, p_n$. Either way, the list $\{p_1, p_2, p_3, \ldots, p_n\}$ could not have been complete. This contradiction completes the proof. ◊

Brute Force, Cases, Exhaustion, Enumeration

Satisfying as it is to dispatch broad claims with elegant proofs like Euclid's, some claims resist such treatment. Sometimes we just argue case-by-case, through several—or even many—different parts of a general assertion. Such "brute force" or "exhaustive" methods can be ugly, but sometimes there is no alternative. And working through special cases, especially with a computer's help, can reveal important patterns.

Sometimes there is an alternative, but it's hard to find.

EXAMPLE 6. Prove Claim VI: Every even integer from 4 to 100 is the sum of two primes.

SOLUTION. Why not just prove Claim VII—every even integer greater than two is the sum of two primes—and be done with it? That would indeed do the job, but Claim VII is *Goldbach's conjecture*, a famous problem dating to the 1740s. Considering that Claim VII has already stumped the likes of Euler, we'll stick with Claim VI. Handling its 49 special cases is easy, although perhaps tedious:

Try some more!

$$4 = 2 + 2; \quad 6 = 3 + 3; \quad 8 = 3 + 5; \quad \ldots \quad 98 = 11 + 87; \quad 100 = 7 + 93.$$

Proof by Induction

Mathematical induction is among every mathematician's favorite power tools. It is a simple, structured, and sometimes astonishingly powerful approach to proving whole families of claims at once. We illustrate the idea first informally, by example.

EXAMPLE 7.   Discuss Claim VIII: A set with n elements has 2^n subsets.

SOLUTION.   Claim VIII is really an infinite collection of propositions $P(1)$, $P(2)$, $P(3)$, ..., one for each positive integer n. To emphasize this structure, we can list the propositions individually (we may as well use $\{1, 2, \ldots, n\}$ as a convenient n-member set):

$$P(1) : \{1\} \text{ has 2 subsets,}$$
$$P(2) : \{1, 2\} \text{ has 4 subsets,}$$
$$\ldots$$
$$P(42) : \{1, 2, \ldots, 42\} \text{ has } 2^{42} \text{ subsets.}$$
$$\ldots$$

It is easy in this case to check $P(n)$ for small n, simply by listing. For $P(1)$, the two subsets are $\{1\}$ and $\emptyset$. For $n = 2$ the list is

Every set has $\emptyset$ as a subset.

$$\emptyset, \quad \{1\}, \quad \{2\}, \quad \{1, 2\}.$$

With $n = 3$ we get eight subsets, as expected: four from the $n = 2$ case plus four more that include a 3:

See the pattern?

$$\emptyset, \quad \{1\}, \quad \{2\}, \quad \{1, 2\}, \quad \{3\}, \quad \{1, 3\}, \quad \{2, 3\}, \quad \{1, 2, 3\}.$$

The basic ingredients of a proof should now be clear:

(i) a set with one element "obviously" has two subsets; and

(ii) adding a new element to any finite set *doubles* the total number of subsets (every "old" subset generates a new one when we throw in the new element).

Hence sets with 1, 2, 3, 4, ... elements have 2, 4, 8, 16, ... subsets, as desired.
$$\Diamond$$

Induction, formally. Example 7 illustrates the usual setting for a proof by induction: a claim that some proposition $P(n)$ holds for every positive integer n. To prove such a claim by induction takes two (named) steps:

• *The base case:* Show that $P(1)$ holds.

- *The inductive step:* Show that $P(k)$ implies $P(k + 1)$ for every positive integer k. (Here $P(k)$ is called the *inductive hypothesis*.)

We illustrate the idea (and the customary shop talk) by formalizing the proof that, for all n, the set $\{1, 2, \ldots, 42\}$ has 2^n subsets.

Proof (of Claim VIII, by induction): The base case $n = 1$ holds because the set $\{1\}$ has only itself and the empty set as subsets.

For the inductive step we assume the inductive hypothesis—that $\{1, 2, \ldots, k\}$ has 2^k subsets—and try to *show* that $\{1, 2, \ldots, k, k + 1\}$ has 2^{k+1} subsets. To this end, note first that all 2^k subsets of $\{1, 2, \ldots, k\}$ are also subsets of $\{1, 2, \ldots, k + 1\}$. Every remaining subset of $\{1, 2, \ldots, k + 1\}$ contains $k + 1$, and so can be formed by adding $k + 1$ to some subset of $\{1, 2, \ldots, k\}$. Thus $\{1, 2, \ldots, k, k + 1\}$ has $2 \times 2^k = 2^{k+1}$ elements, as desired, and the proof is complete. $\square$

After one more simple example we'll draw some conclusions.

EXAMPLE 8. The identity $1 + 2 + 3 + \cdots + n = \frac{n(n+1)}{2}$ holds for all positive integers n.

SOLUTION. We'll write $P(n)$ for the identity above; for instance, $P(10)$ says that $1 + 2 + \cdots + 10 = 10 \cdot 11/2 = 55$. This is easily checked directly, but we'd rather prove $P(n)$ for *all* integers n—just the right job for mathematical induction.

PROOF BY INDUCTION. The base case, $P(1)$, holds by a trivial calculation—both sides of the equation are 1.

For the inductive step we assume equation $P(k)$ and derive $P(k + 1)$. A little algebra is all we need:

adding $k + 1$ to both sides of $P(k)$

by algebra

$$1 + 2 + \cdots + k + (k + 1) = \frac{k(k + 1)}{2} + (k + 1)$$
$$= (k + 1)\left(\frac{k}{2} + 1\right) = \frac{(k + 1)(k + 2)}{2}$$

That the first and last quantities are equal is just what $P(k + 1)$ asserts. The proof is complete. $\Diamond$

Notes on the method. Mathematical induction is so useful and so ubiquitous in mathematics that some general themes deserve mention.

- *Not cheating:* To many conscientious newcomers the inductive step in a proof feels like cheating—assuming the desired result in order to prove it. Caution is always wise, but here we're OK: this part of the proof shows only that $P(k)$ *implies* $P(k + 1)$, not that $P(k)$ itself must hold.

- *Dominoes, ladders, lily pads:* Choose your favorite metaphor—different authors use different real-world images to illustrate mathematical induction. My favorite involves dominoes labeled $P(1)$, $P(2)$, $P(3)$, etc., arranged in the classical childhood manner (if only the child had infinitely many), so that every $P(n)$ falls to the right, toppling $P(n+1)$. The inductive idea is now a domino theory: if (i) the first domino falls; and (ii) each domino fells its right-hand neighbor, then *all* the dominoes fall (to the child's infinite delight).

  Ladders (getting to the first rung is the base step) and frogs crossing lily ponds, pad by pad, lead to similar stories—flesh out your own details.

- *Other bases:* Not every claim for which induction looks promising starts with the $n = 1$ case. For instance, the inequality $n^2 < 2^n$ fails for some small integers, but holds whenever $n \geq 5$. An inductive proof *is* possible, with the small twist that the base case now corresponds to $n = 5$. Claim VII (every even integer from 4 on is the sum of two primes) is another case in point: an inductive proof would have base case $n = 4$. (Whether an inductive proof can be found is another matter entirely.)

  See the exercises.

- *Why induction works:* Inductive proofs themselves may be hard or easy, but the underlying "domino theory" is simple and plausible. A rigorous proof that mathematical induction "works" depends on an axiom about the natural numbers:

  If $S \subseteq \mathbb{N}$ is such that (i) $1 \in S$, and (ii) $n + 1 \in S$ whenever $n \in S$, then $S = \mathbb{N}$.

  We'll take this basic fact, like others about the integers, as an axiom rather than something to be proved.

Exercises

1. In each part of this problem, either prove or disprove the given claim. If you prove the claim, indicate whether your proof is direct, indirect, by contradiction, or something else. If you disprove the claim, use a counterexample. (It's OK to assume known facts, such as that $\sqrt{2}$ is irrational.) In all cases, x and y are real numbers.

   (a) If $x \in \mathbb{Q}$, then $\sqrt{2} + x \notin \mathbb{Q}$.

   (b) If $x \notin \mathbb{Q}$, then $\sqrt{2} + x \notin \mathbb{Q}$.

   (c) If $x + y$ is irrational, then at least one of x and y is irrational.

   (d) If p is a prime number, then $2^p - 1$ is prime.

(e) For all real numbers x and y, $|x - y| \leq |x^2 - y^2|$.

2. In each part following, either prove or disprove the *converse* of the given claim. If you prove the converse, indicate whether your proof is direct, indirect, by contradiction, or something else. If you disprove the converse, use a counterexample. (Assume known facts, such as that $\sqrt{2}$ is irrational.) In all cases, x and y are real numbers.

(a) If $x \in \mathbb{Q}$, then $\sqrt{2} + x \notin \mathbb{Q}$.

(b) If $x \notin \mathbb{Q}$, then $\sqrt{2} + x \notin \mathbb{Q}$.

(c) If $x + y$ is irrational, then at least one of x and y is irrational.

3. Any nonempty finite set $F = \{x_1, x_2, \ldots, x_n\}$ of n real numbers can be arranged from smallest to largest. Doing so depends on an even simpler fact: A finite set F of n real numbers has a largest element. Prove this by mathematical induction.

4. Let S be a nonempty finite set, with n elements.

(a) Explain why S has an *even* number of subsets (including S and $\emptyset$). Hint: Look at results in this section. Or think about complements.

(b) Suppose n is odd. Show that the number of subsets with an even number of elements is the same as the number of subsets with an odd number of elements. Hint: This can be done by induction, but it's quicker to use complements.

(c) The result in the preceding part actually holds for both odd and even n. Prove this, perhaps by induction.

5. Consider the equation $1 + 3 + 5 + \cdots + (2n - 1) = n^2$.

(a) Check directly that the equation holds for $n = 1$ and for $n = 10$.

(b) Prove by induction that the formula holds for all positive integers n.

(c) Use (don't reprove) the formula in Example 8 to give another proof (not involving induction) of the equation above.

6. Guess a formula (in terms of n) for the sum

$$\frac{1}{1 \cdot 2} + \frac{1}{2 \cdot 3} + \frac{1}{3 \cdot 4} + \cdots + \frac{1}{n \cdot (n + 1)}.$$

Prove your answer by induction.

7. Suppose that $x \geq -1$. Prove that $(1 + x)^n \geq 1 + nx$ for all $n \in \mathbb{N}$.

8. Show that

$$(1 + 2 + 3 + \cdots + n)^2 = 1^3 + 2^3 + 3^3 + \cdots + n^3$$

holds for all positive integers n. (Hint: Use the formula for $1 + 2 + \cdots + n$ in Example 8.)

9. Recall the product rule from elementary calculus: If f_1 and f_2 have derivatives f_1' and f_2' then $(f_1 f_2)' = f_1' f_2 + f_1 f_2'$.

 (a) Use the ordinary product rule to show the analogous formula $(f_1 f_2 f_3)' = f_1' f_2 f_3 + f_1 f_2' f_3 + f_1 f_2 f_3'$.

 (b) Let n be any positive integer. Guess a formula for $(f_1 f_2 f_3 \cdots f_n)'$ and prove it by induction.

10. (a) Show that $5^n > n!$ for positive integers $n < 12$.

 (b) Show that $5^n < n!$ for positive integers $n \geq 12$.

11. (a) Guess a formula (in terms of n) for the sum $1 \cdot 2 + 2 \cdot 3 + 3 \cdot 4 + \cdots + n \cdot (n + 1)$. Prove your answer by induction.

 (b) It is well known, and readily proved by induction, that

$$\sum_{j=1}^{n} j^2 = \frac{n(n + 1)(2n + 1)}{6} \quad \text{and} \quad \sum_{j=1}^{n} j = \frac{n(n + 1)}{2}.$$

 Use these facts to give another proof of the result in (a).

12. Let $r \neq 1$ be a real number. Show by induction that for all positive integers n,

$$1 + r + r^2 + r^3 + \cdots + r^n = \frac{r^{n+1} - 1}{r - 1}.$$

13. Show that the inequality $2^n < n!$ holds for all integers $n > 3$.

14. Every calculus student knows that if $f(x) = x^n$ for any positive integer n, then $f'(x) = nx^{n-1}$. Prove this by induction, assuming (i) if $f(x) = x$, then $f'(x) = x$; and (ii) the product rule. (We'll define the derivative, state and prove the product rule, and firm up other ideas later in this book; here the point is to see induction in a familiar setting.)

15. Another familiar calculus formula (see Problem 14) says that

$$\text{if} \quad g(x) = \frac{1}{x^n}, \quad \text{then} \quad g'(x) = -\frac{n}{x^{n+1}}$$

for every positive integer n. Prove this by induction, assuming both the $n = 1$ case and the product rule.

16. A version of Theorem 1.3 says that if n is a positive integer and $\sqrt{n}$ is rational, then n is a perfect square. Prove this using the following outline, the idea of prime factorization, and the fact that a positive integer n is a perfect square if and only if every prime factor of n appears to an even power. See also Example 4, page 45.

    If $\sqrt{n} = a/b$, where a and b are positive integers, then squaring both sides gives $nb^2 = a^2$. Now factor each side of this equation as a product of prime numbers. Because the right side is a square, each prime factor on the right side appears to an *even* power, and so the same must be true on the left. Each prime factor of b^2 appears to an even power, and so (do you see why?) each prime factor of n must *also* appear to an even power.

17. If your wallet contains two $5 bills and an unlimited supply of $3 bills. then you can pay out some amounts, like $8 and $300, but not others, like $2 and $7. Exactly which amounts can you pay out? Guess an answer and prove it by induction.

18. **Claim:** In any group of n kittens, if one is orange, then all are orange. That's absurd, of course, but what's wrong with the following "proof" by induction?

    Proof: The claim is trivial if $n = 1$, so the base case holds. To illustrate the inductive step, assume the claim holds for $n = 42$. Suppose we're given a group of 43 kittens, including at least one—say, Hans—that's orange. Any other kitten—say Fritz—can be put in some 42-member group with Hans, and so Fritz must also be orange by the inductive hypothesis. Thus all 43 kittens are orange, and the proof is done.   □

1.6   Sets 103: Finite and Infinite Sets; Cardinality

How "big" is a set?

Finite sets: few surprises. "Small" sets offer few surprises. It seems clear, for instance, that $S = \{a, b\}$, with two elements, is "smaller" than $T = \{a, b, c\}$, with three. Similarly, $N_{42} = \{1, 2, \ldots, 42\}$ is obviously "smaller" than FIFTYSTATES $= \{$Alabama, Alaska, $\ldots$, Wisconsin, Wyoming$\}$, even though the two sets have completely different types of elements, while N_{26} and ENGLISHALPHABET $= \{$a, b, c, $\ldots$, z $\}$ have the same "size."

"Measuring" sets by counting their elements works well for sets with *finitely* many elements. For such sets, moreover, our intuition is usually reliable. If, say, a set S has 427 elements and $T \subsetneq S$, then T must be "strictly smaller," with 426 elements or fewer.

Infinite sets: many surprises. Matters are very different for *infinite* sets. Consider, for instance,

$$\mathbb{N} = \{1, 2, 3, 4, \ldots \quad \text{and} \quad E = \{2, 4, 6, 8 \ldots\}.$$

In one way E seems obviously "smaller"—it omits all of the (infinitely many!) odd numbers in $\mathbb{N}$. On the other hand, the mapping

$$1 \mapsto 2, \quad 2 \mapsto 4, \quad 3 \mapsto 6, \quad \ldots, \quad n \mapsto 2n, \quad \ldots$$

is a one-to-one correspondence between $\mathbb{N}$ and E, so maybe the two sets have the "same size." Similarly, the interval $(0, 1)$ is "shorter" than the interval $(0, 2)$, but the same mapping, $x \mapsto 2x$, is a one-to-one correspondence. An important aim of this section is to develop useful ways of "measuring" infinite sets.

Cardinality: The Idea

So how *should* we gauge "size" for infinite sets? The answer for present purposes involves *cardinality*. We need some clear definitions, and an excuse to drop all those lawyerly quotation marks:

Definition 1.13. Two nonempty sets A and B have the *same cardinality* if there is a function $f : A \to B$ that is both one-to-one and onto. In this case we write $A \sim B$.

Definition 1.14. A set A is *finite* if either $A = \emptyset$ or $A \sim \{1, 2, \ldots, n\}$ for some positive integer n. Otherwise, A is *infinite*.

Informally speaking, two sets have the same cardinality if there is a *one-to-one correspondence* between them. We saw this above for $\mathbb{N}$ and E and also for $(0, 1)$ and $(0, 2)$; thus, $\mathbb{N} \sim E$ and $(0, 1) \sim (0, 2)$.

EXAMPLE 1.   Show that $(0, 1)$, $(-2, 5)$, and $(0, \infty)$ all have the same cardinality.

SOLUTION.   It is enough to find one-to-one and onto functions $f : (0, 1) \to (-2, 5)$ and $g : (0, \infty) \to (0, 1)$. Figure 1.1 hints at two of the many possibilities.

The function f shown is the *linear* function whose graph passes through $(0, -2)$ and $(1, 5)$; its formula turns out to be $f(x) = 7x - 2$. The function g shown has formula $g(x) = x/(1 + x)$. Graphs of f and g *illustrate* their one-to-one and onto properties; honest proofs require the formulas.

Check this easy calculation.

To prove, for instance, that g is *onto*, we have to find, for each $b \in (0, 1)$, some $a \in (0, \infty)$ with $f(a) = a/(1 + a) = b$. A little symbolic work reveals the recipe:

$$\frac{a}{1 + a} = b \iff a = b + ab \iff a - ab = b \iff a = \frac{b}{1 - b}.$$

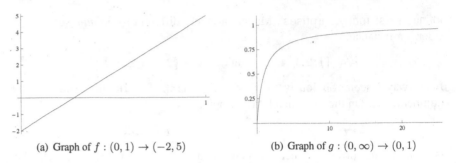

(a) Graph of $f : (0, 1) \rightarrow (-2, 5)$ (b) Graph of $g : (0, \infty) \rightarrow (0, 1)$

Figure 1.1. Showing that sets have the same cardinality.

Thus, for any given $b \in (0, 1)$, the positive number $a = b/(1 - b)$ satisfies $g(a) = b$.   ◇

EXAMPLE 2.   Show that $\mathbb{N}$ and $\mathbb{Z}$ have the same cardinality.

SOLUTION.   The following table suggests one possible one-to-one correspondence:

$\mathbb{N}$:	1	2	3	4	5	6	7	8	9	$\ldots$
$\mathbb{Z}$:	0	1	-1	2	-2	3	-3	4	-4	$\ldots$

The key property should be clear: every $z \in \mathbb{Z}$ appears *once and only once* in the second row. The integer $z = -42$, for instance, corresponds to $n = 85$.   ◇

We can describe this correspondence a little more formally; see the exercises.

We've just seen that $\mathbb{Z}$ can be "paired up with" or "counted off against" $\mathbb{N}$—even though $\mathbb{Z}$ is, in another sense, a bigger set. The following definition captures this possibility:

Definition 1.15. A nonempty set A is *countable* if either A is finite or A has the same cardinality as $\mathbb{N}$. In the latter case, A is *countably infinite*.

We've already seen several examples, including $\mathbb{Z}$, of countably infinite sets. We'll show soon, too, that $\mathbb{Q}$ is countable. (This is probably surprising—after all, between *any* two rationals lie infinitely many others, so $\mathbb{Q}$ might seem "larger" than $\mathbb{N}$.) We might wonder, then, whether *any* set of numbers, even $\mathbb{R}$ itself, is *uncountable*, i.e., too big to be in one-to-one correspondence with $\mathbb{N}$. The answer, as we'll see, turns out to be yes.

Working with Countable Sets

Finite sets. Finite sets have commonsense properties as far as cardinality is concerned, at least if we assume some basic properties of the integers. Following are

As indeed we do in this book.

some properties we will accept, without formal proofs, as intuitively reasonable.

- *The pigeonhole principle:* If a (finite!) hotel has fewer rooms than guests, then at least one room must house at least two guests. This simple but useful idea is known as the *pigeonhole principle* (trade rooms for nest boxes and guests for pigeons). Here is a more formal statement:

  We'll revisit some of these facts in the exercises.

  > Let $f : A \to B$ be any function, where A and B are finite sets, and B has fewer members than A. Then f is *not* one-to-one.

- *Subsets are finite, too:* If A is a finite set and $B \subset A$, then B is finite, too.

- *New finite sets from old:* If A and B are finite sets, then $A \cup B$, $A \cap B$, $A \setminus B$, and $A \times B$ are all finite.

- *Listing and ordering:* If A is any finite set, say with 42 elements, then we can *list* the elements: $A = \{a_1, a_2, a_3, \ldots, a_{42}\}$. If A happens to be a finite set of real numbers, we can *order* our list from smallest to largest:

$$a_1 < a_2 < a_3 < \cdots < a_{40} < a_{41} < a_{42}.$$

Biggest and smallest members. The last property above has a useful form that deserves special mention:

We discuss this further in Problem 3, page 50.

Fact 1.16. *Every* finite *set of real numbers* contains *a maximum and a minimum element.*

The words "finite" and "contains" both matter. Some infinite sets, such as $\mathbb{Z}$, are *unbounded*, and therefore obviously lack maximum and minimum elements. The case of a *bounded* infinite set, such as the open interval $(1, 3)$, is a little subtler. The problem is that 3 and 1—the obvious candidates for biggest and smallest—are not *members* of $(1, 3)$.

The *closed* interval [1,3] does contain maximum and minimum elements.

Countably infinite sets. The set $\mathbb{N}$ is the "model" countably infinite set, but many other sets turn out to have the same cardinality as $\mathbb{N}$. The following proposition says, in effect, that $\mathbb{N}$ is the "smallest" infinite set.

Proposition 1.17. *Let S be countably infinite, and let $T \subset S$ be any nonempty subset. Then T is either finite or countably infinite.*

Notice the possible surprise: Every infinite subset of $\mathbb{N}$—the odd numbers, the primes, the powers of 2, the powers of 123456789—is in the sense of cardinality just as "big" as $\mathbb{N}$ itself.

The idea of the proof is to *list* all the members of T:

$$t_1, t_2, t_3, \ldots, t_{234}, \ldots.$$

Such a list defines a nice one-to-one correspondence: $1 \mapsto t_1$, $2 \mapsto t_2$, ..., $234 \mapsto t_{234}$, and so on. The details follow.

Proof (sketch): By hypothesis, S is in one-to-one correspondence with $\mathbb{N}$, so we may as well assume right away that $S = \mathbb{N}$. and that $T \subset \mathbb{N}$. If T is finite we're done already, so we assume T is infinite.

It's discussed in Section 1.1.

To finish the proof we need a bijective function $f : \mathbb{N} \to T$. To start, let t_1 be the *smallest* element of T (the *well-ordering principle* guarantees that there *is* a smallest element), and set $f(1) = t_1$. Now let t_2 be the smallest element of $T \setminus \{t_1\}$, and set $f(2) = t_2$. Continuing, we set $f(3) = t_3$, where t_3 is the smallest element of $T \setminus \{t_1, t_2\}$, and so on. This process continues indefinitely because T is infinite, and so does indeed construct a function $f : \mathbb{N} \to T$, as desired. The method of construction assures that the t_i are all different from one another, and so f is one-to-one. The function f is also onto. If, say, $t = 42 \in T$, then there are at most 41 *smaller* members of T, so $f(k) = 42$ must hold for some $k \leq 42$. A similar argument works for every $t \in T$, so f is indeed onto, and the proof is done. □

New countable sets from old. Proposition 1.17 gives one way of creating countable sets: start with a given countable set and take *subsets*. Propositions 1.18 and 1.19 describe some other operations that leave countability intact. You may be surprised ....

Proposition 1.18. *Let A and B be countable sets. The Cartesian product*

$$A \times B = \{(a, b) \mid a \in A \text{ and } b \in B\}$$

is countable.

The proof is a little easier if either or both are finite.

Proof: We'll take both A and B to be countably *infinite*. In this case, we have

$$A = \{a_1, a_2, a_3, \dots\} \quad \text{and} \quad B = \{b_1, b_2, b_3, \dots\},$$

The matrix extends infinitely upward and to the right, like the first quadrant in the *xy*-plane.

and we can list all members of $A \times B$ in a "doubly-infinite matrix":

$$
\begin{array}{ccccc}
\cdots & \cdots & \cdots & \cdots & \cdots \\
(a_1, b_3) & (a_2, b_3) & (a_3, b_3) & (a_4, b_3) & \cdots \\
(a_1, b_2) & (a_2, b_2) & (a_3, b_2) & (a_4, b_2) & \cdots \\
(a_1, b_1) & (a_2, b_1) & (a_3, b_1) & (a_4, b_1) & \cdots
\end{array}
$$

Find these entries in the big matrix.

Our last trick is to "count off" all the entries in this gigantic array—without missing any. One way to do this is to start at lower left and proceed along "northwest-pointing" diagonals:

$(a_1, b_1), (a_2, b_1), (a_1, b_2), (a_3, b_1), (a_2, b_2), (a_1, b_3), (a_4, b_1), \ldots.$

Do you see the pattern? What matters is that every point (a_i, b_j) appears once, but only once, in our list. These facts imply that the function $f : \mathbb{N} \to A \times B$ given by $n \mapsto n$th entry in the list is the desired bijection.   □

Proposition 1.19. *Let A_1, A_2, A_3, \ldots, A_n, \ldots be a countable collection of sets, all of them countable. Then the union*

$$A_1 \cup A_2 \cup A_3 \cup \cdots = \bigcup_{i=1}^{\infty} A_i$$

is also countable.

Proof (sketch): Again we assume that each A_i is countably infinite, and list its elements: $A_i = \{a_{i,1}, a_{i,2}, a_{i,3}, \ldots\}$. As in the preceding proof, we can list all elements of *all* the A_i in a big "matrix":

$$\begin{array}{ccccc}
\cdots & \cdots & \cdots & \cdots & \cdots \\
a_{3,1} & a_{3,2} & a_{3,3} & a_{3,4} & \cdots \\
a_{2,1} & a_{2,2} & a_{2,3} & a_{2,4} & \cdots \\
a_{1,1} & a_{1,2} & a_{1,3} & a_{1,4} & \cdots
\end{array}$$

Just as before, this "matrix" has countably many entries, so we're almost done. But a minor subtlety needs attention: the A_i might have some elements in common, so some elements of the union might appear more than once in the "matrix." Luckily, this turns out not to matter, because the *distinct* elements of the union correspond to some *subset* of the matrix entries, and we showed in Proposition 1.17 that such subsets are countable.   □

Corollary 1.20. *The set $\mathbb{Q}$ of rational numbers is countable.*

Proof: It is enough, says Proposition 1.19, to write $\mathbb{Q}$ as a countable union of countable sets. For $i = 1, 2, 3, \ldots$, let

$$A_i = \left\{ \frac{0}{i}, \frac{1}{i}, \frac{-1}{i}, \frac{2}{i}, \frac{-2}{i}, \ldots \right\}$$

be the set of fractions a/i, where $a \in \mathbb{Z}$. Now each A_i is clearly countable—it is in one-to-one correspondence with $\mathbb{Z}$—and $\mathbb{Q}$ is the union of all the A_i.   □

A given rational, say 3/7, appears not just in A_7 but also in A_{14}, A_{21}, and so on.

Uncountable Sets

The preceding propositions imply that many apparently large sets, such as $\mathbb{Q}$, are in fact no larger in cardinality than $\mathbb{N}$. Might $\mathbb{R}$ itself be countable? The answer is hardly obvious, but it turns out to be no, as the German mathematician Georg Cantor showed around 1873.

Theorem 1.21. $\mathbb{R}$ *is uncountable.*

We'll give a version of Cantor's ingenious proof below, but first let's consider some striking implications of the theorem:

Corollary 1.22. *The interval* $(0,1)$ *is uncountable. Every* *nonempty open interval* (a,b) *is uncountable. The* irrational *numbers are uncountable.*

All parts of the corollary say that the sets mentioned are, in cardinality, *much larger* than the comparatively "sparse" set $\mathbb{Q}$. Throw a dart at random at the real line, and you'll almost certainly hit an irrational.

Proofs of the corollaries. In Example 1 we showed that the intervals $(0,1)$, $(-2,5)$, and $(0,\infty)$ all have the same cardinality. Similar methods show that *all* nonempty open intervals, including $(-\infty,\infty)$ (aka $\mathbb{R}$), have the same cardinality. If the set P of irrationals were countable, then $\mathbb{R} = P \cup \mathbb{Q}$ would be the union of two countable sets, and hence countable itself.

We found explicit bijective functions between these intervals.

Cantor's "diagonal" proof. Cantor uses the idea of infinite decimal expansion to show that the interval $(0,1)$ is uncountable. The proof is by contradiction. If $(0,1)$ *were* countable, we could list *all* of its elements in an infinite sequence: x_1, x_2, x_3, ... Now each x_i has an infinite decimal expansion, of the form $x_i = 0.d_{i1}\, d_{i2}\, d_{i3}\, d_{i4}\, d_{i5} \ldots$, where the d_{ij} are decimal digits, ranging from 0 to 9. Thus, we can write

$$x_1 = 0.d_{11}\, d_{12}\, d_{13}\, d_{14}\, d_{15} \cdots$$
$$x_2 = 0.d_{21}\, d_{22}\, d_{23}\, d_{24}\, d_{25} \cdots$$
$$x_3 = 0.d_{31}\, d_{32}\, d_{33}\, d_{34}\, d_{35} \cdots$$
$$x_4 = 0.d_{41}\, d_{42}\, d_{43}\, d_{44}\, d_{45} \cdots$$
$$x_5 = 0.d_{51}\, d_{52}\, d_{53}\, d_{54}\, d_{55} \cdots$$
$$\cdots = \cdots$$

Here comes the clever part. Cantor uses the *diagonal* entries

$$d_{1,1},\, d_{2,2},\, d_{3,3},\, d_{44},\, d_{5,5},\, \ldots$$

in this array to produce a new number $x_0 = 0.e_1\,e_2\,e_3, \ldots$, which lies in $(0,1)$ but *all* of whose digits differ from the corresponding diagonal entry:

$$e_1 \neq d_{11}, \quad e_2 \neq d_{22}, \quad e_3 \neq d_{33}, \quad e_4 \neq d_{44}, \quad \ldots .$$

There are ten choices for each digit, so there are plenty of ways to choose the e_i, and therefore countless possible numbers x_0. What matters is that x_0 differs from x_1 in the first digit, from x_2 in the second digit, from x_3 in the third digit, and so on. Thus x_0 is nowhere among the original x_i, which contradicts our original assumption.

Pun intended.

Exercises

1. Assume (it's true!) that the union $F_1 \cup F_2$ of any *two* finite sets is finite. Prove by induction that $F_1 \cup F_2 \cup \cdots \cup F_n$ is finite for every positive integer n and finite sets F_i.

2. Find a one-to-one correspondence $f : A \to B$ (expressed as a simple formula) in each part below.

   (a) $A = \{1, 2, 3, 4, \ldots, 10\}$; $B = \{2, 4, 8, \ldots, 1024\}$.
   (b) $A = \mathbb{N}$; $B = \{-3, -2, -1, \ldots\}$.
   (c) $A = [0, 1)$; $B = (0, 1]$.
   (d) $A = \mathbb{R}$; $B = (0, \infty)$

3. Consider the functions $f : (0, 1) \to (-2, 5)$ and $g : (0, \infty) \to (0, 1)$ defined in Example 1. (It is shown there that g is onto.)

   (a) Show that f and g are both one-to-one on their domains.
   (b) Show that f is onto.
   (c) Show that $f \circ g$ is one-to-one and onto. What does this mean about cardinality?

4. Let (a, b) and (c, d) be any two bounded intervals. Find a linear function $f : (a, b) \to (c, d)$ that is one-to-one and onto. (Give an explicit formula for f in terms of a, b, c, and d.)

5. Let A be a set with 42 elements and B a set with 43 elements, and let $f : A \to B$ and $g : B \to A$ be functions. Can f be one-to-one? Onto? What about g? Give examples of what can happen, and explain what can't. How is the pigeonhole principle involved?

6. Let S be a *finite* set. Use the pigeonhole principle to show that if $f : S \to S$ is one-to-one, then f is also onto.

7. It can be shown that a set is S is infinite if and only if there exists a function $f : S \to S$ that is one-to-one but not onto. (The "if" part is the claim of Problem 6.)

   (a) Find a function $f : \mathbb{N} \to \mathbb{N}$ that is one-to-one but not onto. (There are many possibilities.)

   (b) Find a function $g : \mathbb{R} \to \mathbb{R}$ that is one-to-one but not onto. (There are many possibilities.)

   (c) Find a function $h : \mathbb{R} \to \mathbb{R} \setminus \{1\}$ that is one-to-one and onto.

8. Suppose that $A \sim B$ and $B \sim C$. Use Definition 1.13 to show that $A \sim C$.

9. Suppose A is infinite, B is finite, and $B \subset A$. Show that $A \setminus B$ is infinite.

10. Let $f : \mathbb{N} \to \mathbb{Z}$ be the function described by the table in Example 2.

    (a) Find a symbolic formula for $f(n)$, perhaps using cases.

    (b) Let $g : \mathbb{Z} \to \mathbb{N}$ be the other function described by the table in Example 2 (read the table from bottom to top). Find a symbolic formula for $g(z)$, perhaps using cases.

    (c) Show that both $g \circ f : \mathbb{N} \to \mathbb{N}$ and $f \circ g : \mathbb{Z} \to \mathbb{Z}$ are identity functions.

11. We proved Proposition 1.19 ($\mathbb{Q}$ is countable) by writing $\mathbb{Q}$ as the union of countably many countably *infinite* sets. We could, instead, write $\mathbb{Q}$ as a *nested* union of countably many *finite* sets.

    (a) For $i \in \mathbb{N}$, let Q_i be the set of rationals p/q (written in reduced form), for which $p^2 + q^2 < i$. How many members does Q_3 have? What about Q_{10}? Can $Q_i = Q_{i+1}$ for some i? (Hints: Members of Q_i correspond to some, but not all, of the points inside the circle $p^2 + q^2$ in the pq-plane. For Q_3 and Q_{10}, just count these points.)

    (b) Find an upper bound in terms of i for the number of elements in Q_i.

    (c) Show that $Q_1 \subseteq Q_2 \subseteq Q_3 \subseteq \dots$ and that $\bigcup_{i=1}^{\infty} Q_i = \mathbb{Q}$.

12. Show that $[0, 1]$ and $(0, 1)$ have the same cardinality by finding a function $f : [0, 1] \to (0, 1)$ that is one-to-one and onto.

    Hints: The claim seems reasonable, but finding a good function f is tricky. Here is one possibility: Set $f(0) = 1/2$, $f(1) = 1/3$, $f(1/2) = 1/4$, $f(1/3) = 1/5$, $f(1/4) = 1/6$, $f(1/5) = 1/7$, etc. For all other x, set $f(x) = x$. Think about it, draw a graph, etc.; show that this function does the job. (We'll show later, by the way, that no *continuous* function f can have the desired properties.)

13. Show that $[0, \infty)$ and $(0, \infty)$ have the same cardinality.

14. Use results from these problems to explain why *all* intervals in $\mathbb{R}$ have the same cardinality.

15. A "book" is a finite string of characters from some finite "alphabet," such as the 128-character ASCII system. Is the set B of all possible "books" finite, countably infinite, or uncountable?

16. A *polynomial with integer coefficients* is an expression of the form $p(x) = a_0 + a_1 x + a_2 x^2 + \cdots + a_n x^n$, where n is a nonnegative integer and all the a_i are integers. The set of all such polynomials is denoted $\mathbb{Z}[x]$. A real number a is called *algebraic* if $p(a) = 0$ for some p in $\mathbb{Z}[x]$. If, say, $p(x) = x^2 - 2$, then $p(\sqrt{2}) = 0$, so $\sqrt{2}$ is algebraic.

   (a) Show that $\mathbb{Z}[x]$ is countable.

   (b) Show that every rational number is algebraic.

   (c) Show that the set of algebraic numbers is countable.

   (d) A real number that is not algebraic is called *transcendental*. Show that the set of transcendental numbers is uncountable. (Note: Showing that a particular number is transcendental is usually difficult. But "nearly every" number is transcendental.)

17. (This extended "problem" could be the basis of a possible project.) Suppose we have two sets A and B and a *one-to-one* function $f : A \to B$. Then it's reasonable to say that the cardinality of A is *not greater than* that of B; in symbols, $|A| \le |B|$. (Think of $|A|$ as the "size" of A, not as an absolute value.) For instance, the one-to-one function $f : \mathbb{N} \to \mathbb{Z}$ given by $f(n) = n$ suggests, correctly, that $|\mathbb{N}| \le |\mathbb{Q}|$. In words, $\mathbb{N}$ is "not greater" than $\mathbb{Z}$.

   If there is *also* a one-to-one function $g : B \to A$, then we can write both $|A| \le |B|$ and $|B| \le |A|$, and it's natural to expect that A and B have the *same* cardinality—i.e., $|A| = |B|$, or $A \sim B$ in our preferred notation.

   The famous Cantor–Schröder–Bernstein theorem says that this is indeed so. For any sets A and B and *injective* functions $f : A \to B$ and $g : B \to A$, we can construct from f and g a *bijective* function $h : A \to B$.

   (a) Find out (from books or online) exactly what the Cantor–Schröder–Bernstein theorem says, and describe a proof in your own words.

   (b) Let $A = [0, 1] = B$, and consider the functions $f : [0, 1] \to [0, 1]$ given by $f(x) = x/2$ and $g : [0, 1] \to [0, 1]$ given by $g(x) = x/2$. Illustrate the proof idea in the previous part by constructing from f and g the advertised bijective function h. (The identity function $h(x) = x$

is obviously bijective, but it's not constructed from f and g as the proof idea describes.)

(c) Try to describe the function h using the binary expansion of points in $[0,1]$.

1.7 Numbers 102: Absolute Values

Recall the familiar *absolute value function*:

$$|x| = \begin{cases} x & \text{if } x \geq 0; \\ -x & \text{if } x < 0. \end{cases}$$

In beginning calculus this function is a favorite example of mildly bad behavior—it's continuous but has no derivative at $x = 0$. We'll use the same example later, too, but the absolute value also plays a much more basic role in real analysis: it measures the *size*, or *magnitude*, of a quantity. Such questions—how big? how close?—are crucial in real analysis; so, therefore, are equalities and inequalities with absolute values.

The absolute value has a nice geometric interpretation:

If x and a are real numbers on a number line, then $|x - a|$ is the *distance* between x and a.

This simple idea is surprisingly helpful in making sense of expressions that involve absolute values.

EXAMPLE 1.   Which real numbers x satisfy (i) $|x - 3| < |x + 2|$? What about (ii) $|3x - 7| < 5$?

SOLUTION.   Both inequalities are slightly awkward algebraically because of the need to handle positive and negative cases separately. Distance interpretations can help—a lot. Inequality (i), recast as $|x - 3| < |x - (-2)|$, now says simply that x is *closer* to 3 than to -2. This means, in turn, that x lies *right* of $1/2$, the midpoint of $(-2, 3)$; equivalently, $x > 1/2$.

Make a sketch.

Is all the algebra OK?

Inequality (ii) can also be interpreted in distance language, after a little rearrangement:

$$|3x - 7| < 5 \iff 3\left|x - \frac{7}{3}\right| < 5 \iff \left|x - \frac{7}{3}\right| < \frac{5}{3}.$$

The last version makes the meaning clear: x lies within distance $5/3$ from $7/3$. Equivalently, $x \in (2/3, 4)$.   ◊

Absolute Truths

We collect some key properties of absolute values in two theorems. The first is easy; proofs are omitted or left as exercises.

Theorem 1.23. *Let x and y denote arbitrary real numbers; assume $k > 0$.*

(a) $-|x| \le x \le |x|$.

(b) $|x| = |-x|$; $|x - y| = |y - x|$.

(c) $|x| < k$ $\Longleftrightarrow$ $-k < x < k$.

(d) $|x| > k$ $\Longleftrightarrow$ $x > k$ *or* $x < -k$.

(e) $|x \cdot y| = |x| \cdot |y|$.

The next theorem is important enough to have its own name.   We explain the name below.

Theorem 1.24 (Triangle inequality). *Let x and y be any real numbers. Then*

$$|x + y| \le |x| + |y| .$$

The reverse triangle inequality *says*

$$|x - y| \ge \big| |x| - |y| \big| .$$

On proofs. The triangle inequality follows from several bits of Theorem 1.23. From (a), we get

$$-|x| \le x \le |x| \quad \text{and} \quad -|y| \le y \le |y| .$$

Adding these inequalities gives

$$-(|x| + |y|) = -|x| - |y| \le x + y < (|x| + |y|) ,$$

which is equivalent, by part (c) of Theorem 1.23, to $|x + y| \le |x| + |y|$, as desired.

The reverse triangle inequality can be derived from the standard triangle inequality; see the exercises.

Variations on a theme. The triangle inequalities are often used in forms slightly different from the "vanilla" versions in the theorem. Following are some examples; observe occasional uses of Theorem 1.23:

(i) $|x - y| = |x + (-y)| \le |x| + |y|$;

(ii) $|x - y| = |x - a + a - y| \le |x - a| + |a - y| = |x - a| + |y - a|$;

(iii) $|x| - |y| \le |x - y| \le |x| + |y|$.

Note especially what (ii) says about distance: the trip from x to y is no longer than the *sum* of the distances from x to a and from a to y. This should sound reasonable—stopping at a enroute shouldn't *shorten* the trip from x to y—and it helps explain the allusion to triangles. Note also (iii), which *traps* $|x - y|$ between upper and lower bounds.

In fact, triangle-type inequalities hold in quite general mathematical settings, some beyond the scope of this book. The basic triangle inequality $|x + y| \leq |x| + |y|$ holds for *complex* numbers x and y, for instance, if $|x|$ denotes the *length* of x considered as a vector in the plane. Generalizing in another direction, related facts like

$$|x_1 + x_2 + \cdots + x_{100}| \leq |x_1| + |x_2| + \cdots + |x_{100}|$$

and

$$\left| \int_0^1 f(x)\, dx \right| \leq \int_0^1 |f(x)|\, dx$$

also hold, and are sometimes called triangle inequalities.

EXAMPLE 2.   Suppose we know that $|x - 7| < 0.01$, $|y - 7| < 0.02$, and $|z - 7| > 0.03$. What can be said about $|x - y|$, $|x - z|$, $|xy|$, and $|x/y|$?

SOLUTION.   Triangle inequalities help. To estimate $|x - y|$ we have

$$|x - y| = |x - 7 + 7 - y| \leq |x - 7| + |y - 7| < 0.01 + 0.02 = 0.03.$$

For $|x - z|$ there is no *upper* bound (why?), but the reverse triangle inequality gives a *lower* bound:

$$|x - z| = |(x - 7) - (z - 7)| \geq |z - 7| - |x - 7| > 0.03 - 0.01 = 0.02.$$

To estimate $|xy|$ and $|x/y|$ we notice first that $6.99 < |x| < 7.01$ and $6.98 < |y| < 7.02$. Thus,

These bounds follow from the triangle inequality; common sense works, too.

$$|xy| = |x||y| \leq 7.01 \cdot 7.02 \approx 49.21 \quad \text{and} \quad \frac{|x|}{|y|} \leq \frac{7.01}{6.98} \approx 1.004.$$

$\Diamond$

Exercises

1. It's well known that $|\sin(x)| \leq 1$ and $|\cos(x)| \leq 1$ for all x. Assume this and other familiar properties of sine and cosine in this problem.

   (a) The triangle inequality says that $|\sin(x) + \cos(x)| \leq 2$? Does a number smaller than 2 actually work in this inequality?

(b) What does the reverse triangle inequality say about $|\sin(x) - \cos(x)|$? Can this inequality be "improved"? Explain.

(c) Use the triangle inequality to show that $\sin(x)^2 + \cos(x)^2 \leq 2$. Can this inequality be "improved"? Explain.

2. This problem is about the reverse triangle inequality (RTI).

(a) Use the ordinary triangle inequality to prove the RTI.

(b) Each side of the RTI can be interpreted as the *distance* between two numbers. What does the RTI say in this language?

(c) Under what conditions on x and y is the RTI actually an *equation*?

3. Explain why the inequalities

$$|x - y| \leq |x| + |y| \quad \text{and} \quad |x - y| \geq |x| - |y|$$

hold for all real numbers x and y.

4. Find all real values of x that satisfy the given inequality; express answers in interval notation.

(a) $|2x + 7| < 11$

(b) $|2x + 7| > 11$

(c) $|2x^2 + 7| < 11$

5. Find all real values of x that satisfy the given inequality; express answers in interval notation.

(a) $|x - 3| < |x + 1|$

(b) $|x^2 - 4| < 0.07$

(c) $|x^2 + 2x + 1| < 4$

6. Suppose that $|x - 3| < 0.07$ and $|y - 5| < 0.04$.

(a) Show that $|5x - 15| < 0.35$.

(b) Show that $|x + y - 8| < 0.11$.

(c) Show that $14.5328 < xy < 15.4728$.

7. Suppose that $|x - 3| < 0.07$ and $|y - 5| < 0.04$.

(a) Show that $|y - 4| > 0.96$.

(b) Find K, as small as possible, such that $|y - x| < K$.

(c) Find L, as large as possible, such that $|y - x| > L$.

8. Suppose that $|x - 1| < 0.03$ and $|y| > 5$.

   (a) Find L, as large as possible, such that $|y - x| > L$.
   (b) Find all the possible values of $|y - x|$.

9. The triangle inequality (TI) for *three* summands follows from the ordinary TI:

$$|x_1 + x_2 + x_3| = |(x_1 + x_2) + x_3| \le |x_1 + x_2| + |x_3| \le |x_1| + |x_2| + |x_3|.$$

   Show by induction that the TI holds for n summands, where $n \ge 2$.

10. The triangle inequality $|\vec{x} + \vec{y}| \le |\vec{x}| + |\vec{y}|$ also holds for *vectors* $\vec{x}$ and $\vec{y}$ in the plane, where $|\vec{x}|$ denotes the *length* of $\vec{x}$. Draw a picture to illustrate this fact; make note of the triangle. When does equality hold?

11. This problem explores a connection between the absolute value and the maximum or minimum of two quantities.

    (a) Show that if x and y are real numbers, then

    $$\max\{x, y\} = \frac{x + y + |x - y|}{2}.$$

    (b) Find a similar formula for the *minimum* of x and y.
    (c) Let $f : I \to \mathbb{R}$ and $g : I \to \mathbb{R}$ be functions defined on an interval I. Find a formula involving the absolute value for the new function $h : I \to \mathbb{R}$ defined by $h(x) = \max\{f(x), g(x)\}$.
    (d) Let $f(x) = \sin x$ and $g(x) = e^x$, and let $h(x)$ be defined as in the preceding part. Use technology to plot f, g, and h together in an interval that shows clearly what's happening.

12. Let ϵ be a positive number, and suppose $|x - 7| < \epsilon$ and $|y - 7| < \epsilon$. Show that $|x - y| < 2\epsilon$.

13. (a) Suppose that $|x - 1| < 0.5$. Show that $x > 0.5$.
    (b) Suppose that $c \ne 0$ and $|x - c| < \frac{|c|}{2}$. Show that $|x| > \frac{|c|}{2}$.

14. Yet another triangle-type inequality has the form

$$\left| \int_0^1 f(x)\, dx \right| \le \int_0^1 |f(x)|\, dx.$$

Discuss, perhaps with a picture, why this is plausible. How are Riemann sums involved? (No formal proof is possible until we define the integral!)

1.8 Bounds

Theorems and proofs in real analysis often turn on *boundedness*. We'll prove later, for instance, that a continuous function defined on a closed and *bounded* interval $I = [a, b]$ is itself *bounded* on I. To make sense of such a claim requires clear meanings for all its words—which include *two* somewhat different instances of the b-word. In this section we start to unpack the idea and language of boundedness in several settings.

We'll also need to define "continuous," of course.

Definition 1.25. Let S be a nonempty set of real numbers.

- S is *bounded above* if there exists a number M with $s \leq M$ for all $s \in S$; here M is an *upper bound* for S.

- S is *bounded below* if there exists a number m with $s \geq m$ for all $s \in S$; in this case m is a *lower bound* for S.

- S is *bounded* if it is bounded both above and below.

It is clear, for instance, that all three of the sets $\{1, 2, 3\}$, $[1, 3]$, and $(1, 3)$ are bounded above by 3 and below by 1, while the interval $(1, \infty)$ has the same lower bound, but *no upper bound*.

Here are some basic notes on the idea and language of boundedness:

- *Many choices:* Upper and lower bounds are far from unique. For $\{1, 2, 3\}$, $[1, 3]$, and $(1, 3)$, for instance, -42 and 42 also work as lower and upper bounds. These may seem unlikely choices, but in practice we sometimes care more about the *existence* (or absence) of bounds than about particular numerical values.

- *Unboundedness:* A set is *unbounded* if it is not bounded. The definition is far from surprising, but deciding whether a set is bounded or unbounded can be challenging. What do you think about

And similarly for sets unbounded above or unbounded below.

$$\left\{ 1, 1 + \frac{1}{2}, 1 + \frac{1}{2} + \frac{1}{3}, 1 + \frac{1}{2} + \frac{1}{3} + \frac{1}{4}, \ldots \right\}$$

and

$$\left\{ 1, 1 + \frac{1}{2}, 1 + \frac{1}{2} + \frac{1}{4}, 1 + \frac{1}{2} + \frac{1}{4} + \frac{1}{8}, \ldots \right\},$$

for example?

See the exercises.

- *Boundedness and absolute values:* Absolute values—which measure magnitude— have a natural connection to boundedness. Indeed:

Fact 1.26. *Let $S \subset \mathbb{R}$ be a nonempty set. S is bounded if and only if there exists $M > 0$ such that $|s| \leq M$ for all $s \in S$.*

We outline the (straightforward) proof in the exercises.

EXAMPLE 1.   Let S and T be nonempty sets of real numbers. Prove some basic properties of bounded sets:

(a) If $S \subset T$ and T is bounded, then S is bounded, too.

(b) If S is finite, then S is bounded.

(c) Let $|S| = \{|s| \mid s \in S\}$. Then S is bounded if and only if $|S|$ is bounded.

(d) Let $S + T = \{s + t \mid s \in S,\, t \in T\}$ and $ST = \{st \mid s \in S,\, t \in T\}$. If S and T are bounded, then so are $S + T$ and ST.

SOLUTION.   These facts follow directly from the definition of boundedness. Claims (a) and (c) are left to the exercises; claim (b) follows from the fact that every finite set of numbers has a largest and a smallest member. For (d), note first that, by assumption, there are positive numbers M_S and M_T with $|s| \leq M_S$ and $|t| \leq M_T$ for all $s \in S$ and $t \in T$. Hence we have

See the triangle inequality at work?

$$|s + t| \leq |s| + |t| \leq M_S + M_T \quad \text{and} \quad |st| = |s||t| \leq M_S M_T$$

for all s and t in question, and the claims about boundedness follow. $\Diamond$

Getting Edgy: sup, lub, inf, and glb

The interval $I = (-3, 42]$, like every bounded set, has infinitely many upper and lower bounds. But there is something obviously special about -3 and 42: the former is the *greatest lower bound* and the latter the *least upper bound* for I. The following definition formalizes these properties—and introduces some impressive Latin synonyms.

Definition 1.27 (Supremum and infimum). Let $S \subset \mathbb{R}$ be a nonempty set. A number α is the *infimum* of S, and we write $\alpha = \inf(S)$ (or $\alpha = \mathrm{glb}(S)$) if

(i) α is a lower bound for S; and

(ii) if α' is any lower bound for S, then $\alpha \geq \alpha'$.

A number β is the *supremum* of S, and we write $\beta = \sup(S)$ (or $\alpha = \mathrm{lub}(S)$) if similar inequalities hold for upper bounds.

This may all seem, for now, like a lot of fuss over not much. Indeed, there is little interesting to be said about sups and infs of simple sets, like bounded intervals. We will see soon, however, that the existence of real sups and infs for more general bounded sets—a property called *completeness*—is essential in the theory of real analysis. Here, for the moment, are some simpler notes on the definition.

- *Do they exist?* If S is unbounded above (or below), then, obviously, S has no supremum (or infimum). Every *bounded* set of real numbers, by contrast, *has* a real supremum β and a real infimum α. We discuss this subtle property and its implications carefully in the next section.

- *In or out?* The interval $I = (-3, 42]$ contains its supremum but not its infimum; including or excluding intervals' endpoints shows that all such combinations are possible. If a set S contains its supremum β (or infimum α), then β is the *maximum* (or minimum) element of S.

- *Only one?* We said "the" in the preceding definition, and this is justified: a set S has at most *one* supremum and *one* infimum. The proof is easy. If β_1 and β_2 are *both* suprema for S, then we'd have both $\beta_1 \le \beta_2$ and $\beta_2 \le \beta_1$, and so $\beta_1 = \beta_2$.

Elbow room. Every member of $I = (0, 1)$ is positive, but I pushes right up against 0 in that members of I can be found within any given distance of 0. By contrast, I leaves 1.003 a little "elbow room": no member of I is within 0.003 of 1.003. This distinction, which comes up often in real analysis, can be described efficiently in terms of boundedness.

Definition 1.28. If a is a number and $S \subset \mathbb{R}$ is a nonempty set, then S is *bounded away from* a if, for some $\delta > 0$, $|s - a| > \delta$ for all $s \in S$.

For $I = (0, 1)$ and $a = 1.003$, for example, we can use $\delta = 0.003$ (or any smaller value of δ). Just as clearly, *no* positive δ works for $I = (0, 1)$ and $a = 0$.

EXAMPLE 2.   The interval $I = (2, 3)$ is bounded away from π because $|\pi - x| > 0.14$ for all $x \in (2, 3)$. By contrast, $3 \notin I$, but I is *not* bounded away from 3, since I has members within *any* small distance δ from 3. The set $\mathbb{N}$ is bounded away from π, but $\mathbb{Q}$ is not, since for any $\delta > 0$, no matter how small, there are rational numbers within δ of π. ◊

EXAMPLE 3.   Let $S = \{s_1, s_2, \ldots, s_n\}$ be a *finite* set of real numbers, with $a \notin S$. Show that S is bounded away from a.

SOLUTION.   The idea is that, since S is finite, one of the s_i must be *closest* to a. More precisely, let $d_i = |s_i - a|$ be the distance from s_i to a. Then $D = \{d_1, d_2, \ldots, d_n\}$ is a finite set of positive numbers, and so has a *minimum* member, say d_1. (Every finite set of numbers has largest and smallest elements; see Fact 1.16, page 55.) Now it is clear that $\delta = d_1/2$ works in Definition 1.28. ◊

Any positive δ smaller than d_1 would work here.

Functions and Bounds

Boundedness for *functions* is much like that for *sets*, but the jargon is slightly different:

Definition 1.29. Let f be a real-valued function and A a subset of the domain of f. If $f(a) \leq M$ for all $a \in A$, then f is *bounded above by M on A*. If $f(a) \geq m$ for all $a \in A$, then f is *bounded below by m on A*. If f is bounded above and below on A, then f is *bounded on A*.

EXAMPLE 4.   Let

$$f(x) = \sin x, \quad g(x) = x^3 - 5x^2 + 7x + 13, \quad \text{and} \quad h(x) = 1/x.$$

On which sets A are f, g, and h bounded or unbounded?

SOLUTION.   The function f is least interesting. Because $-1 \leq \sin(x) \leq 1$ for *all* real x, f is bounded (above by 1 and below by -1) on *every* set $A \subset \mathbb{R}$. Sharper bounds can be found on some smaller sets, of course. If, say, $A = [0.36, 1.27]$ and $a \in A$, then $0.353 \approx \sin 0.36 \leq \sin a \leq \sin 1.27 \approx 0.955$.

sin *x* increases on this interval.

We know (informally for now) from elementary calculus that $g(x)$—like every odd-degree polynomials—"blows up" for large (positive or negative) values of x. Thus, g is *unbounded* on unbounded sets like $[0, \infty)$, $(-\infty, 42)$, and $\mathbb{R}$ itself. On the other hand, g is bounded on *every* bounded set A. If, say, A is the interval $[-K, K]$ and $a \in A$, then we have

Thanks, triangle inequality.

$$|g(a)| = |a^3 - 5a^2 + 7a + 13| \leq |a^3| + 5|a^2| + 7|a| + 13$$
$$\leq K^3 + 5K^2 + 7K + 13;$$

the last quantity is an upper bound. (With more work we might find a smaller upper bound.)

The function h behaves somewhat differently: although bounded on some infinite intervals, such as $[0.00017, \infty)$ and $(-\infty, -0.24)$, h is *unbounded* on some *finite* intervals, like $(0, 1)$ and $(-0.24, 0)$.   ◇

Exercises

1. Prove parts (a) and (c) of Example 1, page 68.

2. Show that if $S \subset \mathbb{R}$ is a *finite* set, then S contains both $\sup(S)$ and $\inf(S)$. (Hint: See Fact 1.16, page 55.)

3. In each part, give an example of an interval $I \subseteq \mathbb{R}$ with the given properties.

(a) I has a minimum but no supremum

(b) I has a supremum but no infimum

(c) I has a supremum and an infimum, but no minimum

(d) Can any set S have a maximum but no supremum? Why?

4. We said in this section that a set S of real numbers is bounded if and only if there is some $M > 0$ such that $|s| \leq M$ for all $s \in S$. Equivalently, $S \subseteq [-M, M]$.

   (a) Suppose S has upper and lower bounds -5 and 3, respectively. Find an M that satisfies the condition above.

   (b) Suppose S has upper and lower bounds a and b, respectively. Find an M that satisfies the condition above.

   (c) Suppose $M = 42$ satisfies the condition above. Find upper and lower bounds for S.

   (d) Prove the statement at the beginning of this exercise.

5. Consider the functions $f(x) = e^x$ and $g(x) = \sin(x)$ and the set $A = [0, 10]$.

   (a) Find upper and lower bounds for f and g on A. Is either f or g bounded above or below on $\mathbb{R}$?

   (b) Find upper and lower bounds for the composite functions $f \circ g$ and $g \circ f$ on A. Is either of these function bounded on $\mathbb{R}$?

6. Consider the function $f : \mathbb{R} \to \mathbb{R}$ given by $f(x) = x^2 + 2x$.

   (a) Explain briefly why f is bounded below but not bounded above on $\mathbb{R}$.

   (b) Find a set $A \subseteq \mathbb{R}$ (as large as possible) such that f is bounded above on A by $M = 120$.

   (c) Find a set $A \subseteq \mathbb{R}$ (as large as possible) such that f is bounded below on A by $m = -1$.

   (d) Find a set $A \subseteq \mathbb{R}$ (as large as possible) such that f is bounded above on A by $M = 0$.

   (e) Let $A \subset \mathbb{R}$ be any set with 1234 points. Explain briefly why f is bounded above and below on A.

7. Show that if $A \subseteq \mathbb{R}$ is bounded and $f(x) = 3x + 5$, then f is bounded on S.

8. Let $f : \mathbb{R} \to \mathbb{R}$ and $g : \mathbb{R} \to \mathbb{R}$ be functions, with $-2 \leq f(x) \leq 1$ and $-3 \leq g(x) < 4$ for all $x \in \mathbb{R}$.

   (a) Find upper and lower bounds for the function $f + g$ on $\mathbb{R}$.

   (b) Find upper and lower bounds for the function $|fg|$ on $\mathbb{R}$.

   (c) Find upper and lower bounds for the function $f \circ g$ and $g \circ f$ on $\mathbb{R}$.

9. For each set S following find—if possible—$\inf(S)$ and $\sup(S)$. Are any of these maxima or minima? No proofs needed.

   (a) $S = \{n \in \mathbb{N} \mid n^2 < 10\}$

   (b) $S = \{p \in \mathbb{N} \mid p \text{ is prime}\}$

   (c) $S = \{p \in \mathbb{N} \mid p \text{ is an even prime}\}$

   (d) $S = \{\sin(x) + 42 \mid x \in \mathbb{R}\}$

   (e) $S = \{r \in \mathbb{Q} \mid r^2 \leq 2\}$

   (f) $S = \{x \in \mathbb{R} \mid x^2 \leq 2\}$

10. Is each of the following sets bounded? If so, give upper and lower bounds. If not, why not? (No proofs needed; it's OK to use ideas from elementary calculus.)

   (a) $\left\{ 1, 1 + \dfrac{1}{2}, 1 + \dfrac{1}{2} + \dfrac{1}{3}, 1 + \dfrac{1}{2} + \dfrac{1}{3} + \dfrac{1}{4}, \cdots \right\}$

   (b) $\left\{ 1, 1 + \dfrac{1}{2}, 1 + \dfrac{1}{2} + \dfrac{1}{4}, 1 + \dfrac{1}{2} + \dfrac{1}{4} + \dfrac{1}{8}, \cdots \right\}$

   (c) $\left\{ \dfrac{\ln 1}{1}, \dfrac{\ln 2}{2}, \dfrac{\ln 3}{3}, \dfrac{\ln 4}{4}, \cdots \right\}$

   (d) $\left\{ \dfrac{2^1}{1^2}, \dfrac{2^2}{2^2}, \dfrac{2^3}{3^2}, \dfrac{2^4}{4^2}, \cdots \right\}$

   (e) $\{\tan(x) \mid x \in \mathbb{R}\}$

   (f) $\{x \in \mathbb{R} \mid x^3 - 5x^2 + 7x - 1234 = 0\}$

11. Consider the sets EW of all non-hyphenated English words and WITP of all words in this problem. Let f be the function with rule

$$f(\text{word}) = \text{number of letters in word}.$$

   (a) Is f bounded above and/or below on EW? Can you give good upper and lower bounds? Explain.

   (b) Find upper and lower bounds for f on the set WITP.

12. Consider the real-valued function g whose domain is the set GCES of grammatically correct English sentences, and has rule

$$g(\text{sentence}) = \text{number of words in sentence}.$$

Is g bounded on GCES? Can you give upper and lower bounds? What bounds apply g on the subset SITP of sentences in this problem?

13. In each part following, describe the sets $S \subseteq \mathbb{R}$ with the given property.

(a) $\forall s \in S \, \exists M \in \mathbb{R}$ such that $|s| \leq M$

(b) $\exists M \in \mathbb{N}$ such that $|s| \leq M \quad \forall s \in S$

(c) $\exists M \in \mathbb{N}$ such that $|s| \geq M \quad \forall s \in S$

14. Consider the interval $I = [-10, 10]$. Find good upper and lower bounds for each of the following functions defined on I. Use the triangle inequality or ideas from elementary calculus.

(a) $f(x) = \dfrac{1}{1 + x^2}$.

(b) $g(x) = 3x^2 + 2x - 7$.

(c) $h(x) = \sin^2 x + \cos^2 x + x$.

15. Do the preceding problem, but replace the interval I with the interval $J = [-K, K]$, where $K > 0$. (Answers may depend on K, of course.)

16. Let $f(x) = Ax + B$, where A and B are any real constants, and let I be the interval $[-K, K]$, where $K > 0$. Find sharp (i.e., best possible) upper and lower bounds for f on I. (Hint: Handle the cases $A > 0$, $A = 0$, and $A < 0$ separately.)

17. (This problem alludes to Definition 1.28, page 69.) Let $S \subset \mathbb{R}$ be a nonempty set with $0 \notin S$.

(a) Can S be both (i) unbounded; and (ii) bounded away from 0? Either give an example or explain why none is possible.

(b) Show that S is bounded away from 0 if and only if the set $T = \{1/s \mid s \in S\}$ is bounded.

(c) Show that S is bounded away from 0 if and only if there is an open interval $I = (a, b)$ with $0 \in I$ and $S \cap I = \emptyset$.

18. (This problem alludes to Definition 1.28, page 69.) In each of the following cases either give an example or say why none can exist.

(a) A set of positive numbers that is bounded away from one but not from zero.

(b) A number a such that $\mathbb{Q}$ is bounded away from a.

(c) A set S that is bounded away from $a = 42$ but not bounded away from any other integer.

19. This problem is about the intervals $I = [0, 1]$ and $J = (2, 3)$.

(a) Show that if $a \notin I$, then I is bounded away from a.

(b) If $a \notin J$, must J be bounded away from a? Why or why not?

(c) If $a \notin \mathbb{N}$, must $\mathbb{N}$ be bounded away from a?

20. Let $S \subset \mathbb{R}$ be bounded, with $\beta = \sup(S)$. Show that S is *not* bounded away from β.

1.9   Numbers 103: Completeness

There are many, many rational numbers. Indeed, between *any* two particular rationals lie infinitely many others. Between zero and one, for instance, we find

$$\frac{1}{2}, \frac{1}{3}, \frac{1}{4}, \ldots, \frac{1}{507}, \frac{1}{508}, \ldots$$

Between any two of *these*, say 1/507 and 1/508, lie infinitely many others; and between any two of *those* .... Who could ask for more?

We could, and we do. There are just not enough rational numbers out there to permit the sort of analysis—or even algebra—that we want to do. We have seen, for instance, that there is no rational, exact solution to the simple-looking equation $x^2 = 2$, even though rational numbers like 1.4142135 (too small) and 1.4142136 (too big) come very, very close.

The *real* numbers have no such gaps. Somewhere between all those too-small rational numbers x (with $x^2 < 2$) and the too-big rational numbers y (with $y^2 > 2$) is a real number r that deserves the name $\sqrt{2}$ because $r^2 = 2$ *exactly*. We will take this "gap-free" property, known technically as *completeness*, as an *axiom*: a defining property of the real numbers rather than a claim that requires proof.

> *The Completeness Axiom for* $\mathbb{R}$. Every nonempty set of real numbers that is bounded above has a supremum. In symbols: If $S \subset \mathbb{R}$, $S \neq \emptyset$, and S is bounded above, then there is a real number β such that $\beta = \sup(S)$.

Like other succinct mathematical statements, this one can use some unpacking:

- *What about infs?* For brevity, the completeness axiom mentions only sups. But the story is similar for infs: Every nonempty set that is bounded *below* has an *infimum*. The inf version, moreover, follows easily from the sup version.

  See the exercises.

- *The real advantage:* The completeness axiom guarantees that every bounded set of *rationals* has a supremum—which may or may not be a rational number. By contrast, the supremum of a bounded set of reals must be real. In this sense $\mathbb{R}$ is complete, while $\mathbb{Q}$ is incomplete.

  Rational numbers are also real.

- *In or out?* We know already that the sup and the inf of a given set may or may not lie within the set. It is pretty obvious, for instance, that the half-closed interval $(1, 2]$ contains its supremum but not its infimum. For more complicated sets, things may be less clear. Consider, for example, the set $S = \{r \in \mathbb{Q} \mid r^2 \le 152399024\}$. A little thought shows that S is bounded, so the completeness axiom guarantees that a sup and an inf exist. We might also guess (correctly!) that $\sup(S) = \sqrt{152399024}$ and $\inf(S) = -\sqrt{152399024}$, but do these numbers lie inside or outside S? This is far from clear at a glance—and the completeness axiom is no help at all. To decide, we'd need to discover whether $\sqrt{152399024}$ is rational, and this takes a little effort.

  The question boils down to whether 152399024 is a perfect square. It isn't ... quite.

Using Completeness: Order Properties of $\mathbb{R}$

The following theorem collects some useful properties of the real numbers as an ordered set. Most are familiar; the surprise, if any, is the role that completeness plays in their proofs.

Theorem 1.30 (Order properties of numbers).

- $\mathbb{N}$ is unbounded: *Given any real number M, there exists a positive integer n with $n > M$.*

- Squeezing in: *For every positive number ϵ, no matter how small, there exists a positive integer n such that $0 < \frac{1}{n} < \epsilon$.*

- The Archimedean principle: *Given real numbers a and b with $0 < a < b$, there exists a positive integer n with $na > b$.*

More on the Archimedean principle. The Archimedean principle can be thought of geometrically. The challenging case occurs when a is small and b is large, as suggested in Figure 1.2.

The principle says that steps of size a, no matter how short, will eventually complete a trip of length b, no matter how long. As Archimedes (around 250 BCE) might have put it, a positive number a, no matter how small, is not "infinitesimal": successive multiples a, $2a$, $3a$, ... will eventually exceed any proposed bound b.

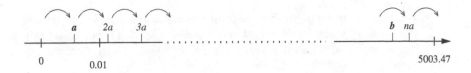

Figure 1.2. Archimedes in motion: a long trip in short steps.

Proof: To show that $\mathbb{N}$ is unbounded, assume toward contradiction that *no* integer n exceeds M. Then M is an upper bound for $\mathbb{N}$, and so, by the completeness axiom, $\mathbb{N}$ must have a real supremum, say β. Because β is the least upper bound, $\beta - 1$ is not an upper bound, and so $n_0 > \beta - 1$ must hold for some positive integer n_0. But then $n_0 + 1 > \beta$, which contradicts our assumption that *no* integer exceeds β. Thus $\mathbb{N}$ is unbounded, as claimed.

The remaining claims turn out to be lightly disguised versions of the result just shown. For the Archimedean principle, note that, for positive a and b,

$$na > b \quad \Longleftrightarrow \quad n > \frac{b}{a}.$$

But we've just shown that $\mathbb{N}$ is unbounded, so the right-hand inequality must hold for sufficiently large n.

Proof of the "squeezing-in" principle is left as an exercise. $\square$

Nested Intervals

A collection of intervals $I_1, I_2, I_3, \ldots$ is called *nested* if

$$I_1 \supseteq I_2 \supseteq I_3 \supseteq I_4 \supseteq \cdots.$$

Here is a nested collection of *open* intervals:

$$(0, 1) \supseteq (0, 1/2) \supseteq (0, 1/3) \supseteq (0, 1/4) \supseteq \cdots.$$

Observe that, although each interval I_n contains infinitely many points, the intersection of the full collection is *empty*.

The story is different for *closed* intervals. The nested collection

$$[0, 1] \supseteq [0, 1/2] \supseteq [0, 1/3] \supseteq [0, 1/4] \supseteq \cdots$$

has exactly one point of intersection—the number 0. The following theorem describes the situation in general; completeness is the key.

Theorem 1.31 (The nested intervals theorem). *Consider a nested infinite collection*

$$I_1 \supseteq I_2 \supseteq I_3 \supseteq I_4 \supseteq \cdots$$

of closed and bounded intervals. The intersection

$$I_1 \cap I_2 \cap I_3 \cap I_4 \cap \dots$$

contains at least one point. If the intervals' lengths shrink to zero, then the intersection is a single point.

Proof (sketch): We sketch the main idea, leaving some details to exercises. If we write $I_n = [a_n, b_n]$ for all positive integers n, then the nesting condition means

$$a_1 \leq a_2 \leq a_3 \leq \dots \leq b_3 \leq b_2 \leq b_1.$$

In particular, the "left endpoint set" $A = \{a_1, a_2, a_3, \dots\}$ is bounded above. By completeness, we can set $\alpha = \sup A$, and it is readily shown that α lies in *all* the I_n. A similar argument shows that $\beta = \inf\{b_1, b_2, b_3, \dots\}$ lies in all the I_n. If the lengths of the I_n shrink to zero, then $\alpha = \beta$. $\qquad\square$

Rationals and Irrationals: Tightly Packed

We mentioned casually at the beginning of this section that between any two given rational numbers lie infinitely many other rational numbers. A little more is true, and now we can prove it.

Theorem 1.32. *Let x and y be any two real numbers, with $x < y$.*

(a) *The interval (x, y) contains at least one rational and one irrational number.*

(b) *The interval (x, y) contains infinitely many rationals and infinitely many irrationals.*

Proof: Claim (b) looks much stronger than (a), but showing that (a) implies (b) is surprisingly easy. Let's suppose, toward contradiction, that (a) holds but (x, y) contains only *finitely* many rationals r_1, r_2, ..., r_n. Then one of these, say r_n, is largest. But then the interval (r_n, y) must contain *no* rationals, which contradicts (a). Exactly the same proof applies to irrational numbers, so we conclude that (a) implies (b).

To prove (a) we'll use Theorem 1.30. For convenience, we'll handle here only the special case $0 \leq x < y$, and mop up remaining cases as exercises. If we set $\epsilon = y - x$, then the squeezing-in part of Theorem 1.30 says that for some integer n the *rational* number $r = 1/n$ lies in the interval $(0, \epsilon)$—as does the *irrational* number $p = r/\sqrt{2}$.

To finish the proof, we show that at least one of r, $2r$, $3r$, $4r$, ... (all are rational) and at least one of p, $2p$, $3p$, $4p$, ... (all are irrational) lie inside (x, y). This is intuitively reasonable—for both sequences, the "jumps" are too small to miss (x, y) entirely. To put it formally, say for p, $2p$, $3p$, $4p$, ..., consider the

Recall: Every nonempty finite set of reals has a maximum element.

Note that $0 < p < r < y - x$.

set $S = \{np \mid n \in \mathbb{N} \text{ and } np < y\}$. By the Archimedean principle, $np > y$ for sufficiently large n, which amounts to saying that S is finite. Hence S has a largest element, say $n_0 p$, with $n_0 p < y$ but $(n_0 + 1)p \geq y$. But now

$$(n_0 + 1)p \geq y \quad \Longrightarrow \quad n_0 p \geq y - p > y - (y - x) = x,$$

and so $x < n_0 p < y$, as desired. The proof that $x < m_0 r < y$ for some positive integer m_0 is almost identical. $\square$

Exercises

1. The list $\frac{1}{2}, \frac{1}{3}, \frac{1}{4}, \ldots, \frac{1}{n}, \ldots$ suggests an explicit "recipe" for an endless collection of rational numbers between 0 and 1.

   (a) Give a similar recipe for an endless list of rationals between 0 and $\frac{1}{507}$.

   (b) Give a similar recipe for an endless list of rationals between $\frac{1}{2}$ and $\frac{1}{3}$.

   (c) Let a and b be any two rationals, with $a < b$. Give a recipe (involving a and b) for an endless list of rationals between a and b.

2. In the spirit of Problem 1:

   (a) Give a recipe for an endless list of *irrationals* between 0 and $\frac{1}{507}$.

   (b) Let a and b be any two rationals, with $a < b$. Give a recipe (involving a and b) for an endless list of irrationals between a and b.

3. Use the unboundedness of $\mathbb{N}$ (the first part of Theorem 1.30) to prove the squeezing-in principle (the second part of Theorem 1.30).

4. Our statement of the Archimedean principle (part of Theorem 1.30) includes the hypothesis $0 < a < b$.

   (a) Is the claim true if $0 < b \leq a$? Give a proof or counterexample.

   (b) Is the Archimedean claim true or false if we replace the hypothesis $0 < a < b$ with $a < b$? What if $0 \neq a < b$?

5. (a) Show with a counterexample that the following statement (which resembles the Archimedean principle) is *false*: If a and b are real numbers with $a < b$, then there exists a positive integer n such that $na > b$.

   (b) Prove (using the Archimedean principle) or disprove (with a counterexample) the following statement: If a and b are nonzero real numbers with $a < b$, then there exists an integer n such that $na > b$.

(c) Prove *without* using the Archimedean principle: If a and b are positive numbers with $a < b$, then there exists a real number n such that $na > b$.

6. Part (a) of Theorem 1.32 says that every interval (x, y) contains at least one rational and one irrational. The proof given there assumed that $0 \leq x < y$.

   (a) Show that the same result holds if $x < 0 < y$. (Hint: Apply the result already proved to the new interval $(X, Y) = (0, y)$.)

   (b) Show that the result still holds if $x < y \leq 0$. (Hint: Look at the new interval $(X, Y) = (-y, -x)$.)

7. The completeness axiom says this: *If $S \subset \mathbb{R}$ is bounded above, then S has a supremum β, and $\beta \in \mathbb{R}$.*

   (a) Is the italicized statement true or false if $\mathbb{R}$ is replaced (in both places) by $\mathbb{Z}$? Explain.

   (b) Is the italicized statement true or false if $\mathbb{R}$ is replaced (in both places) by $\mathbb{Q}$? Explain.

8. Use the completeness axiom for sups to prove the analogous result for infs: If $S \subset \mathbb{R}$, $S \neq \emptyset$, and S is bounded below, then there is $\alpha \subset \mathbb{R}$ such that $\alpha = \inf(S)$. (Hint: Given a nonempty set S, look at the new set $-S$ defined by $-S = \{-s \mid s \in S\}$. Apply the completeness axiom to $-S$ and interpret the result.)

9. In each case, find a nested collection $I_1 \supseteq I_2 \supseteq \ldots$ of intervals as described.

   (a) Each I_n is open, the I_n are nested, and the intersection is the single point 3.

   (b) Each I_n is closed, no two I_n are equal, and the intersection is the interval $[-3, 42]$.

   (c) Each I_n is open, no two I_n are equal, and the intersection is the interval $[-3, 42]$.

10. Show that every *closed* interval $[a, b]$ is the intersection of a nested $I_1 \supset I_2 \supset \ldots$ of *open* intervals.

11. Show that no *open* interval (a, b) is the intersection of a nested $I_1 \supset I_2 \supset \ldots$ of *closed* intervals. Hint: It's OK to assume the (true) fact that if I is a closed interval and $(a, b) \subseteq I$, then $[a, b] \subseteq I$, too.

12. This problem is about some details in the proof of Theorem 1.31, page 76; see the notation there.

    (a) Prove that, for all m and n, $a_m \leq b_n$.

    (b) Prove that $\alpha \leq b_n$ for all n.

    (c) Prove that, for all n, $a_n \leq \alpha \leq \beta \leq b_n$.

13. We said in this section that completeness of $\mathbb{R}$ guarantees that 2 has a *real* square root.

    (a) Let $S = \{x \mid x^2 < 2\}$. Explain why S has a least upper bound; call it β. We'll show that $\beta^2 = 2$.

    (b) Suppose toward contradiction that $\beta^2 < 2$. Now choose any $h > 0$ such that (i) $0 < h < 1$ and (ii) $h \leq (2 - \beta^2)/(2\beta + 1)$. Show that $(\beta + h)^2 < 2$. Hint: Use the fact that $h^2 < h$.

    (c) The preceding calculation leads to a contradiction. Identify it, and conclude $\beta^2 < 2$ is impossible.

    (d) Prove by contradiction that $\beta^2 > 2$ is also impossible. (Hint: Set $k = (\beta^2 - 2)/(2\beta)$ and consider $\beta - k$.)

14. Imitate Problem 13 to show that *every* positive number a has a unique positive square root.

15. Use the result of Problem 14 to show that every positive number a has unique positive roots $\sqrt{a}$, $\sqrt[4]{a}$, $\sqrt[8]{a}$, $\sqrt[16]{a}$, etc.

16. In the spirit of Problem 14, it's true that every positive number a has a unique positive *cube* root. We can show this by defining $S = \{x \mid x^3 < a\}$, setting $\beta = \sup(S)$, and proving that $\beta^3 = a$. Complete details below.

    (a) Suppose toward contradiction that $\beta^3 < a$. Choose h with $0 < h < 1$ and $h < (a - \beta^3)/(3\beta^2 + 3\beta + 1)$. (Why is this possible?) Show that $(\beta + h)^3 < a$, a contradiction.

    (b) Show that $\beta^3 > a$ also leads to a contradiction. To do so, choose an appropriate positive value of k and show that $(\beta - k)^3 > a$.

17. Use results of problems above to show that that every positive number a has unique positive roots $\sqrt[6]{a}$, $\sqrt[12]{a}$, and $\sqrt[18]{a}$.

18. It is well known that every number has an *infinite decimal expansion*. For instance, we can write

$$\frac{1}{3} = 0.3333333\ldots; \quad \pi = 3.1415926\ldots; \quad 42 = 42.0000000\ldots.$$

Here is a slightly subtler fact: Every infinite string of decimal digits corresponds to a unique real number β. Use the completeness axiom to explain why. (Hint: For convenience, consider only strings of the form $0.d_1d_2d_3d_4d_5\ldots$, where each d_i is a decimal digit. Use these digits to construct a bounded set with supremum β.)

CHAPTER 2

Sequences and Series

2.1 Sequences and Convergence

An *infinite sequence* is a list $a_1, a_2, a_3, \ldots$ of real numbers, indexed by the positive integers. Here are some examples:

$$
\begin{aligned}
a_1, a_2, a_3, a_4, \ldots, a_{42}, \ldots &= 1, \frac{1}{2}, \frac{1}{3}, \frac{1}{4}, \ldots, \frac{1}{100}, \ldots; \\
b_1, b_2, b_3, b_4, \ldots, b_{42}, \ldots &= 1, -1, 1, -1, \ldots, -1, \ldots; \\
c_1, c_2, c_3, c_4 \ldots, c_{42}, \ldots &= \sin 1, \sin 2, \sin 3, \sin 4, \ldots, \sin 42, \ldots; \\
f_1, f_2, f_3, f_4, \ldots, f_{42}, \ldots &= 1, 1, 2, 3, \ldots, 267914296, \ldots; \\
g_1, g_2, g_3, g_4, \ldots, g_{42}, \ldots &= \frac{1}{1}, \frac{2}{1}, \frac{3}{2}, \frac{5}{3}, \ldots, \frac{433494437}{267914296}, \ldots.
\end{aligned}
$$

Observe:

- *Words and notations:* A particular entry a_n in a sequence is called its nth *term*. Notice that the subscript, or *index variable* name, is arbitrary: a_n and a_i and even a_x can all mean the same thing. For the entire sequence we write $\{a_n\}_{n=1}^{\infty}$, or just $\{a_n\}$; the braces emphasize the fact that a sequence is a *set*.

  The a_x notation is legal but unwise, since x usually denotes a "continuous" variable.

- *Sequences as functions:* A sequence $\{a_n\}$ is, among other things, a real-valued *function*, with domain $\mathbb{N}$. Writing, say, $a(42)$ rather than a_{42} can help emphasize this viewpoint.

- *Sequence rules:* Like other functions, sequences are often defined by symbolic rules. Among the sequences above, for example, we have the rules

$$
a_n = 1/n, \quad b_n = (-1)^{n+1}, \quad \text{and} \quad c_n = \sin n.
$$

Not every sequence (or function, for that matter) has such a simple recipe. The sequence $\{f_n\}$ above is the famous *Fibonacci sequence*, which is defined *recursively*:

$$
f_1 = 1; \quad f_2 = 1; \quad f_n = f_{n-2} + f_{n-1} \quad \text{if } n \geq 3.
$$

The sequence $\{g_n\}$ can be built (can you guess how?) from $\{f_n\}$.

- *Converge or diverge?* It is clear at a glance that $\{a_n\}$ *converges* to the limit zero. Sequences $\{b_n\}$ and $\{f_n\}$ presumably *diverge*, because the former never settles on any single limit, while the latter blows up in size. Precise definitions, which we'll give in a moment, will formalize these intuitions and lead to rigorous proofs. Whether $\{c_n\}$ and $\{g_n\}$ converge or diverge may be harder to guess.

Calculating some decimal values might help us guess.

Convergence and Divergence

Informally speaking, a sequence $\{a_n\}$ converges to a limit, say 3, if the numbers a_n "approach" 3 as n "tends to infinity." Such an intuitive view can be useful, but it's too vague to permit careful proofs. The words in quotes are especially important—but especially unclear as written. We need a formal definition:

Definition 2.1. Let $\{a_n\}$ be a sequence and L a number. We say $\{a_n\}$ *converges* to L if for every $\epsilon > 0$ there exists a number N so that

$$|a_n - L| < \epsilon \quad \text{whenever } n > N.$$

In this case L is the *limit*, and we write $\lim_{n\to\infty} a_n = L$, or just $a_n \to L$. If no such L exists, we say $\{a_n\}$ *diverges*.

The definition is subtle, and deserves some unpacking:

- *Greek letters:* The Greek letter ϵ (lower-case epsilon) is traditionally used in mathematics to denote small positive quantities. Later we'll use δ (lower-case delta), as well, for similar purposes.

- *ϵ measures "nearness":* The inequality $|a_n - L| < \epsilon$ means that a_n is within ϵ of L. If ϵ is very small then a_n is very near to L.

- *N measures "waiting time":* The definition requires that *every* term a_n with $n > N$ be within ϵ of L. The number N says how long we need to wait for this happy outcome. If, say, $\epsilon = 0.001$ and $N = 234.7$, then we're guaranteed that a_{235}, a_{236}, a_{237}, and *all* following terms are within 0.001 of the limit.

- *Symbolic shorthand:* The heart of the definition can be written compactly in symbols:

$$\forall \epsilon > 0 \quad \exists N \in \mathbb{N} \quad \text{such that} \quad n > N \implies |a_n - L| < \epsilon$$

Using the definition: notes on proofs. Our concise definition can lead, ideally, to equally concise proofs. But such polished products can seem both impressive and mysterious, like a Ferrari with the hood up. To raise the hood a bit, we'll often precede formal proofs with informal discussion—but keep the two separate. We start with some positive and negative examples.

EXAMPLE 1. Prove that the sequence $\{a_n\}$ with $a_n = \frac{1}{n}$ converges to zero. In symbols, $\lim_{n \to \infty} \frac{1}{n} = 0$.

Pre-proof discussion. The *fact* that $a_n \to 0$ seems obvious; how could it be otherwise? To invoke the definition, we work with the desired inequality, $|a_n - L| < \epsilon$. Here we have

$$|a_n - L| = \left| \frac{1}{n} \right| = \frac{1}{n}, \quad \text{and} \quad \frac{1}{n} < \epsilon \iff n > \frac{1}{\epsilon}.$$

(The first calculation works because—but only because—both n and ϵ are positive.) Now we're getting somewhere: $N = 1/\epsilon$ does what the definition requires, and we're ready for a slick and seamless proof.

Clearing out pesky absolute value bars is a big help.

Proof: Let $\epsilon > 0$ be given. Set $N = 1/\epsilon$. This N "works" because if $n > N$, then

$$|a_n - L| = \left| \frac{1}{n} \right| = \frac{1}{n} < \frac{1}{N} = \epsilon,$$

which is what the definition requires.   □

◊

EXAMPLE 2. Let $x_n = (3n + 2)/(n + 5)$. Show that $\{x_n\}$ converges.

Discussion. First we need a candidate for L; a little thought or plugging in large numbers suggests $L = 3$. As in Example 1, we manipulate the desired inequality in search of a suitable N. This time we need a little more algebra:

Again the annoying absolute value eventually vanishes.

$$|x_n - L| - \left| \frac{3n + 2}{n + 5} - 3 \right| = \left| \frac{3n + 2 - 3(n + 5)}{n + 5} \right| = \left| \frac{-13}{n + 5} \right| = \frac{13}{n + 5}$$

and

$$\frac{13}{n + 5} < \epsilon \iff \frac{13}{\epsilon} < n + 5 \iff \frac{13}{\epsilon} - 5 < n.$$

So $N = \frac{13}{\epsilon} - 5$ does the job, and we're ready for another brief proof.

Proof. Let $\epsilon > 0$ be given. Set $N = \frac{13}{\epsilon} - 5$. This N "works" because if $n > N$, then (omitting some routine algebra in the second equality) we have

$$|x_n - L| = \left| \frac{3n + 2}{n + 5} - 3 \right| = \frac{13}{n + 5} < \frac{13}{N + 5} = \frac{13}{\frac{13}{\epsilon} - 5 + 5} = \epsilon,$$

just as the definition requires.   ◊

EXAMPLE 3.   Show that the oscillating sequence $b_1, b_2, b_3, b_4, \cdots = 1, -1, 1,$ $-1, \ldots$ diverges.

Discussion.   As in Example 1 the claim is hardly surprising. The new twist is to prove a negative: *No* number L works in the definition. One approach involves the fact that no single number L can be close (within 0.01, say) to both 1 and -1.

Proof.   Assume, toward contradiction, that $\{b_n\}$ converges to any number L. Let $\epsilon = 0.01$ and choose N as in the definition. If n_0 is any *odd* integer with $n_0 > N$, then we have

$$|b_{n_0} - L| = |1 - L| < 0.01 \implies L \in (0.99, 1.01).$$

Similarly, if n_1 is an *even* integer with $n_1 > N$, then

$$|b_{n_1} - L| = |-1 - L| < 0.01 \implies L \in (-1.01, -0.99).$$

Thus L lies in two disjoint intervals, which is absurd.   ◇

Visualizing Sequences and Convergence

Like ordinary graphs in a calculus course, geometric views of sequences and convergence help us visualize their behavior—good and bad. Different views are possible.

Points on a line.   Selected terms of a sequence $\{a_n\}$ can be plotted as labeled points on a line. The resulting diagrams are crude and limited, but they can illuminate properties of sequences and convergence. Figure 2.1, for example, illustrates the ϵ–N condition for convergence: All terms a_n with $n > N$ lie within ϵ of L.

Points in a plane.   Terms of $\{a_n\}$ can also be viewed as points (n, a_n) on the graph of the function $a : \mathbb{N} \to \mathbb{R}$. Figure 2.2 offers glimpses of two sequences, $\{1/n\}$ and $\{\sin n\}$. One converges to zero (as we've proved!); the other looks unlikely to converge to anything. Such graphs can reveal a lot, but never everything—a sequence has infinitely many terms, and a picture shows only a tiny sample.

Proving this is another matter.

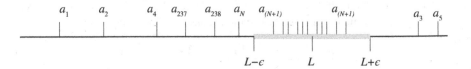

Figure 2.1. Convergence: a one-dimensional view.

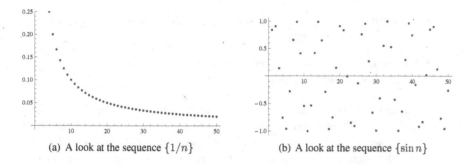

(a) A look at the sequence $\{1/n\}$ (b) A look at the sequence $\{\sin n\}$

Figure 2.2. Graphical views of two sequences.

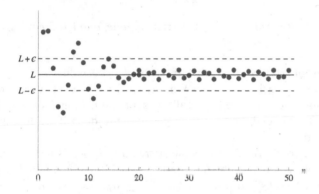

Figure 2.3. What convergence means graphically.

Sequence graphs can also show how ϵ and N interact. The idea is to express the key definition in graphical language: for any $\epsilon > 0$, no matter how small, there is some N on the (horizontal) n-axis so that all graph points *to the right* of N lie *inside* an "ϵ-band" around the horizontal line $y = L$, as Figure 2.3 suggests.

Properties of Sequences

Sequences, like other real-valued functions, may or may not have various behavioral properties. Here are some typical definitions:

Definition 2.2. Let $\{a_n\}$ be a sequence of real numbers.

- *Boundedness:* $\{a_n\}$ is *bounded above* if there exists M such that $a_n \leq M$ for all $n \in N$; it is *bounded below* if there exists m such that $a_n \geq m$ for all n.

- *Monotonicity:* $\{a_n\}$ is *increasing* if $a_n \leq a_{n+1}$ for all $n \in N$; $\{a_n\}$ is *decreasing* if $a_n \geq a_{n+1}$ for all $n \in N$. $\{a_n\}$ is *monotone* if it is either increasing or decreasing.

- *Strictness:* $\{a_n\}$ is *strictly increasing* if $a_n < a_{n+1}$ for all $n \in N$. $\{a_n\}$ is *strictly decreasing* if $a_n > a_{n+1}$ for all $n \in N$.

The sequence $\{1/n\}$, for example, is strictly decreasing and bounded (above and below), while the Fibonacci sequence is increasing and unbounded (above).

The following two theorems link these properties to convergence and divergence. Neither statement is surprising, but the formal proofs are nice exercises—and *left* as exercises—in using the definition. Some pre-proof discussion follows each result.

Theorem 2.3. *If a sequence $\{a_n\}$ is (i) monotone, and (ii) bounded, then $\{a_n\}$ converges.*

Almost-a-proof discussion. Let's assume that $\{a_n\}$ is *increasing*; a similar proof works for the decreasing case. Since the set $\{a_n\}$ is bounded, it has a *supremum*, by the completeness axiom. Let's call the supremum L, and show that L is the limit, too. To this end, let $\epsilon > 0$ be given; we're done if we find any N that "works" for this ϵ.

Here's the trick. Because L is the supremum, $L - \epsilon$ is *not* an upper bound for $\{a_n\}$, and so there must be some term, a_N, with $L - \epsilon < a_N$. The punchline is that this N "works": If $n > N$, then $a_n \geq a_N$ because $\{a_n\}$ is increasing; therefore, we have

$$L - \epsilon < a_N \leq a_n \leq L < L + \epsilon,$$

as desired.

Nice to see it turn up again.

Theorem 2.4. *If a sequence $\{a_n\}$ converges, then $\{a_n\}$ is bounded both above and below.*

Discussion. Suppose $\{a_n\}$ converges to L. If we set $\epsilon = 1$, then—by the definition—there is some N such that $L - 1 < a_n < L + 1$ for all $n > N$. In particular, the "upper tail" $\{a_{N+1}, a_{N+2}, a_{N+3}, \dots\}$ is bounded (above by $L + 1$ and below by $L - 1$). Now the "front end" $\{a_1, a_2, \dots, a_N\}$ is a *finite* set, and hence also bounded. Thus the entire sequence $\{a_n\}$ is the union of two bounded sets, and is therefore bounded.

We could play the same game with smaller ϵ, but there's no need.

Beyond the basics: inequalities. The sequences in Examples 1 and 2 led to relatively simple inequalities and, in turn, to relatively simple recipes for N in terms of ϵ. Sequences with more complicated definitions may require more cleverness. Fortunately, a little creativity with *inequalities* can dramatically simplify the work. Following are two examples.

In Example 2, we found $N = 13/\epsilon - 5$.

EXAMPLE 4.   Show that the sequence $\{z_n\}$ with

$$z_n = \frac{1}{n + \sqrt{n+1} + 5}$$

converges to zero.

Discussion.   But for the annoying denominator $\{z_n\}$ resembles $\{1/n\}$, which converges to zero (see Example 1). To show that $z_n \to 0$, too, for given $\epsilon > 0$, we need to find N for which

$$|z_n - 0| = \left| \frac{1}{n + \sqrt{n+1} + 5} \right| = \frac{1}{n + \sqrt{n+1} + 5} < \epsilon$$

whenever $n > N$. This looks clumsy, but a simple inequality brings radical improvements:

$$|z_n - 0| = \frac{1}{n + \sqrt{n+1} + 5} < \frac{1}{n}.$$

Now it is easy to see that

$$\frac{1}{n} < \epsilon \iff n > \frac{1}{\epsilon},$$

which means that $N = 1/\epsilon$ "works." We assemble the parts in the concise proof.

Proof.   Let $\epsilon > 0$ be given. Set $N = \frac{1}{\epsilon}$. If $n > N$ then

Check each step; note the key inequality.

$$|z_n - 0| = \left| \frac{1}{n + \sqrt{n+1} + 5} \right| = \frac{1}{n + \sqrt{n+1} + 5} < \frac{1}{n} < \frac{1}{N} = \epsilon$$

as the definition requires.   ◊

EXAMPLE 5.   Let $w_n = 4n/(2n - 85)$; show that $\{w_n\}$ converges to 2.

Discussion.   This time the key quantity $|w_n - L|$ takes the form

Check the final calculation.

$$|w_n - L| = \left| \frac{4n}{2n - 85} - 2 \right| = \left| \frac{170}{2n - 85} \right|;$$

we want

$$|w_n - L| = \left| \frac{170}{2n - 85} \right| < \epsilon$$

to hold for large n. It would be nice to drop the absolute value—alas, the denominator, $2n - 85$, may be *negative*. The good news is that this happens only for $n \leq 42$. If $n > 42$, then we can indeed drop the absolute value in good conscience, and solve our inequality without undue fuss:

But check all algebra carefully.

$$\left| \frac{170}{2n - 85} \right| = \frac{170}{2n - 85} < \epsilon \iff \frac{170}{2\epsilon} + \frac{85}{2} < n.$$

The last inequality is what we wanted—a value of n beyond which $|w_n - L| < \epsilon$. Here comes the proof.

Proof. Let $\epsilon > 0$ be given. Set $N = \frac{170}{2\epsilon} + \frac{85}{2}$. (Note that $N > 42$.) This N works, since if $n > N$, then we have

since $n \geq 43$

$$|w_n - 2| = \left| \frac{170}{2n - 85} \right| = \frac{170}{2n - 85}$$

a nice collapse!

$$< \frac{170}{2N - 85} = \frac{170}{\frac{170}{\epsilon} + 85 - 85} = \epsilon,$$

as the definition requires. $\Diamond$

Exercises

1. If a sequence $\{x_n\}$ converges, then for any given $\epsilon > 0$ there is some N that "works" in the sense of the definition. For the sequence $\{1/n\}$ the following table shows values of N associated with values of ϵ:

ϵ	1.000	0.100	0.010	0.001
N	1	10	100	1000

Make a similar table (same values of ϵ) for each sequence following. Try to choose N as small as possible.

   (a) $\left\{ \frac{1}{n^2} \right\}$

   (b) $\left\{ \frac{1}{\sqrt{n}} \right\}$

   (c) $\left\{ \frac{1}{1 + \ln n} \right\}$

2. This problem is about ϵ–N tables for convergent sequences, like those in Problem 1, with $\epsilon = 1, 0.1, 0.01, 0.001$ in the first row.

   (a) Explain why all entries in the second row can be the same.

   (b) Consider the sequence $\{x_n\}$ with $x_n = 0$ for all n. What goes in the second row?

(c) Consider the sequence $\{y_n\}$ with $y_n = 1/d(n)$, where $d(n)$ is the number of base-ten digits in n. (For example, $y_{123456} = 1/6$.) What goes in the second row?

3. Suppose $\{x_n\}$ converges to L, and $a \leq x_n \leq b$ for all n. Show that $a \leq L \leq b$.

4. Prove or disprove: If $x_n < 17$ for all n and $\{x_n\}$ converges to L, then $L < 17$.

5. Suppose that $\{x_n\}$ converges to 5.

   (a) Show that there is some N such that $x_n \in (4.9, 5.1)$ for all $n > N$.
   (b) Show that there is some N such that $x_n > 4.999$ for all $n > N$.
   (c) Show that the set $\{x_n \mid x_n > 6\}$ is finite.

6. Suppose that $\{x_n\}$ does *not* converge.

   (a) Is it possible that $x_n = 3$ for infinitely many n? If so, give an example. If not, explain why not.
   (b) Is it possible that $x_n = 3$ for all but finitely many n? If so, give an example. If not, explain why not.

7. Suppose that $\{x_n\}$ converges to 4.

   (a) Is it possible that $x_n \neq 4$ for infinitely many n? If so, give an example. If not, explain why not.
   (b) Is it possible that $x_n > 4$ for infinitely many n *and* $x_n < 4$ for infinitely many n? If so, give an example. If not, explain why not.

8. Suppose that $\{x_n\}$ converges to a number L, with $L > 4$.

   (a) Can $x_n = 4$ for exactly 1234 values of n? If so, give an example. If not, explain why not.
   (b) Can $x_n = 4$ for infinitely many values of n? If so, give an example. If not, explain why not.

9. Prove that each of the following sequences converges. (Example 2, page 85, is similar.)

   (a) $\{a_n\}$, with $a_n = \dfrac{2n}{3n + 5}$.
   (b) $\{b_n\}$, with $b_n = \dfrac{2n + 300000}{3n + 5}$.

(c) $\{c_n\}$, with $c_n = \dfrac{2n}{3n^2 + 5}$.

10. Guess and then prove a limit for each sequence following. (Examples 4 and 5 may be helpful.)

(a) $\{a_n\}$, with $a_n = \dfrac{2n}{3n - 5}$.

(b) $\{b_n\}$, with $b_n = \dfrac{2n}{3n + \sin n + 5}$.

(c) $\{c_n\}$, with $c_n = (-1)^n \dfrac{2n}{3n^2 + 5}$.

11. Suppose that $\{a_n\}$ converges to 1. Prove (use ϵ and N, not theorems) that $\{17a_n\}$ converges to 17.

12. Suppose $\{x_n\}$ is monotone decreasing and $\inf\{x_n\} = 17$. Show that $x_n \to 17$.

13. Convert the informal discussion after Theorem 2.3, page 88, into a concise proof that a bounded, *decreasing* sequence converges.

14. Convert the informal discussion after Theorem 2.4, page 88, into a concise proof.

15. Let $\{x_n\}$ be a sequence. Define a new sequence $\{y_n\}$ by $y_n = x_n - 5$. Use the definition of convergence to show that $x_n \to 5 \iff y_n \to 0$.

16. From any given sequence $\{x_n\}$ we can form the related sequence $\{y_n\} = \{5x_n + 2\}$. Use the definition of convergence to show that if $\{x_n\}$ converges to 42, then $\{y_n\}$ converges to _____. (First fill in the blank.)

17. Here is a second, "unofficial" definition of convergence to a number L: $\{x_n\}$ converges to L if, for every $\epsilon > 0$, we have $|x_n - L| < \epsilon$ for all but finitely many n.

(a) Show that $\{1/n\}$ converges to zero in the sense above.

(b) Show that if $x_n \to 0$ in the sense above, then $x_n \to 0$ in the "official" sense defined in this section.

(c) What does it mean for a sequence $\{x_n\}$ *not* to converge to zero in the sense above? (*Negate* the unofficial definition.)

(d) What does it mean for a sequence $\{x_n\}$ *not* to converge to zero in the official sense? (*Negate* the official definition.)

18. Show that every real number β is the limit of an increasing sequence of *rational* numbers. Hint: One approach uses infinite decimal expansion; see Exercise 18, page 80.

19. Some sequences $\{x_n\}$ have ϵ–N tables (in the sense of Exercise 1) of the following form:

ϵ	1.0	0.1	0.01	0.001	0.0001	0.00001	$\ldots$
N	0	0	0	0	0	0	$\ldots$

Which sequences are these?

20. Some sequences $\{x_n\}$ have ϵ–N tables (in the sense of Exercise 1) of the following form:

ϵ	1.0	0.1	0.01	0.001	0.0001	0.00001	$\ldots$
N	5	5	5	5	5	5	$\ldots$

Which sequences are these?

21. Consider the sequence $\{x_n\}$ given by $x_n = 1/n$. Decide whether each of the following sentences is true; explain answers briefly.

    (a) The sequence $\{x_n\}$ is bounded above by 2.

    (b) The sequence $\{x_n\}$ is bounded below.

    (c) If $n > 42$, then $|x_n| < .01$.

    (d) There is some integer N such that $|x_n| < .0001$ for all n such that $n > N$.

    (e) There is some integer N such that $|x_n - x_{n+1}| < .0001$ for all n such that $n > N$.

    (f) There is some integer N such that $|x_n| > 0.001$ for all n such that $n > N$.

    (g) For every positive number ϵ there is some integer N such that $|x_n| < \epsilon$ for all n such that $n > N$.

22. Like Problem 21, but with the sequence $\{x_n\}$ given by $x_n = \dfrac{(-1)^{n+1}}{n}$.

    (a) The sequence $\{x_n\}$ is bounded above by 2.

    (b) The sequence $\{x_n\}$ is bounded below.

    (c) If $n > 42$, then $|x_n| < .01$.

(d) There is some integer N such that $|x_n| < .0001$ for all n such that $n > N$.

(e) There is some integer N such that $|x_n - x_{n+1}| < .0001$ for all n such that $n > N$.

(f) There is some integer N such that $|x_n| > 0.001$ for all n such that $n > N$.

(g) For every positive number ϵ there is some integer N such that $|x_n| < \epsilon$ for all n such that $n > N$.

23. Like Problem 21, but with the sequence $\{x_n\}$ given by $x_n = \dfrac{\sin(n)}{n}$.

(a) The sequence $\{x_n\}$ is bounded above by 2.

(b) The sequence $\{x_n\}$ is bounded below.

(c) If $n > 42$, then $|x_n| < .01$.

(d) There is some integer N such that $|x_n| < .0001$ for all n such that $n > N$.

(e) There is some integer N such that $|x_n - x_{n+1}| < .0001$ for all n such that $n > N$.

(f) There is some integer N such that $|x_n| > 0.001$ for all n such that $n > N$.

(g) For every positive number ϵ there is some integer N such that $|x_n| < \epsilon$ for all n such that $n > N$.

24. Suppose that $\{x_n\}$ is unbounded above. Use the definition of convergence to show that $\{x_n\}$ does not converge to $L = 1000$.

25. For each sequence, guess a limit L (e.g., try large values of n), if you think one exists. Then complete an associated ϵ–N table like the one below. (There are many possible values for N.)

ϵ	1.00000	0.10000	0.01000	0.00100	10^{-10}
N					

(a) If $a_n = \dfrac{1}{\sqrt{n}}$, then $L = \boxed{}$.

(b) If $a_n = \dfrac{1}{n^2}$, then $L = \boxed{}$.

(c) If $a_n = \dfrac{\sin n}{n^2}$, then $L = \boxed{}$.

(d) If $a_n = \dfrac{3n}{n+2}$, then $L = \boxed{}$.

(e) If $a_n = \min\{n, 42\}$, then $L = \boxed{}$.

26. (Do Problem 25 first.) Write a brief sentence or two in each part.

   (a) Look at your *rightmost* N-entry in any table above. Could you correctly use this same entry in *every* N-position in that table?

   (b) Suppose you've found "good" N-entries for a table like these. Would your table still be OK if you *double* every N-entry but leave the ϵ-entries alone? What if you double the ϵ-entries and leave the N-entries alone?

27. In this section we mentioned the Fibonacci sequence $\{f_n\}$, defined by $f_1 = f_2 = 1$ and $f_n = f_{n-2} + f_{n-1}$ for $n \geq 3$. It is clear that $\{f_n\}$ is unbounded, but how fast does $\{f_n\}$ increase? We explore this question in this problem.

   Let's show first, by induction, that $f_n < 2^n$ for $n \geq 1$. Because the Fibonacci sequence is defined using *two* previous terms, it is convenient to take both $n = 1$ and $n = 2$ as base cases.

   The claim is obvious for $n = 1$ and $n = 2$. For the inductive step, we suppose the claim holds for all n up through k (with $k \geq 2$) and show, as follows, that it holds for $n = k + 1$:

   $$f_{k+1} = f_{k-1} + f_k < 2^{k-1} + 2^k < 2^k + 2^k = 2^{k+1},$$

   as desired.

   (a) Show by induction that $f_n > 1.5^n$ for all $n \geq 11$. (It is easy to check with technology that the inequality is false for smaller n.)

   (b) It can also be shown by induction (as in the preceding problem) that $f_n > 1.6^n$ for all "sufficiently large" n. Use technology to decide which n have this property.

28. Consider the sequence $\{g_n\}$ defined by $g_1 = 1$ and $g_{n+1} = 1 + \frac{1}{g_n}$.

   (a) Write out the first few terms of the sequence to see the Fibonacci numbers pop up.

   (b) Prove by induction that $g_n \leq 2$ for all n.

   (c) Observe (with technology) that $\{g_n\}$ appears to hop back and forth across the number $\phi = (1+\sqrt{5})/2 \approx 1.618$ (this is the famous *golden ratio*). Prove this by showing (algebraically; no induction needed) that (i) if $g_n > \phi$ then $g_{n+1} < \phi$; and (ii) if $g_n < \phi$ then $g_{n+1} > \phi$.

29. Following are several meaningful (if possibly clumsy) sentences about a sequence $\{a_n\}$. Describe in your own words what each sentence means about $\{a_n\}$. Do any of these sentences imply any others?

    (a) $\forall \epsilon > 0 \, \exists N \in \mathbb{N}$ such that $n > N \implies |a_n - \pi| < \epsilon$

    (b) $\exists N \in \mathbb{N}$ such that $\forall \epsilon > 0, \, n > N \implies |a_n - \pi| < \epsilon$

    (c) $\forall \epsilon > 0$ and $\forall N \in \mathbb{N}, \, n > N \implies |a_n - \pi| < \epsilon$

2.2   Working with Sequences

New Sequences from Old

As with other functions, "old" sequences can be combined in various ways to form "new" ones. Naturally enough, convergence and divergence properties of a built-up sequence reflect properties of the building blocks.

Theorem 2.5 (Algebra with convergent sequences). *Let $\{a_n\}$ and $\{b_n\}$ be convergent sequences. Let a, b, and c be real numbers, with $a_n \to a$ and $b_n \to b$.*

- Sums and differences: *The sequence $\{a_n \pm b_n\}$ converges to $a \pm b$.*

- Constant multiples: *The sequence $\{ca_n\}$ converges to ca.*

- Products: *The sequence $\{a_n \cdot b_n\}$ converges to $a \cdot b$.*

- Quotients: *If $b \neq 0$ and $b_n \neq 0$ for all n, then the sequence $\{a_n/b_n\}$ converges to a/b.*

Let's use the theorem first; proofs come later.

EXAMPLE 1.   In Example 2, page 85, we used the definition of convergence to show that $\left\{ \frac{3n+2}{n+5} \right\}$ converges to three. Could Theorem 2.5 have helped?

SOLUTION.   Yes. Basic algebra gives

$$\frac{3n+2}{n+5} = \frac{3+2/n}{1+5/n} = \frac{3 \cdot 1 + 2 \cdot \frac{1}{n}}{1 + 5 \cdot \frac{1}{n}},$$

which shows the original sequence as a combination of the much simpler sequences $\{1\}$, and $\{1/n\}$, which converge to 1 and 0, respectively. Now Theorem 2.5 implies that

$$\frac{3 \cdot 1 + 2 \cdot \frac{1}{n}}{1 + 5 \cdot \frac{1}{n}} \to \frac{3 \cdot 1 + 2 \cdot 0}{1 + 5 \cdot 0} = 3,$$

as expected.

We just *assumed*, of course, that $\{1\}$ and $\{1/n\}$ do indeed converge to one and zero. These claims need proof, too, but the proofs are easy—and need to be done just once.   ◊

EXAMPLE 2.   Let $\{a_n\}$ and $\{b_n\}$ be *divergent* sequences, and let $\{c_n\}$ be a *convergent* sequence. Can the sum $\{a_n + b_n\}$ converge? Can it diverge? What about $\{a_n + c_n\}$?

SOLUTION.   Theorem 2.5 says nothing about sums of *divergent* sequences. But easy examples show that $\{a_n\}$ and $\{b_n\}$ can either converge or diverge. With $a_n = n$ and $b_n = \pi - n$, for example, we get $a_n + b_n = \pi$ for all n, so $\{a_n + b_n\}$ converges.

Theorem 2.5 *does* help with $\{a_n + c_n\}$; let's call it $\{d_n\}$ for the moment. If $\{d_n\}$ were convergent, then the difference $\{d_n - c_n\} = \{a_n\}$ would converge, too, contradicting our assumption. Thus, $\{a_n + c_n\}$ *diverges*.   ◊

> Find your own example where $\{a_n + b_n\}$ diverges.

EXAMPLE 3.   A sequence $\{s_n\}$ is defined recursively by

$$s_1 = 2.0 \quad \text{and} \quad s_n = \frac{s_{n-1}}{2} + \frac{1}{s_{n-1}} \quad \text{for} \quad n > 1.$$

The first few terms (rounded) are $2.0, 1.5, 1.41667, 1.41422, 1.41421, \ldots$. What is happening here? Why?

SOLUTION.   The numbers suggest convergence to $\sqrt{2}$, but can we be certain? The answer is yes—if we invoke appropriate theorems. The definition implies that $s_n > 0$ for all n. It's also true (and proved in exercises) that $\{s_n\}$ is *decreasing*, and so Theorem 2.3, page 88, implies that $\{s_n\}$ converges to *some* limit L. Invoking different parts of Theorem 2.5 lets us find L. Because $s_n \to L$, we know

$$s_n = \frac{s_{n-1}}{2} + \frac{1}{s_{n-1}} \quad \Longrightarrow \quad L = \frac{L}{2} + \frac{1}{L},$$

which implies, in turn, that $L^2 = 2$, or $L = \sqrt{2}$.   ◊

> $L = -\sqrt{2}$ is clearly impossible since all s_n are positive.

Proving Theorem 2.5.   Theorem 2.5 contains few surprises. Concerning sums, for instance, it should seem reasonable that if $a_n \approx a$ and $b_n \approx b$ for large n, then also $a_n + b_n \approx a + b$ for large n. Rigorous *proofs* are another matter, of course.

The definition of convergence is all about inequalities; therefore, so are convergence proofs. Here, we'll need to show that inequalities like

$$|(a_n + b_n) - (a + b)| < \epsilon, \quad |a_n b_n - (a \pm b)| < \epsilon, \quad \text{and} \quad \left| \frac{a_n}{b_n} - \frac{a}{b} \right| < \epsilon$$

hold for large n. What we have to work with are

$$|a_n - a| < \epsilon \quad \text{and} \quad |b_n - b| < \epsilon,$$

which *do* hold for large n, by hypothesis. The trick is to parlay these simpler inequalities into proofs of their more complicated cousins. The triangle inequality will come in very handy; watch for it in the following proof.

Watch for a technical trick, too.

Proof (for sums): Let $\epsilon > 0$; we need to choose N so that $|(a_n + b_n) - (a + b)| < \epsilon$ whenever $n > N$. Set $\epsilon' = \epsilon/2$; note $\epsilon' > 0$. Because $a_n \to a$, there is a number N_1 that "works" for ϵ':

$$n > N_1 \implies |a_n - a| < \epsilon'.$$

Similarly, there exists N_2 such that

$$n > N_2 \implies |b_n - b| < \epsilon'.$$

Now let N be the *larger* of N_1 and N_2. If $n > N$, then both $n > N_1$ and $n > N_2$, and so

$$
\begin{aligned}
|(a_n + b_n) - (a + b)| &= |(a_n - a) + (b_n - b)| \\
&\le |a_n - a| + |b_n - b| < \epsilon' + \epsilon' = \epsilon,
\end{aligned}
$$

where we used the triangle inequality at the line break. This shows that N works in the desired sense, and completes the proof. $\qquad\square$

Working with products. To prove our claim about products we will *use* the corresponding properties for sums (just proved) and for constant multiples (left as an exercise). The proof starts with a standard analyst's trick: subtracting and adding the same quantity:

We'll see this again.

$$a_n b_n = a_n b_n - a_n b + a_n b = a_n(b_n - b) + a_n b.$$

The result for constant multiples shows that $a_n b \to ab$; we'll be done if we can also show that $a_n(b_n - b) \to 0$. The key insight, as we will see in the formal proof, is that the $\{a_n\}$ are *bounded*.

Proof (for products): First we write $a_n b_n = a_n(b_n - b) + a_n b$. By the result for sums, it's enough to show that

$$a_n b \to ab \quad \text{and} \quad a_n(b_n - b) \to 0.$$

And left as an exercise.

That $a_n b \to ab$ follows from the property of constant multiples that we assumed. To show $a_n(b_n - b) \to 0$, observe first that the convergent sequence $\{a_n\}$ is bounded (Theorem 2.4, page 88, says so); in other words, there exists $M > 0$

with $|a_n| \leq M$ for all n. Now let $\epsilon > 0$ be given; note that $\epsilon/M > 0$, too. Since $b_n \to b$, there exists N such that $|b_n - b| < \epsilon/M$ when $n > N$. This N works for $\{a_n(b_n - b)\}$, because if $n > N$, then

$$|a_n(b_n - b) - 0| = |a_n| \, |b_n - b| \leq M \, |b_n - b| < M\frac{\epsilon}{M} = \epsilon,$$

as desired. $\square$

Quotients. To show $a_n/b_n \to a/b$ requires—even to make sense—that $b \neq 0$ and $b_n \neq 0$ for all n. If we assume this, it is enough to show that $1/b_n \to 1/b$, for then the result for *products* gives $a_n \cdot 1/b_n \to a \cdot 1/b$, as desired. We will outline a proof that $1/b_n \to 1/b$ in the exercises.

Squeezing. The algebraic methods of Theorem 2.5 don't immediately help with the sequence $\{\sin n/n\}$, which has a trigonometric ingredient. Still, we expect the sequence to converge to zero because the numerator remains tamely bounded, while the denominator "blows up." More precisely, we know

$$-\frac{1}{n} \leq \frac{\sin n}{n} \leq \frac{1}{n}$$

for all n. Both the left- and right-hand sequences converge to zero, and we expect them to "squeeze" the middle sequence to the same limit. That intuition is correct:

Theorem 2.6 (The squeeze principle). *Let $\{a_n\}$, $\{b_n\}$, and $\{c_n\}$ be sequences such that $a_n \leq b_n \leq c_n$ for all n. If $a_n \to L$ and $c_n \to L$, then $b_n \to L$, too.*

Proof: For given $\epsilon > 0$, we need to find N such that $n > N$ implies $|b_n - L| < \epsilon$, or, equivalently, $L - \epsilon < b_n < L + \epsilon$.

Because $a_n \to L$ we can choose N_1 so

$$n > N_1 \implies L - \epsilon < a_n < L + \epsilon.$$

Similarly, there is N_2 so

$$n > N_2 \implies L - \epsilon < c_n < L + \epsilon.$$

Now let $N = \max\{N_1, N_2\}$. This N works for $\{b_n\}$, since if $n > N$ then

$$L - \epsilon < a_n \leq b_n \leq c_n < L + \epsilon,$$

as desired. $\square$

Theorem 2.6 suggests additional "squeeze-like" properties of sequences.

Proposition 2.7 (More squeezing). *Let $\{a_n\}$ and $\{b_n\}$ be sequences and L a number.*

(a) $a_n \to 0 \iff |a_n| \to 0$

(b) If $a_n \to L$, *then* $|a_n| \to |L|$.

(c) If $|a_n| \to 0$ *and* $\{b_n\}$ *is* bounded, *then* $a_n b_n \to 0$.

Proof: Claim (a), left as an exercise, is a basic application of the definition.

Claims (b) and (c) follow from (a) by judicious squeezing. For (b), notice first that $a_n - L \to 0$ by hypothesis, and so $|a_n - L| \to 0$, according to (a). The *reverse* triangle inequality now gives our squeezing inequality:

Here one sequence is constant.

$$0 \leq \big|\, |a_n| - |L| \,\big| \leq |a_n - L|$$

Since the right-hand sequence tends to zero, so must the middle one, as we aimed to show.

For (c), we first use boundedness of $\{b_n\}$ to choose $B > 0$ with $|b_n| \leq B$ for all n. This leads to another squeezing inequality:

$$0 \leq |a_n b_n| \leq B\,|a_n|\,.$$

Again, the left- and right-hand sequences tend to zero (the constant multiple B does no harm) and therefore squeeze the middle sequence to the same limit. □

Sequences to order. Sequences with specified properties can often be made to order, as the following two (similar) samples suggest. We'll call them lemmas because the sequences involved here are often used in proofs of other results. (Lemmas are junior-grade theorems, normally used to prove something else.)

The German *Hilfsatz*, or "helping sentence," is more descriptive.

Lemma 2.8. *Let L be any number. There exist sequences $\{r_n\}$ and $\{p_n\}$, both converging to L, with $r_n \in \mathbb{Q}$ and $p_n \notin \mathbb{Q}$ for all n. If desired, $\{r_n\}$ and $\{p_n\}$ and can be chosen to be either strictly increasing or strictly decreasing.*

Proof: If L is rational, then the sequences defined by

$$r_n = L + \frac{1}{n} \quad \text{and} \quad p_n = L + \frac{\sqrt{2}}{n}\,.$$

are strictly decreasing and converge to L. Other cases are similar, and left as exercises. □

Lemma 2.9. *Let $C \subseteq \mathbb{R}$ be any nonempty bounded set, with $\alpha = \inf(C)$ and $\beta = \sup(C)$. There exist sequences $\{a_n\}$ and $\{b_n\}$ contained in C with $a_n \to \alpha$ and $b_n \to \beta$.*

The $\{a_n\}$ case is similar.

Proof: Let's find $\{b_n\}$. For each $n \in N$ we know $\beta - 1/n$ is *not* an upper bound for C, so there is some $b_n \in C$ with

$$\beta - \frac{1}{n} \leq b_n \leq \beta \quad \text{and so} \quad b_n \to \beta$$

by the "squeezing" principle, Theorem 2.6. Further details are in an exercise. □

Divergence to Infinity

Sequences like $\{n\}$ and $\{-n^3\}$ clearly have no finite limit, but they misbehave less seriously than, say, the zig-zag sequence $\{1, -2, 3, -4, \ldots\}$. The following definition formalizes these ideas.

Definition 2.10. A sequence $\{x_n\}$ *diverges to infinity* if for every $M > 0$ there exists a number N such that

$$x_n > M \quad \text{whenever } n > N.$$

In this case, we write $\lim_{n \to \infty} x_n = \infty$, or just $x_n \to \infty$.

Observe:

- *Divergence to* $-\infty$*:* A similar definition holds for divergence to $-\infty$: For given $M > 0$ there must exist N with $x_n < -M$ whenever $n > N$.

- *Big M, small ϵ:* For ordinary convergence we focus mainly on *small* positive values of ϵ; here the challenging values of M are *large*.

- *Convergence or divergence?* Some authors refer to *convergence* to ∞; we prefer to reserve "convergence" for *finite* limits.

- *Unboundedness and divergence to infinity:* An unbounded sequence need not diverge to $\pm\infty$, as $\{1, -2, 3, -4, \ldots\}$ illustrates. An unbounded *monotone* sequence, on the other hand, *must* diverge to $\pm\infty$; see Example 5.

- *Zero or infinity?* A sequence $\{x_n\}$ of positive terms diverges to ∞ if and only if $\{1/x_n\}$ converges to 0. (Proofs follow directly from the definitions; see the exercises.) A similar result holds for *negative* sequences.

Swept away: a comparison test. Here's an unsurprising claim about comparing one sequence to another:

If $a_n \leq b_n$ for all n and $a_n \to \infty$, then $b_n \to \infty$, too.

(A similar claim holds for sequences tending to $-\infty$.) Proofs follow easily from the definitions; see the exercises. The surprise is how powerful this straightforward idea can be in applications.

EXAMPLE 4. The sequences $\{h_n\}$ and $\{s_n\}$ given by

$$h_n = \frac{1}{1} + \frac{1}{2} + \frac{1}{3} + \cdots + \frac{1}{n}, \qquad s_n = \frac{1}{1} + \frac{1}{\sqrt{2}} + \frac{1}{\sqrt{3}} + \cdots + \frac{1}{\sqrt{n}}$$

are positive and increasing. Do they diverge to infinity?

If using unproved calculus
results seems like cheating, see
the exercises for another
approach.

SOLUTION. In a word, yes. Observe first that $s_n \geq h_n$ for all n, so it is enough to show that $h_n \to \infty$. One approach is to recall from elementary calculus that

$$h_n = \frac{1}{1} + \frac{1}{2} + \frac{1}{3} + \cdots + \frac{1}{n} > \int_1^n \frac{dx}{x} = \ln n.$$

It is well known that $\ln n \to \infty$, and so both $\{s_n\}$ and $\{h_n\}$, being larger, are also "swept away" to ∞.   ◊

EXAMPLE 5. Prove concisely that if $\{x_n\}$ is unbounded and increasing, then $x_n \to \infty$.

SOLUTION. Let $M > 0$ be given. Since $\{x_n\}$ is unbounded, there exists N with $x_N > M$. This N works in the definition: if $n > N$ then, since $\{x_n\}$ is increasing, we have $x_n \geq x_N > M$, as desired.   ◊

Exercises

1. Prove the following part of Theorem 2.5, page 96: If $a_n \to a$ and c is any constant, then $ca_n \to ca$. (Hint: The result is trivial if $c = 0$, so assume $c \neq 0$.)

2. Another part of Theorem 2.5, page 96, says that $a_n/b_n \to a/b$ if $b \neq 0$ and $b_n \neq 0$ for all n. To complete the proof sketched in this section, we need to show that $1/b_n \to 1/b$.

   (a) Explain why $|\frac{1}{b_n} - \frac{1}{b}| = \frac{1}{|b_n b|}|b_n - b|$.

   (b) We will show just below that the sequence $\{1/|b_n b|\}$ is bounded, say by $M > 0$. Assuming this for the moment, prove that $1/b_n \to 1/b$. (Hint: The preceding identity gives $|\frac{1}{b_n} - \frac{1}{b}| < M|b_n - b|$. Now argue as in the proof for products of sequences.)

   (c) Show in two steps that $\{1/|b_n b|\}$ is bounded.

      (i) Show that the set $\{b_n\}$ is *bounded away from zero*—i.e., there is some $\delta > 0$ such that $|b_n| \geq \delta$ for all n. (Hints: Let $\epsilon = |b|/2 > 0$. Since $b_n \to b$ there is an integer N such that $|b_n - b| < \epsilon$ when $n > N$; note that $|b_n| > |b|/2$ for all such n. The remaining set $\{|b_1|, |b_2|, \ldots, |b_N|\}$ is a *finite* set of positive numbers, and so has a *positive* minimum. Use these facts to find a suitable value of δ.)

      (ii) Show that $1/|b_n b| \leq 1/(\delta|b|)$ for all n.

3. Let $\{x_n\}$ be a sequence and L a number. Show that $x_n \to L \iff (x_n - L) \to 0 \iff |x_n - L| \to 0$.

4. This problem is about Lemma 2.8, page 100.

   (a) What does the lemma say if $L = \sqrt{2}$?

   (b) Find a strictly decreasing sequence $\{p_n\}$ with $p_n \notin \mathbb{Q}$ for all n and $p_n \to \sqrt{2}$.

   (c) Describe how to find a strictly increasing sequence $\{r_n\}$ with $r_n \in \mathbb{Q}$ for all n and $r_n \to \sqrt{2}$.

5. As in Lemma 2.9, page 100, let $C \subseteq \mathbb{R}$ be a nonempty set with $\alpha = \inf(C)$. Explain carefully how to construct a sequence $\{a_n\}$ with $a_n \in C$ for all n and $a_n \to \alpha$.

6. Let C be the set of irrational numbers between 0 and 1. Describe the sequences $\{a_n\}$ and $\{b_n\}$ mentioned in Lemma 2.9, page 100.

7. Let $C \subseteq \mathbb{R}$ be a bounded set, with $\sup(C) = \beta$. Complete the details following to show that there is a sequence $\{b_n\} \subseteq C$ with $b_n \to \beta$.

   (a) Find such a sequence if $C = [0, 1)$.

   (b) Find such a sequence if $C = \{0, 1\}$.

   (c) Explain why the claim is trivial if $\beta \in C$.

   (d) Explain: For each $n \in N$ we can choose some $b_n \in C$ with $\beta - \frac{1}{n} < b_n \leq \beta$.

   (e) Show: The sequence $\{b_n\}$ defined in the preceding part converges to β. Hint: Squeeze.

8. Let $S \subseteq \mathbb{R}$ be a bounded set, with $\sup S = \beta$. Show that there is a *monotone* sequence $\{s_n\} \subseteq S$ with $s_n \to \beta$.

9. Consider the sequence $\{s_n\}$ in Example 3, page 97. Show by induction that $\{s_n\}$ is strictly decreasing; i.e., show $s_{n+1} < s_n$ for all $n \geq 1$.

10. Consider the sequence $\{s_n\}$ given by $s_n = \frac{1}{1} + \frac{1}{\sqrt{2}} + \cdots + \frac{1}{\sqrt{n}}$. We showed in Example 4, page 101, that $\{s_n\}$ diverges to infinity. Give another comparison proof, this time with the sequence $\{\sqrt{n}\}$.

11. Consider again the sequence $\{h_n\}$ in Example 4, page 101. Observe that $h_{2n} = h_n + \frac{1}{n+1} + \cdots + \frac{1}{2n} \geq h_n + n \cdot \frac{1}{2n} = h_n + \frac{1}{2}$. Use this idea to prove (without calculus) that $\{h_n\}$ diverges to infinity.

12. The sequence $\{S_n\}$ given by

$$S_n = \frac{1}{1} + \frac{1}{2^2} + \frac{1}{3^2} + \cdots + \frac{1}{n^2}$$

is clearly increasing, so it either converges or diverges to infinity. Which happens? To decide we work with the similar series $\{T_n\}$ given by

$$T_n = \frac{1}{1+1} + \frac{1}{2^2+2} + \frac{1}{3^2+3} + \cdots + \frac{1}{n^2+n}.$$

(a) Use technology to evaluate some terms of both $\{S_n\}$ and $\{T_n\}$. Do they appear to converge or diverge?

(b) Guess a simple formula for T_n; prove your guess by induction.

(c) Conclude from the preceding part that $T_n \to 1$.

(d) Prove that $\{S_n\}$ converges to some limit $L \le 2$. (Hint: Show first that $S_n \le 2T_n$ for all $n \ge 1$.)

13. Let $\{x_n\}$ be a sequence of *positive* terms.

(a) Prove that if $x_n \to \infty$, then $-x_n \to -\infty$. (The converse is also true, and the proof is almost identical.)

(b) Show that $x_n \to \infty$ if and only if $1/x_n \to 0$.

(c) State and prove a similar claim for sequences of *negative* terms. (Use the first two parts.)

14. Suppose $a_n > 0$ for all n and $a_n \to 3$. Show that $\sqrt{a_n} \to \sqrt{3}$. (Hint: Show first by algebra that $|\sqrt{a_n} - \sqrt{3}| < |a_n - 3|$.)

15. Use theorems from this section (not the definitions of convergence) to discuss each of the following limits. It is OK to assume such basic facts as $1/n \to 0$, but say what you're assuming.

(a) $\left\{ \dfrac{2n + 3\sin n}{5n + \cos n} \right\}$

(b) $\left\{ \sqrt{n^2 + 1} \right\}$

(c) $\left\{ \sqrt{n^2 + n} - n \right\}$

(d) $\left\{ \dfrac{n^2 + \arctan n}{n + 2} \right\}$

16. For any sequence $\{a_n\}$ we can form a new sequence $\{b_n\}$ by the rule $b_n = \max\{|a_1|, |a_2|, \ldots, |a_n|\}$. Show that b_n converges if and only if $\{a_n\}$ is bounded.

17. Given any two sequences $\{a_n\}$ and $\{b_n\}$, we can construct a new sequence $\{c_n\}$ with the rule $c_n = \max\{a_n, b_n\}$.

   (a) Give an example (as simple as possible; there are many possibilities) to show that $\{c_n\}$ may converge even if both $\{a_n\}$ and $\{b_n\}$ diverge.

   (b) Suppose $\{a_n\}$ and $\{b_n\}$ converge to a and b respectively. Show that $\{c_n\}$ converges, too. (Guess the limit first.)

18. A sequence $\{x_n\}$ is *quasi-positive* if $\exists N \in \mathbb{N}$ such that $\forall n > N$, $x_n > 0$.

   (a) Is the sequence $x_n = 2n - 50$ quasi-positive?

   (b) Is the sequence $x_n = (-1)^n \sqrt{n} + 100$ quasi-positive?

   (c) Suppose $a > 0$ and $x_n \to a$. Prove or disprove that $\{x_n\}$ is quasi-positive.

19. It's well known that the golden ratio, φ, satisfies $\varphi \approx 1.61803$. Let $x_1 = 1$ and $x_{n+1} = 1 + \dfrac{1}{x_n}$.

   (a) Find (first as rational numbers and then as decimals) x_1 through x_8. What patterns do you see? Is this sequence monotone?

   (b) Assume (it's true) that $\{x_n\}$ converges to some number L. Explain why $L = 1 + \dfrac{1}{L}$. Then use the equation to find an exact formula for L, involving a square root.

20. As every calculator knows, $\sqrt{3} \approx 1.732051$. Consider the sequence defined by $x_1 = 2$ and $x_{n+1} = \dfrac{x_n + 3/x_n}{2}$ for $n \geq 1$.

   (a) Find x_2, x_3, x_4 by hand, as rational numbers. Then find decimal approximate values.

   (b) Show that if $x_n > 0$, then $x_{n+1} \geq \sqrt{3}$. (It's OK to use a little calculus if you like.) Conclude that $x_n \geq \sqrt{3}$ for all n.

   (c) Explain why $x_{n+1} < x_n$ for all n.

   (d) We know now that $x_n \to L$ for some number L. Which theorem(s) say this?

   (e) We also know now that $L = \dfrac{L + 3/L}{2}$. Which theorems say *this*?

   (f) What's the exact value of L? Why?

21. Suppose $\{x_n\}$ is defined by $x_1 = 3$ and $x_{n+1} = \dfrac{x_n^2 + 4}{5}$.

(a) Prove that $\{x_n\}$ is a decreasing sequence.

(b) It is clear that $x_n \geq 0$ for all $n \in \mathbb{N}$. Use this fact and the result from part (a) to prove that $\{x_n\}$ converges. (Hint: Your proof should be quite short.)

(c) Find the limit $L = \lim_{n \to \infty} x_n$.

22. A sequence $\{x_n\}$ *pseudo-diverges* to ∞ if $\forall M, N \in \mathbb{N}$ there exists $n > N$ such that $x_n > M$. Prove or disprove that $\{x_n\} = (-1)^n n$ pseudo-diverges to ∞.

23. Show that if $\{x_n\}$ is bounded, all $y_n \neq 0$, and $y_n \to \infty$, then $x_n/y_n \to 0$.

2.3   Subsequences

Basic Ideas

Any given sequence $x_1, x_2, x_3, x_4, \ldots$ has many, many *subsequences*. Here are three examples:

$$x_1, \ x_3, \ x_5, \ x_7, \ \ldots \qquad x_1, \ x_4, \ x_9, \ x_{16}, \ \ldots \qquad x_{123}, \ x_{124}, \ x_{125}, \ \ldots$$

Subsequences are formed by choosing—in order—any infinite subset of the original sequence $\{x_n\}$. The last example above could be called an *upper tail subsequence*, formed simply by skipping over an initial string. Order matters, and repetitions aren't allowed, so

$$x_2, \ x_1, \ x_4, \ x_3, \ x_6, \ x_5, \ \ldots \quad \text{and} \quad x_1, \ x_1, \ x_2, \ x_2, \ x_3, \ x_3, \ \ldots$$

are *not* considered subsequences of the "parent" $\{x_n\}$.

Subsequences may or may not behave much like their parents. If, say,

$$x_1, \ x_2, \ x_3, \ x_4, \ \cdots = 1, \ 2, \ 1, \ 2, \ \ldots,$$

then $\{x_n\}$ has no single limit, and therefore diverges. But the subsequences

$$x_1, \ x_3, \ x_5, \ \cdots = 1, \ 1, \ 1, \ \ldots \quad \text{and} \quad x_2, \ x_4, \ x_6, \ \cdots = 2, \ 2, \ 2, \ \ldots$$

obviously converge to 1 and 2, respectively. Still other subsequences, such as

$$x_3, \ x_6, \ x_9, \ x_{12}, \ \ldots \quad \text{and} \quad x_1, \ x_{10}, \ x_{100}, \ x_{1000}, \ \ldots$$

may or may not converge. By contrast, the sequence $\{y_n\}$ with

$$y_1, \ y_2, \ y_3, \ y_4, \ \cdots = 1.0, \ 0.1, \ 0.01, \ 0.001, \ \ldots$$

converges to zero—and so does *every one* of its subsequences, including

$$y_1, \ y_3, \ y_5, \ \cdots \ y_{83}, \ \ldots \quad \text{and} \quad y_1, \ y_{10}, \ y_{100}, \ \ldots, \ y_{10^{41}}, \ \ldots.$$

Stranger examples are possible.

EXAMPLE 1. (A walk along the rationals.) Because $\mathbb{Q}$ is countably infinite, it can be listed as a sequence $r_1, r_2, r_3, r_4, \ldots$. Is there a subsequence consisting entirely of positive integers? Must some subsequence converge to zero?

SOLUTION. Yes and yes. Lacking rigorous definitions, we'll argue (very!) informally.

Coming soon.

To find a subsequence of positive integers, imagine walking from left to right along the rational sequence, underlining each term that happens to be a positive integer. The result might look like this:

$$r_1, \underline{r_2}, \underline{r_3}, r_4, r_5, r_6, \ldots, r_{23}, \underline{r_{24}}, r_{25}, \ldots, r_{1234}, \underline{r_{1235}}, \ldots.$$

The process never stops because the supply of positive integers is infinite, so we have our desired subsequence:

$$r_2, r_3, r_{24}, r_{1235} \ldots.$$

To find a subsequence that converges to zero, recall first that for every positive ϵ, no matter how small, the interval $(-\epsilon, \epsilon)$ contains infinitely many rationals. To find our desired subsequence, therefore, we walk again from left to right along the original sequence. At some position, say a_9, we find $a_9 \in (-1, 1)$. Continuing our rightward walk, we find, say,

$$a_{17} \in (-0.1, 0.1), \quad a_{137} \in (-0.01, 0.01), \quad a_{2988} \in (-0.001, 0.001),$$

and so on. We never need to stop because, at any stage, we've left behind only finitely many of the infinite family of residents of $(-\epsilon, \epsilon)$. $\Diamond$

Similar reasoning shows that any listing $\{r_n\}$ of $\mathbb{Q}$ has subsequences consisting entirely of prime numbers or entirely of fractions with denominator 424242. It is less obvious—but true—that $\{r_n\}$ has an increasing subsequence with limit π, a decreasing subsequence with limit e, a decreasing subsequence of integers diverging to $-\infty$, and countless other subsequences of interest. We sort out such possibilities in this section.

Formalities. The idea of subsequences is simple enough but the notation takes some getting used to. To create a subsequence from a given sequence $\{a_n\}$ means to choose a *strictly increasing* sequence of subscripts

$$n_1 < n_2 < n_3 < n_4 < n_5 < \ldots,$$

and use them to form the subsequence

$$a_{n_1}, a_{n_2}, a_{n_3}, a_{n_4}, a_{n_5}, \ldots.$$

If, say, $n_1 = 4$, $n_2 = 7$, $n_3 = 11$, $n_4 = 29, \ldots$, then the subsequence has the form

$$a_4, \, a_7, \, a_{11}, \, a_{29}, \, \ldots.$$

The entire subsequence is denoted $\{a_{n_k}\}$; note the double subscript.

The process can be described precisely in the language of functions. Recall, first, that a sequence $\{a_n\}$ *is* a function $a : \mathbb{N} \to \mathbb{R}$, with $a(1) = a_1$, $a(2) = a_2$, etc. From this viewpoint a subsequence is formed by *composition*, as the formal definition says:

Definition 2.11 (Subsequence). Let $\{a_n\}$ be a sequence, regarded as a function $a : \mathbb{N} \to \mathbb{R}$. A *subsequence* $\{a_{n_k}\}$ of $\{a_n\}$ is a composite function $a \circ n : \mathbb{N} \to \mathbb{R}$, where $n : \mathbb{N} \to \mathbb{N}$ is any *strictly increasing* function.

Observe:

- *A useful inequality:* The definition implies a simple inequality that is useful in proofs: $n_k \geq k$ for all k. A formal proof might involve induction or the pigeonhole principle, but the idea is mainly common sense—the kth term of a subsequence can't appear before the kth term of the parent sequence.

- *Another subsequence:* The subsequence indices n_1, n_2, n_3, n_4, $\ldots$ for $\{a_{n_k}\}$ form still another subsequence—a *strictly increasing* subsequence of $\{1, 2, 3, \ldots\}$.

EXAMPLE 2.   Let $\{a_n\}$ be a sequence. Interpret the subsequences

$$a_1, \, a_3, \, a_5, \, a_7, \, \ldots \quad \text{and} \quad a_2, \, a_4, \, a_6, \, a_8, \, \ldots$$

in functional language and notation. What can be said about $\{a_n\}$ if both subsequences converge to three?

SOLUTION.   For the "odd" subsequence we have $n(1) = 1$, $n(2) = 3$, and $n(42) = 83$; in general, $n(k) = n_k = 2k - 1$. Thus,

$$a \circ n(k) = a(n(k)) = a(2k - 1)$$

for all k. In ordinary sequence notation, we can write $\{a_{n_k}\} = \{a_{2k-1}\}$; notice that k, not n, is now the index variable. For the "even" subsequence similar reasoning gives $\{a_{n_k}\} = \{a_{2k}\}$.

A bit laboriously.

If *both* subsequences $\{a_{2k-1}\}$ and $\{a_{2k}\}$ converge to three, then it is reasonable to expect the parent sequence $\{a_n\}$ to do so as well. To prove this let $\epsilon > 0$ be given. By hypothesis, there exist numbers K_1 and K_2 such that

$$k > K_1 \implies |a_{2k-1} - 3| < \epsilon \quad \text{and} \quad k > K_2 \implies |a_{2k} - 3| < \epsilon.$$

Now it turns out that $N = \max\{2K_1, 2K_2\}$ "works" for ϵ in the parent sequence.

To see why, suppose $n > N$. If $n = 2k$ happens to be *even*, then $n = 2k > 2K_2$, and so $k > K_2$, which implies that

$$|a_n - 3| = |a_{2k} - 3| < \epsilon.$$

If $n = 2k - 1$ is *odd*, then $n = 2k - 1 > 2K_1$, and so $k > K_1$, and we see

$$|a_n - 3| = |a_{2k-1} - 3| < \epsilon.$$

Thus we have $|a_n - 3| < \epsilon$ for *all* $n > N$, and the proof is complete.   ◊

Properties of Subsequences

Every infinite sequence $\{x_n\}$ has countless subsequences. What can be said about such an enormous set? The answer is, "quite a lot": subsequences often "inherit" their parents' basic properties.

Uncountably many, in fact.

Theorem 2.12. *Let $\{x_n\}$ be a sequence and L a number.*

(a) *If $\{x_n\}$ converges to L, then every subsequence $\{x_{n_k}\}$ converges to L, too.*

(b) *If $\{x_n\}$ diverges to $\pm\infty$, then every subsequence $\{x_{n_k}\}$ diverges to $\pm\infty$, too.*

(c) *If $\{x_n\}$ has subsequences converging to* different *limits, then $\{x_n\}$ diverges.*

About proofs. Key to both proofs is the fact, mentioned above, that, for *any* subsequence, $n_k \geq k$ for all k. To prove (a), let $\epsilon > 0$ be given. By hypothesis there exists N that works for $\{x_n\}$ in the sense that $|x_n - L| < \epsilon$ whenever $n > N$. The key fact above implies that *the same N works* for $\{x_{n_k}\}$; if $k > N$ then $n_k \geq k > N$, and so $|x_{n_k} - L| < \epsilon$, as desired.

The proof of (b) is similar, and (c) follows immediately from (a). Further details are left to exercises.

Monotone subsequences. Following is a striking property of *every* sequence; we'll make good use of it below in proving this section's marquee result, the celebrated Bolzano–Weierstrass theorem.

Proposition 2.13. *Every sequence has a monotone subsequence.*

Proposition 2.13 is easy to state, but it's a bit tricky to prove *anything* about all sequences—convergent or divergent, bounded or unbounded, tame or wild. Here,

"Nontrivial," in math-speak.

for instance, is a randomly-chosen stretch (from $n = 125$ to $n = 130$) of terms from the sequence $\{n\cos(n^2)\}$:

$$\ldots, 35.617, -4.847, 126.746, -106.052, -128.983, -25.531, \ldots.$$

It is hard to see much pattern there, but Proposition 2.13 guarantees that *somewhere* in there is an increasing subsequence, a decreasing subsequence, or maybe both.

We'll sneak up on a proof of Proposition 2.13 through two technical lemmas about special cases.

Lemma 2.14. *Every unbounded sequence $\{x_n\}$ has a monotone subsequence that diverges to $\pm\infty$.*

Proof (sketch): For a sequence $\{x_n\}$ unbounded *above*, we'll find an *increasing* sequence $\{x_{n_k}\}$ with $x_{n_k} > k$ for all k. Since $\{x_n\}$ is unbounded above, we can choose n_1 so $x_{n_1} > 1$. Now the "upper tail"

$$x_{n_1+1}, \ x_{n_1+2}, \ x_{n_1+3}, \ x_{n_1+4}, \ \ldots$$

Do you see why?

is *also* unbounded above, and so we can choose n_2 with $n_2 > n_1$ and $x_{n_2} > \max\{2, x_{n_1}\}$. Then we choose $n_3 > n_2$ with $x_{n_3} > \max\{3, x_{n_2}\}$. Continuing this process produces the desired subsequence. A similar construction produces a strictly *decreasing* subsequence if $\{x_n\}$ is unbounded *below*.   □

Lemma 2.15. *Let $\{x_n\}$ be a bounded sequence, with infimum α and supremum β. If $\alpha \notin \{x_n\}$, then there is a decreasing subsequence $\{x_{n_k}\}$ with $x_{n_k} \to \alpha$. If $\beta \notin \{x_n\}$, then there is an increasing subsequence $\{x_{n_k}\}$ with $x_{n_k} \to \beta$.*

Proof (sketch): To save a little labor we'll *use* Lemma 2.14 and some fancy algebra. In the case $\beta \notin \{x_n\}$, we define a new sequence $\{y_n\}$ by the rule

$$y_n = \frac{1}{\beta - x_n}.$$

Do you see why?

Now $\{y_n\}$ is unbounded above, and so Lemma 2.14 guarantees that there is an *increasing* subsequence $\{y_{n_k}\}$ with $y_{n_k} \to \infty$. But this implies that

$$\frac{1}{y_{n_k}} = \beta - x_{n_k} \to 0.$$

This implies, in turn, that $\{x_{n_k}\}$ is increasing, with limit β. A similar argument applies if $\alpha \notin \{x_n\}$.   □

At last we can prove Proposition 2.13. Watch for the completeness axiom to pop up, and for Lemma 2.15 to provide the key technical insight.

Proof (of Proposition 2.13): Let $\{x_n\}$ be our sequence. If $\{x_n\}$ is unbounded, we're done, by Lemma 2.14. So we'll assume that

(i) $\{x_n\}$ is bounded,   and   (ii) $\{x_n\}$ has no increasing subsequence,

and use these assumptions to construct a *decreasing* subsequence.

To get started, let $\beta_1 = \sup\{x_1, x_2, x_3, \dots\}$. (Assumption (i) and the completeness axiom guarantee that β_1 exists.) The key observation is that β_1 is *in* the set $\{x_1, x_2, x_3, \dots\}$. Otherwise, says Lemma 2.15, some *increasing* subsequence would converge to β_1. Thus, we can find n_1 with $x_{n_1} = \beta_1$; this is the first term of our desired subsequence.

To continue, we set

$$\beta_2 = \sup\{x_{n_1+1}, x_{n_1+2}, x_{n_1+3}, \dots\},$$

Again, β_2 exists by the completeness axiom, and Lemma 2.15, applied this time to the upper tail sequence

$$x_{n_1+1},\ x_{n_1+2},\ x_{n_1+3},\ x_{n_1+4} \cdots,$$

implies that $\beta_2 = x_{n_2}$ for some $n_2 > n_1$. It is clear, too, that $\beta_2 \le \beta_1$, as β_2 bounds a smaller set than does β_1.

Continuing this process indefinitely produces the desired decreasing subsequence $\{x_{n_1}, x_{n_2}, x_{n_3}, \dots\}$, and the proof is complete.   □

A big theorem. It is now easy to prove a famous theorem, which we'll use repeatedly. We've done all the hard work.

Starting in the very next section.

Theorem 2.16 (Bolzano–Weierstrass). *Every bounded sequence $\{x_n\}$ has a convergent subsequence.*

Proof: By Proposition 2.13, $\{x_n\}$ has a *monotone* subsequence $\{x_{n_k}\}$, which is also *bounded* because $\{x_n\}$ is. By Theorem 2.3, page 88, every monotone and bounded sequence converges.   □

Exercises

1. Give an example in each part as indicated; no proofs needed.

   (a) A sequence with subsequences converging to 1, 2, and 3.

   (b) A convergent sequence with a subsequence of positive terms and another subsequence with negative terms.

   (c) A sequence with subsequences converging to 0 and to ∞.

   (d) A positive sequence with monotone subsequences converging to 0 and to ∞.

(e) A sequence of positive integers with subsequences that converges to *every* positive integer.

2. The Bolzano–Weierstrass theorem (BWT) says that if a sequence $\{x_n\}$ is bounded, then $\{x_n\}$ has a convergent subsequence.

   (a) State the converse and the contrapositive of the BWT. Disprove the false one.

   (b) Consider the sequences $\{y_n\}$ and $\{z_n\}$ defined by $y_n = \sin n$ and $z_n = \frac{\sin n}{n}$. What can be said about their subsequences? Is the BWT helpful? Is it needed?

   (c) Show (assuming the BWT) that *every* sequence $\{x_n\}$ has either (i) a convergent subsequence, or (ii) a subsequence that diverges to $\pm\infty$. Can both (i) and (ii) occur?

3. Because $\mathbb{Q}$ is countably infinite, we know that it can be written as a sequence.

   (a) Explain why the set of *nonnegative* rationals can also be written as a sequence, say, $\{p_n\}$.

   (b) Give a reason why $\{p_n\}$ cannot be monotone.

   (c) Although $\{p_n\}$ itself is not monotone, it must have a monotone subsequence of positive integers. Explain why.

4. Let $\{x_n\}$ be a sequence and x_0 a number.

   (a) Show that $x_n \to x_0$ if and only if for every $\epsilon > 0$ the set $\{n \mid x_n \notin (x_0 - \epsilon, x_0 + \epsilon)\}$ is finite.

   (b) Show that some subsequence $\{x_{n_k}\}$ converges to x_0 if and only if for every $\epsilon > 0$ the set $\{n \mid x_n \in (x_0 - \epsilon, x_0 + \epsilon)\}$ is infinite.

5. Consider the subsequence $x_{4242}, x_{4243}, x_{4244}, \ldots$ formed from a parent sequence $x_1, x_2, x_3, \ldots$. If the subsequence converges to L, then so does the parent sequence.

6. Suppose $x_n \to \infty$. Then every subsequence $\{x_{n_k}\}$ diverges to ∞, too.

7. Following are several statements about a sequence $\{a_n\}$. They belong in two groups. What are the groups?

   (a) $\{a_n\}$ converges to 3

   (b) a subsequence of $\{a_n\}$ converges to 3

(c) $\forall \epsilon > 0$ only finitely many a_n are *not* in $(3 - \epsilon, 3 + \epsilon)$

(d) $\forall \epsilon > 0$, infinitely many of the a_n are in $(3 - \epsilon, 3 + \epsilon)$

(e) $\forall \epsilon > 0 \quad \exists N \in \mathbb{N}$ such that $n > N \implies |a_n - 3| < \epsilon$

(f) $\forall \epsilon > 0 \quad \exists N \in \mathbb{N}$ such that $a_{N+1}, a_{N+2}, a_{N+3}, \ldots$ are all in $(3 - \epsilon, 3 + \epsilon)$

8. Consider the sequence $\{x_n\}$ defined by $x_n = (-1)^n \frac{n}{n+1}$.

   (a) Find the subsequences $\{x_{2k}\}$ and $\{x_{2k-1}\}$. What are their limits? (No proofs needed.) What does the answer imply about convergence of $\{x_n\}$?

   (b) Let $\{x_{n_k}\}$ be any subsequence of $\{x_n\}$. Show that if $\{x_{n_k}\}$ converges, then it must converge either to 1 or to -1.

9. Let $\{x_n\}$ be a sequence that does *not* converge to 3.

   (a) Show by an example that $x_n = 3$ can hold for infinitely many n.

   (b) Show that there exists some $\epsilon > 0$ and a subsequence $\{x_{n_k}\}$ such that $|x_{n_k} - 3| \geq \epsilon$ for all k.

10. Show that if $\{x_n\}$ diverges and x_0 is any number, then there exists $\epsilon > 0$ and a subsequence $\{x_{n_k}\}$ such that $|x_{n_k} - x_0| \geq \epsilon$ for all k. (In other words, some subsequence $\{x_{n_k}\}$ is bounded away from x_0.)

11. This problem is about Theorem 2.12, page 109.

    (a) Prove Theorem 2.12(b).

    (b) State the contrapositive of Theorem 2.12(a). How is this related to Theorem 2.12(c)?

12. Let $\{x_n\}$ be a sequence and x_0 a number. Show that $x_n \to x_0$ if and only if $x_{n_k} \to x_0$ for every *monotone* subsequence $\{x_{n_k}\}$.

13. The sequence $\{x_n\}$ given by $x_n = 1/n$ converges, relatively slowly, to $L = 0$. Find a subsequence $\{x_{n_k}\}$ that converges rapidly to 0 in the sense that $|x_{n_k} - L| < 1/10^k$ for all k.

14. Show that for any sequence $\{x_n\}$ with $x_n \to L$, there's a subsequence $\{x_{n_k}\}$ for which $|x_{n_k} - L| < 1/10^k$ for all k.

15. Given any two sequences $\{a_n\}$ and $\{b_n\}$, we can construct a new sequence $\{z_n\}$ by "zipping" $\{a_n\}$ and $\{b_n\}$ together: $\{z_n\}$ is the new sequence $a_1, b_1, a_2, b_2, \ldots, a_{42}, b_{42}, \ldots$. Show that $\{z_n\}$ converges to L if and only if both $\{a_n\}$ and $\{b_n\}$ converge to L.

2.4 Cauchy Sequences

The obvious way to prove that a sequence $\{x_n\}$ converges is to know or guess a limit, say L, and then show that $x_n \approx L$ for large n (in the precise sense of the definition). For some sequences no obvious candidate for L presents itself, but we may know instead that the x_n are *close to each other* for large n. Consider, for instance, the sequence $\{s_n\}$:

$$1, \ 1 - \frac{1}{3}, \ 1 - \frac{1}{3} + \frac{1}{5}, \ 1 - \frac{1}{3} + \frac{1}{5} - \frac{1}{7}, \ \ldots.$$

Here are approximate numerical values for the first few s_n:

$$1.00, \ 0.667, \ 0.867, \ 0.724, \ 0.835, \ 0.744, \ 0.821, \ 0.754, \ 0.813, \ \ldots.$$

Successive s_n appear to jump back and forth across smaller and smaller intervals. Does $\{s_n\}$ approach some limit?

   In this particular case the answer is classical. Our sequence $\{s_n\}$ is derived from the famous *Leibniz series*

$$1 - \frac{1}{3} + \frac{1}{5} - \frac{1}{7} + \frac{1}{9} - \frac{1}{11} + \frac{1}{13} - \cdots,$$

which has been known for at least 300 years to converge to $\pi/4 \approx 0.7854$. More generally, $\{s_n\}$ illustrates the notion of a *Cauchy sequence*, named for the French mathematician Augustin–Louis Cauchy (1789–1857), a key figure in the development of real analysis.

Cauchy Basics

The formal definition describes precisely what we mean by "close to each other for large n."

Definition 2.17. A sequence $\{x_n\}$ is *Cauchy* if, for every $\epsilon > 0$, there exists N such that

$$|x_n - x_m| < \epsilon \quad \text{whenever } n > m > N.$$

   Observe:

- *Similar, but different:* The definition resembles that for convergence, but with a key difference: no limit L is mentioned.

- *Not just consecutive terms:* The definition requires that *all* terms with large index be close to each other. This applies not only to *consecutive* terms, like x_{12345} and x_{12346}, but also to *any* two terms with large index, like x_{12345} and x_{98765}.

- A *minor convenience:* The condition $n > m$ in the definition is sometimes convenient in proofs. This apparent asymmetry is actually harmless, because the case $n = m$ is trivial, and we lose no generality in using n to denote the larger of two indices.

EXAMPLE 1.   As we know, the sequence $\{1/n\}$ converges to zero, while $\{\sqrt{n}\}$ diverges to infinity. Is either sequence Cauchy?

SOLUTION.   The sequence $\{1/n\}$ *is* Cauchy. Intuitively, we expect that, since terms approach zero, they must also stay close to each other. To prove this formally, let $\epsilon > 0$ be given, and set $N = 1/\epsilon$. This N "works," since if $n > m > N$, then

$$|x_n - x_m| = \left| \frac{1}{n} - \frac{1}{m} \right| = \frac{1}{m} - \frac{1}{n} < \frac{1}{m} < \frac{1}{1/\epsilon} = \epsilon,$$

Notice where we use $n > m$.

as the definition requires.

The sequence $\{\sqrt{n}\}$ is *not* Cauchy—even though the difference between *successive terms* (like $\sqrt{123456} \approx 351.363$ and $\sqrt{123457} \approx 351.364$) tends to zero. We can prove this formally by showing that if $\epsilon = 1$, then *no* N works in the sense of the definition. Indeed, let N be *any* positive number, and let k be any positive integer with $k^2 > N$. If we set $m = k^2$ and $n = (k+1)^2$, then

$$n > m > N \quad \text{but} \quad \sqrt{n} - \sqrt{m} = 1 = \epsilon,$$

and so the definition is not satisfied.   ◊

EXAMPLE 2.   The sequence $\{s_n\}$ at the beginning of this section *is* Cauchy. Explain why.

SOLUTION.   The key idea, on which a formal proof can be based, is the sequence's back-and-forth behavior:

$$s_2 < s_4 < s_6 < s_8 < \cdots < s_7 < s_5 < s_3 < s_1,$$

combined with the fact that the distance between successive terms tends to zero. If, say, $\epsilon = 0.001$, we could set $N = 1000$, and observe that if $n > m > 1000$, then

$$s_{1000} < s_n, s_m \leq s_{1001} = s_{1000} + \frac{1}{1001},$$

which implies that s_n and s_m lie within $1/1001$ of each other.   ◊

Morals from the examples. The examples suggest, correctly, a close connection between sequence convergence and the Cauchy property. Indeed, the two properties turn out to be equivalent. As we'll see, it is easy to show that every convergent sequence is also Cauchy. The fact that every Cauchy sequence converges is deeper, and proving it takes a little more work. The Bolzano–Weierstrass theorem—another convergence guarantee—will be useful.

Properties of Cauchy Sequences

First we pick two low-hanging fruits.

Proposition 2.18. *If $\{x_n\}$ converges, then $\{x_n\}$ is Cauchy.*

The basic idea is that if both x_n and x_m are eventually close to a limit L, then they must also be close *to each other*. The proof makes this precise.

Why is it OK to use $\epsilon/2$, not ϵ?

Proof: Suppose $x_n \to L$. For given $\epsilon > 0$, we can choose N such that $|x_n - L| < \epsilon/2$ whenever $n > N$. This N works in the Cauchy sequence definition, because if $n > m > N$, then

$$|x_n - x_m| \le |x_n - L| + |x_m - L| < \frac{\epsilon}{2} + \frac{\epsilon}{2} = \epsilon;$$

we used the triangle inequality in the first step. $\square$

Proposition 2.19. *If $\{x_n\}$ is Cauchy, then $\{x_n\}$ is bounded.*

Proof: For $\epsilon = 1$, choose N as in the Cauchy definition. Now fix any integer m_0 with $m_0 > N$. If $n > m_0$, then we have $|x_n - x_{m_0}| < 1$, or, equivalently, $x_{m_0} - 1 < x_n < x_{m_0} + 1$. In particular, the set $\{x_{m_0}, x_{m_0+1}, x_{m_0+2}, \dots\}$ is bounded (above by $x_{m_0} + 1$ and below by $x_{m_0} - 1$). Since the remaining terms $\{x_1, x_2, \dots x_{m_0-1}\}$ form a *finite* set, the entire sequence $\{x_n\}$ is bounded. $\square$

Watch for another $\epsilon/2$ trick.

Combining Proposition 2.19 with the Bolzano–Weierstrass theorem lets us prove our main theoretical result.

Theorem 2.20. *A sequence $\{x_n\}$ converges if and only if $\{x_n\}$ is Cauchy.*

Proof: Proposition 2.18 is the "only if" part. To prove the "if" part, suppose that $\{x_n\}$ is Cauchy. Proposition 2.19 implies that $\{x_n\}$ is bounded; by the Bolzano–Weierstrass theorem, $\{x_n\}$ has a subsequence $\{x_{n_k}\}$ that converges to some limit, say, L. To complete the proof, we'll show that the full sequence $\{x_n\}$ also converges to L.

Let $\epsilon > 0$ be given. Since $\{x_n\}$ is Cauchy, we can choose N_1 such that

$$|x_n - x_m| < \epsilon/2 \quad \text{whenever} \quad n > m \ge N_1.$$

Because $\{x_{n_k}\}$ converges to L, we can choose a member of the subsequence, say x_N, with $N > N_1$ and $|x_N - L| < \epsilon/2$. Now this N works in the definition of convergence of $\{x_n\}$ to L, because if $n \geq N$, then

Infinitely many possible x_N exist; any one will do.

$$|x_n - L| \leq |x_n - x_N| + |x_N - L| < \frac{\epsilon}{2} + \frac{\epsilon}{2} = \epsilon,$$

as desired. $\square$

EXAMPLE 3. A "random walk" sequence $a_1, a_2, a_3, \ldots$ is produced iteratively, with help from a fair coin. First we flip the coin and set $a_1 = 1$ if heads show and $a_1 = -1$ otherwise. Then we flip again and set $a_2 = a_1 + 1/2$ for heads and $a_2 = a_1 - 1/2$ for tails. We flip again at each stage, and set

$$a_{n+1} = \begin{cases} a_n + \frac{1}{2^n} & \text{if heads,} \\ a_n - \frac{1}{2^n} & \text{if tails.} \end{cases}$$

Particular values of a_k depend, of course, on random coin flips. Here's what happened on one try:

Some decimals are rounded.

$$-1, \qquad -1.5, \qquad -1.25, \qquad -1.125, \qquad -1.1875, \qquad -1.1563,$$
$$-1.1406, \qquad -1.1484, \qquad -1.1445, \qquad -1.1426, \qquad -1.1436, \qquad \ldots$$

Must *every* such sequence converge?

SOLUTION. Yes, thanks to Theorem 2.20. The sequence $\{a_n\}$ is Cauchy—no matter how the coin falls. To prove this formally observe first that, for any a_n and a_m with $n > m$, we have

$$|a_n - a_m| = \left| \pm\frac{1}{2^m} \pm \frac{1}{2^{m+1}} \pm \cdots \pm \frac{1}{2^{n-1}} \right|$$
$$\leq \frac{1}{2^m} + \frac{1}{2^{m+1}} + \cdots + \frac{1}{2^{n-1}}$$
$$= \frac{1}{2^{m-1}} \left(\frac{1}{2} + \frac{1}{2^2} + \cdots + \frac{1}{2^{n-m}} \right) < \frac{1}{2^{m-1}}.$$

Now for given $\epsilon > 0$, we can choose N with $1/2^{N-1} < \epsilon$. This N works in the Cauchy definition, because if $n > m \geq N$, then, as just shown,

$$|a_n - a_m| < \frac{1}{2^{m-1}} \leq \frac{1}{2^{N-1}} < \epsilon,$$

as desired. $\Diamond$

Cauchy, convergence, completeness: crucial C-words. This section's main result, that Cauchy sequences converge, is not obvious; proving it took planning and effort. The work is worthwhile because it guarantees that sequences like those of Example 3, for which the terms become close together in a strong sense, do indeed have specific numerical limits. Without such guarantees it would be difficult, for example, to define the *integral* of a function (another kind of limit) unambiguously.

As we'll do later.

Another C-word—completeness—is at the heart of Theorem 2.20. Completeness appears implicitly in the proof, by way of the Bolzano–Weierstrass theorem, which itself required the completeness axiom. (We invoked it in proving Proposition 2.13, page 109, for example.) Indeed, completeness is sometimes *defined* not in terms of the supremum (as was done earlier; see page 74) but rather in terms of Cauchy sequence convergence. We pursue this connection briefly in the exercises.

Exercises

1. Show that each of the following sequences either is or is not Cauchy. (Use the definition, not theorems.)

    (a) $\{x_n\}$, where $x_n = \frac{(-1)^n}{n}$.

    (b) $\{y_n\}$, where $y_n = \frac{(-1)^n}{1234}$.

    (c) $\{z_n\}$, where $z_n = \frac{n}{n+1}$.

    (d) $\{w_n\}$, where $w_n = \frac{\sin n}{n^2+1}$.

2. Let $\{x_n\}$ and $\{y_n\}$ be Cauchy sequences. Use Definition 2.17 to show that $\{x_n + y_n\}$ is Cauchy, too.

3. Let $\{x_n\}$ be a Cauchy sequence such that $x_1 = 0 = x_{10} = x_{100} = x_{1000} = \ldots$. Use Definition 2.17 (not theorems) to show that $\{x_n\}$ converges to zero.

4. Let $\{x_n\}$ be a Cauchy sequence, with $x_n \in \mathbb{Z}$ for all n. Show that $\{x_n\}$ is "eventually constant"—i.e., there exists $N > 0$ such that $x_n = x_m$ whenever $n > m > N$.

5. As we said at the end of this section, completeness can be *defined* using Cauchy sequences: A subset $A \subseteq \mathbb{R}$ is *complete* if every Cauchy sequence $\{x_n\}$ contained in A has a limit in A. Use this definition below.

    (a) The set $\mathbb{Q}$ of rational numbers is not complete. Show this by finding a Cauchy sequence $\{r_n\}$ of rational numbers that has no rational limit.

    (b) Is the set $\mathbb{Z}$ of integers complete? (Hint: See Problem 4.)

(c) Show that $[0, 1]$ is complete but $(0, 1]$ is not.

6. From any given sequence $\{x_n\}$ we can form the related sequence $\{y_n\} = \{5x_n + 2\}$.

   (a) Use the definition of a Cauchy sequence (not theorems) to show that if $\{x_n\}$ is Cauchy, then so is $\{y_n\}$.

   (b) Use any convenient theorems to give a shorter (not necessarily better) proof that if $\{x_n\}$ is Cauchy, then so is $\{y_n\}$.

7. Let $\{a_n\}$ be a Cauchy sequence. Show that if $\{x_n\}$ is a sequence such that $|x_n - x_m| \leq 42 \, |a_n - a_m|$ for all n and m, then $\{x_n\}$ is Cauchy, too.

8. This problem is about the sequence $\{a_n\}$ in Example 3, page 117.

   (a) Under what circumstances is $\{a_n\}$ monotone?

   (b) Suppose the first six coin tosses show $HTHTHT$. Find the smallest interval in which the limit can lie.

   (c) Explain why $-2 < a_n < 2$ for all n (regardless of the coin tosses).

9. Consider the sequence $\{\pi_n\}$ defined by $\pi_1 = 3.1$, $\pi_2 = 3.14$, $\pi_3 = 3.141$, $\ldots$; for each n, π_n shows the first n decimal digits of π. Show (use the definition, not theorems) that $\{\pi_n\}$ is Cauchy.

10. This problem is about the fact that if $\{x_n\}$ and $\{y_n\}$ are Cauchy sequences, then the product sequence $\{x_n y_n\}$ is Cauchy, too.

    (a) Use Theorem 2.20, page 116, and any other already-proved results to explain why the fact holds.

    (b) (Harder.) Prove the same fact *without* using Theorem 2.20, page 116. (Hints: An algebraic trick and the triangle inequality give

    $$|x_n y_n - x_m y_m| = |x_n y_n - x_n y_m + x_n y_m - x_m y_m|$$
    $$\leq |x_n y_n - x_n y_m| + |x_n y_m - x_m y_m|.$$

    Now use boundedness of both $\{x_n\}$ and $\{y_n\}$ (Proposition 2.19) to complete the proof.)

11. Let $\{a_n\}$ be a sequence for which *successive* terms are "very close" in the sense that $|a_{n+1} - a_n| \leq 1/2^{n+1}$ for all n. Show that $\{a_n\}$ is Cauchy. Hint: For any n and m with $n > m$, we have

    $$|a_n - a_m| \leq |a_{m+1} - a_m| + |a_{m+2} - a_{m+1}| + \cdots + |a_n - a_{n-1}|$$
    $$\leq \frac{1}{2^{m+1}} + \frac{1}{2^{m+2}} + \cdots + \frac{1}{2^n}.$$

    How large can the last expression be?

12. In each part, either give an example of the given type or say briefly (perhaps by citing an appropriate theorem) why no such example can exist. To describe a sequence $\{a_n\}$, either give a formula for a_n or just write enough terms to make the pattern clear.

   (a) a sequence $\{a_n\}$ that is **not monotone** but **diverges** to ∞

   (b) a **divergent** sequence $\{a_n\}$ such that $\left\{ \dfrac{a_n}{1717} \right\}$ converges

   (c) two **divergent** sequences $\{a_n\}$ and $\{b_n\}$ such that $\{a_n + b_n\}$ **converges** to 17

   (d) two **convergent** sequences $\{a_n\}$ and $\{b_n\}$ such that such that $\{a_n/b_n\}$ **diverges**

   (e) a sequence with **no convergent subsequence**

   (f) a **Cauchy** sequence with an **unbounded** subsequence

2.5   Series 101: Basic Ideas

An *infinite series* is a sum with infinitely many summands, or *terms*:

$$\sum_{k=1}^{\infty} a_k = a_1 + a_2 + a_3 + \dots.$$

Here are two important examples, each of which we will revisit:

$$\sum_{k=1}^{\infty} \frac{1}{2^{k-1}} = 1 + \frac{1}{2} + \frac{1}{4} + \dots; \qquad \sum_{k=1}^{\infty} \frac{1}{k} = 1 + \frac{1}{2} + \frac{1}{3} + \frac{1}{4} + \dots.$$

As usual for infinite processes, some obvious questions arise. Is there some limit (or "infinite sum") to which the series converges? How do we decide? If a series has a limit, how do we find it?

A first observation is that series are close kin to sequences—about which we already know a lot. Indeed, every *series* $\sum a_k$ corresponds in a natural way to a certain *sequence* $\{A_n\}$ whose properties tell us "everything" about the series. The flaw in this happy scenario is that, in practice, getting one's hands directly on the sequence $\{A_n\}$ can be difficult. Much of what follows can be thought of as strategy for getting around this problem.

We'll define it in a moment.

Enough generalities.

Definition 2.21. For a given series $\sum_{k=1}^{\infty} a_k = a_1 + a_2 + a_3 + \dots$, the *sequence of partial sums* is given by

$$A_1 = a_1, \quad A_2 = a_1 + a_2, \quad \dots, \quad A_n = a_1 + a_2 + \dots + a_n, \quad \dots.$$

In general, $A_n = \sum_{k=1}^{n} a_k$.

EXAMPLE 1.   Describe the partial sum sequences for the two series illustrated above. Do they have limits?

SOLUTION.   The series

$$\sum_{k=1}^{\infty} a_k = \sum_{k=1}^{\infty} \frac{1}{2^{k-1}} = 1 + \frac{1}{2} + \frac{1}{4} + \cdots$$

comes from the *geometric* family, in which successive terms differ by a constant multiple. Here we have

Much more about geometric series soon.

$$A_1 = 1, \quad A_2 = 1 + \frac{1}{2} = \frac{3}{2}, \quad A_3 = 1 + \frac{1}{2} + \frac{1}{4} = \frac{7}{4}, \quad A_4 = A_3 + \frac{1}{8} = \frac{15}{8}, \quad \ldots$$

Now it is easy to guess—and to prove by induction—that

$$A_n = \frac{2^n - 1}{2^{n-1}} = 2 - \frac{1}{2^{n-1}}.$$

Clearly, $A_n \to 2$, so it makes sense to write $1 + \frac{1}{2} + \frac{1}{4} + \frac{1}{8} + \cdots = 2$. We'll sanctify this below with a formal definition.

The series

$$\sum_{k=1}^{\infty} h_k = \sum_{k=1}^{\infty} \frac{1}{k} = 1 + \frac{1}{2} + \frac{1}{3} + \frac{1}{4} + \cdots$$

is the famous *harmonic series*. Here are the first few partial sums H_n, also known as *harmonic numbers*:

More soon about this, too.

$$H_1 = 1, \quad H_2 = 1 + \frac{1}{2} = \frac{3}{2}, \quad H_3 = 1 + \frac{1}{2} + \frac{1}{3} = \frac{11}{6}, \quad H_4 = H_3 + \frac{1}{4} = \frac{25}{12}.$$

No pattern seems obvious, so let's try some numerical experiments:

Thanks, technology; values are rounded.

$$H_{10} = 2.929, \quad H_{433} = 6.649, \quad H_{7881} = 9.549, \quad H_{12345} = 9.998.$$

The H_n are clearly increasing, but slowly. And mysteries remain: is the sequence $\{H_n\}$ bounded, and therefore convergent, or do the harmonic numbers diverge to infinity?

By Theorem 2.3 page 88

It turns out—as you probably know, and as we will review in exercises—that $\{H_n\}$ diverges to infinity. This has been known for at least 650 years, thanks to the French philosopher and bishop Nicole Oresme. The moral for now is that this partial sum sequence *has no simple formula*, and so requires some work-around effort to analyze. This state of affairs is typical—most series don't have partial sum sequences with nice, "closed form" expressions. Exceptions, like the first series above, are precious.   ◇

Series: The Basics

Our systematic discussion starts, as usual, with a formal definition.

Definition 2.22. Let $\sum_{k=1}^{\infty} a_k$ be a series of numbers, and $\{A_n\}$ its sequence of partial sums. If $\{A_n\}$ converges to a finite limit A, then we say $\sum_{k=1}^{\infty} a_k$ *converges* to the *sum A*, and write $\sum_{k=1}^{\infty} a_k = A$. Otherwise, $\sum_{k=1}^{\infty} a_k$ *diverges.*

Convince yourself by writing out some terms.

Notes on notation. Sigma notation packs a lot of information into a small suitcase—and so may need careful unpacking. For instance, all of the following expressions mean exactly the same thing:

$$\sum_{k=1}^{\infty} \frac{1}{k}; \quad \sum_{j=1}^{\infty} \frac{1}{j}; \quad \sum_{j=1}^{103} \frac{1}{j} + \sum_{j=104}^{\infty} \frac{1}{j}; \quad \sum_{j=0}^{\infty} \frac{1}{j+1}; \quad \sum_{z=13}^{\infty} \frac{1}{z-12}.$$

With due care taken, such cosmetic differences won't cause trouble. For typographical economy, moreover, we will write just $\sum a_k$, not $\sum_{k=1}^{\infty} a_k$, when confusion seems unlikely.

As another example of notational alternatives, observe that the equation

$$\sum_{k=1}^{\infty} a_k = \lim_{n \to \infty} \sum_{k=1}^{n} a_k = A$$

Efficiency isn't everything.

is a highly compressed form of the definition of convergence of a series $\sum a_k$ to the sum A. There is no shame—and often added clarity—in expanding things a bit, so we might write, instead,

$$a_1 + a_2 + a_3 + \cdots = \lim_{n \to \infty} (a_1 + a_2 + a_3 + \cdots + a_n) = A.$$

EXAMPLE 2. Does the series $\sum_{k=1}^{\infty} 1/(k^2 + k) = \frac{1}{2} + \frac{1}{6} + \ldots$ converge? If so, to what limit?

Try some by hand.

SOLUTION. By a happy coincidence, the partial sums follow a nice pattern:

$$S_1 = \frac{1}{2}, \quad S_2 = \frac{2}{3}, \quad S_3 = \frac{3}{4}, \quad \ldots, \quad S_{100} = \frac{100}{101}, \quad \ldots$$

Now it is easy to guess a formula for S_n, and to see that $S_n \to 1$. (See the exercises for details.)   ◇

Nonnegative terms, increasing sums. Getting one's hands directly on partial sums can be difficult, as we've said. Things are simpler for a *nonnegative* series

$\sum a_k$: one for which $a_k \geq 0$ for all k. In this case the partial sum sequence $\{A_1, A_2, A_3, \dots\}$ is *increasing*:

$$a_1 \leq a_1 + a_2 \leq a_1 + a_2 + a_3 \leq a_1 + a_2 + a_3 + a_4 \leq \dots.$$

By Theorem 2.3, page 88, there are only two possibilities: Either $\{A_n\}$ converges, or it diverges to infinity. If we can somehow rule out either alternative, the other must apply.

Algebra with convergent series. Convergent series, like convergent sequences, allow basic algebraic operations; the following theorem has a counterpart for sequences in Theorem 2.5, page 96. Proofs of the following theorem, too, follow easily from those for sequences.

Theorem 2.23 (Algebra with convergent series). *Let $\sum a_n$ and $\sum b_n$ be convergent sequences, with respective sums A and B.*

- Sums and differences: $\sum (a_n \pm b_n)$ *converges to* $A \pm B$; *equivalently,*

$$\sum (a_n \pm b_n) = \sum a_n \pm \sum b_n.$$

- Constant multiples: $\sum ca_n$ *converges to* cA; *equivalently,*

$$\sum ca_n = c \sum a_n.$$

Proof: We will leave the proof for sums as an exercise. For constant multiples we want to show, in effect, that the distributive law applies to (convergent) *infinite* sums. To do this, we need to show that the sequence of partial sums

$$ca_1, \ ca_1 + ca_2, \ ca_1 + ca_2 + ca_3, \ \dots$$

converges to cA. This follows from the distributive law for *finite* sums, and from what we already know about sequences. By the distributive law, we can rewrite the preceding sequence as

$$ca_1, \ c(a_1 + a_2), \ c(a_1 + a_2 + a_3), \ \dots, \ c(a_1 + a_2 + \dots a_n), \dots,$$

or, equivalently, as

$$cA_1, \ cA_2, \ cA_3, \ \dots, \ cA_n, \ \dots,$$

where $\{A_n\}$ is the sequence of partial sums of $\sum a_n$. But we know by hypothesis that $A_n \to A$, and the constant multiple rule for *sequences* implies that $cA_n \to cA$, as desired. $\qquad \square$

EXAMPLE 3. Does the series

$$\sum_{k=1}^{\infty} \left(\frac{1}{2^{k-1}} + \frac{1}{k} \right) = 1 + 1 + \frac{1}{2} + \frac{1}{2} + \frac{1}{4} + \frac{1}{3} + \cdots$$

converge or diverge? Why?

SOLUTION. It diverges. The given series is the sum of two simpler series—one convergent and one divergent—discussed in Example 1. In fact, *every* such sum diverges. To see why, suppose that $\sum a_n$ converges and $\sum b_n$ diverges. If $\sum (a_n + b_n)$ were convergent, then, by Theorem 2.23, the series

$$\sum b_n = \sum \left((a_n + b_n) - a_n \right)$$

would also converge, against our assumption.   ◇

Detecting Convergence and Divergence

Theorem 2.23 is nice to know. But it is of little use unless we know that some particular series converge or diverge. To this end, we need reliable methods for deciding whether a given series converges or diverges. We address that deficit next, starting with a simple test for *divergence*.

Proposition 2.24 (*n*th term test). *If a series $\sum_{n=1}^{\infty} a_n$ converges, then the sequence $\{a_n\}$ converges to zero.*

In practice, the *contrapositive* is especially useful:

if $\{a_n\}$ does *not* converge to zero, then $\sum a_n$ diverges.

The *converse*,

if $a_n \to 0$ then $\sum a_n$ converges,

is tempting but false: a series $\sum a_n$ may diverge even if $a_n \to 0$.

Don't give in. (The harmonic series is one counterexample.)

Proof: Observe first that, for all $n > 1$,

$$A_n - A_{n-1} = (a_1 + \cdots + a_{n-1} + a_n) - (a_1 + \cdots + a_{n-1}) = a_n.$$

Watch the difference "telescope."

Now convergence of $\sum a_n$ means that $A_n \to A$ for some number A. But then $A_{n-1} \to A$, too; therefore,

$$\lim_{n \to \infty} a_n = \lim_{n \to \infty} (A_n - A_{n-1}) = \lim_{n \to \infty} A_n - \lim_{n \to \infty} A_{n-1}$$
$$= A - A = 0,$$

as claimed.   □

Geometric series: the nicest examples. Proposition 2.24 sometimes helps us identify *divergent* series. But we would still like a good source of *convergent* series; the geometric family fits that bill.

A *geometric series* is one of the form

$$\sum_{k=1}^{\infty} ar^{k-1} = a + ar + ar^2 + ar^3 + \cdots,$$

where a and r are any fixed numbers; notice the common ratio r of each summand to its predecessor. The first series discussed above, for instance, is geometric, with ratio $r = 1/2$:

$$\sum_{k=1}^{\infty} \frac{1}{2^{k-1}} = 1 + \frac{1}{2} + \frac{1}{4} + \frac{1}{8} \cdots = \sum_{k=1}^{\infty} 0.5^{k-1}.$$

Here is another example:

$$3 + 3.3 + 3.63 + 3.993 + 4.3923 + 4.83153 + 5.314683 + \cdots.$$

The geometric form may be slightly hidden in the numbers, but it is easy to check that successive terms have common ratio $r = 1.1$, and so the standard geometric template $\sum_{k=1}^{\infty} 3 \times 1.1^{k-1}$ fits the pattern.

Geometric series satisfy a useful algebraic equation:

It's readily proved by induction.

$$a + ar + ar^2 + ar^3 + \cdots + ar^{n-1} = a\,\frac{1 - r^n}{1 - r} \qquad (*)$$

holds if $r \neq 1$, n is a positive integer, and a is any number. This pretty formula tells us, almost instantly, whether *any* geometric series converges and, if so, to what limit.

Proposition 2.25. *Let a and r be any real constants, with $a \neq 0$.*

(a) *If $|r| < 1$, then $\sum_{k=1}^{\infty} a\,r^{k-1}$ converges to $\frac{a}{1-r}$.*

(b) *If $|r| \geq 1$, then $\sum_{k=1}^{\infty} a\,r^{k-1}$ diverges.*

We can infer, for instance, that $\sum_{k=1}^{\infty} 0.5^{k-1}$ converges to 2—as we proved in Example 1. By contrast, $\sum_{k=1}^{\infty} 3 \times 1.1^{k-1}$ diverges—as the numerical evidence suggests, and as the nth term test also implies.

Proof (sketch): Convergence and divergence of series are all about partial sums, and Equation $(*)$ above tells almost the whole story. If $r \neq \pm 1$ we have

$$\sum_{k=1}^{n} a\,r^{k-1} = a\,\frac{1 - r^n}{1 - r}.$$

What happens to the right side as $n \to \infty$? The answer depends entirely on the r^n in the numerator—and *that* depends, in turn, on the value of r. If $|r| < 1$, then $r^n \to 0$, and so

$$a\frac{1 - r^n}{1 - r} \to a\frac{1}{1 - r},$$

as claimed. If $|r| > 1$, then $\{ar^n\}$ is unbounded, and so the series (badly!) fails the nth term test, and so diverges.

The remaining cases ($r = \pm 1$) are much simpler, and left to the exercises. $\square$

EXAMPLE 4. Proposition 2.25 and Theorem 2.23 justify calculations like the following:

Watch the tricky index change from *k* to *j*.

$$\sum_{k=0}^{\infty} \frac{3^k + 4^k}{5^k} = \sum_{k=0}^{\infty} \frac{3^k}{5^k} + \sum_{k=0}^{\infty} \frac{4^k}{5^k} = \sum_{j=1}^{\infty} \left(\frac{3}{5}\right)^{j-1} + \sum_{j=1}^{\infty} \left(\frac{4}{5}\right)^{j-1}$$

$$= \frac{1}{1 - 3/5} + \frac{1}{1 - 4/5} = \frac{15}{2} = 7.5.$$

Similarly,

$$\pi + e + \frac{1}{3} - \frac{1}{6} + \frac{1}{12} - \frac{1}{24} + \cdots = \pi + e + \frac{1}{3}\left(1 - \frac{1}{2} + \frac{1}{4} - \frac{1}{8} + \cdots\right)$$

$$= \pi + e + \frac{1}{3}\sum_{k=1}^{\infty} (-1/2)^{k-1} = \pi + e + \frac{1}{3}\frac{1}{1 + 1/2} \approx 6.082.$$

Numerical experiments support these results. According to *Mathematica*,

$$\sum_{k=0}^{50} \frac{3^k + 4^k}{5^k} \approx 7.4993; \quad \pi + e + \frac{1}{3}\sum_{k=1}^{50} (-1/2)^{k-1} \approx 6.0821. \qquad \Diamond$$

Comparison and comparison testing. Compare these two series:

$$\sum_{k=1}^{\infty} \frac{1}{2^{k-1}} = \frac{1}{1} + \frac{1}{2} + \frac{1}{4} + \frac{1}{8} + \frac{1}{16} + \cdots;$$

$$\sum_{k=1}^{\infty} \frac{1}{2^{k-1} + 1} = \frac{1}{2} + \frac{1}{3} + \frac{1}{5} + \frac{1}{9} + \frac{1}{17} + \cdots.$$

The first series we recognize—it is geometric and converges to two—but what about the second? Does it also converge, or might its partial sums diverge to infinity?

Comparing the two series suggests some common-sense guesses. Because each term of the second series is *less than* its counterpart in the first, we expect the second series, like the first, to converge, and to some limit less than two. Sampling some partial sums $\{S_n\}$ from this series adds some credence:

$$S_{10} \approx 1.2625479, \quad S_{20} \approx 1.2644979, \quad S_{30} \approx 1.2644998, \quad \ldots .$$

Comparing series for size in this simple way *is* valid—and useful. Details are in the following theorem and proof.

Theorem 2.26 (Comparison test). *Let $\sum a_k$ and $\sum b_k$ be series, with $0 \leq a_k \leq b_k$ for all k.*

(a) *If $\sum b_k$ converges to a number B, then $\sum a_k$ converges to a number A, with $A \leq B$.*

(b) *If $\sum a_k$ diverges, then $\sum b_k$ diverges, too.*

Proof: Let $\{A_n\}$ and $\{B_n\}$ be the partial sum sequences for $\sum a_k$ and $\sum b_k$. Because $\{A_n\}$ and $\{B_n\}$ are increasing sequences, each converges if—but only if—it is bounded above. By hypothesis,

$$A_n = a_1 + a_2 + a_3 + \cdots + a_n \leq b_1 + b_2 + b_3 + \cdots + b_n = B_n$$

for all n. If $\{B_n\}$ happens to be bounded above, then certainly $\{A_n\}$ is also bounded above, and we have

$$A = \sup\{A_n\} \leq \sup\{B_n\} = B.$$

In this case $A_n \to A$ and $B_n \to B$, as (a) claims. Part (b) follows from part (a). $\square$

EXAMPLE 5. Does each of the following series converge or diverge? What can be said about limits?

(a) $\displaystyle\sum_{k=1}^{\infty} \frac{2^k - 1}{3^k}$ (b) $\displaystyle\sum_{k=1}^{\infty} \frac{1}{2k - 1}$ (c) $\displaystyle\sum_{k=1}^{\infty} \frac{\sin k}{2^k}$

SOLUTION. Series (a) converges by comparison to a convergent geometric series:

$$\frac{2^k - 1}{3^k} < \frac{2^k}{3^k}$$

holds for all $k \geq 1$, and so

$$\sum_{k=1}^{\infty} \frac{2^k - 1}{3^k} < \sum_{k=1}^{\infty} \frac{2^k}{3^k} = 2.$$

We can evaluate (a) *exactly* if we recognize it as the difference of two convergent geometric series:

$$\sum_{k=1}^{\infty} \frac{2^k - 1}{3^k} = \sum_{k=1}^{\infty} \frac{2^k}{3^k} - \sum_{k=1}^{\infty} \frac{1}{3^k} = 2 - \frac{1}{2}.$$

Series (b) sums the reciprocals of *odd* integers:

$$\sum_{k=1}^{\infty} \frac{1}{2k - 1} = \frac{1}{1} + \frac{1}{3} + \frac{1}{5} + \frac{1}{7} + \cdots.$$

The harmonic series $\frac{1}{1} + \frac{1}{2} + \frac{1}{3} + \cdots$ comes to mind, but term-by-term comparison goes the wrong way. A better strategy is to multiply the harmonic series by $1/2$:

$$\sum_{k=1}^{\infty} \frac{1}{2k} = \frac{1}{2} + \frac{1}{4} + \frac{1}{6} + \frac{1}{8} + \cdots.$$

The resulting series diverges (why?), and *is* comparable to series (b), which therefore also diverges.

For series (c), comparison with $\sum_{k=1}^{\infty} 1/2^k$ is tempting, but some terms of (c) may be negative, and so Theorem 2.26 doesn't apply. Still, numerical evidence suggests convergence; here are some (approximate) partial sums S_n:

$$s_5 \approx 0.588433, \; s_{10} \approx 0.59333, \; s_{15} \approx 0.59285, \; s_{20} \approx 0.59284.$$

The case for convergence looks strong, but we don't (quite yet) have a proof.   ◊

Absolute Convergence

It's a triangle inequality, but for infinitely many summands.

Series (c) from Example 5 stumped us. The next theorem comes just in time.

Theorem 2.27. *If* $\sum |a_k|$ *converges then* $\sum a_k$ *converges too, and*

$$\left| \sum_{k=1}^{\infty} a_n \right| \leq \sum_{k=1}^{\infty} |a_n|.$$

Let's polish off series (c) from Example 5 before proving this. Clearly,

$$\frac{|\sin k|}{2^k} < \frac{1}{2^k}$$

We don't know exactly where.

for all k, and so Theorem 2.26 guarantees that the left-hand series converges, to some positive limit less than one. Theorem 2.27, in turn, says that series (c) converges too, to a limit somewhere in the interval $(-1, 1)$.

The proof, informally. Suppose $\sum |a_n|$ converges. To show that $\sum a_n$ converges, too, we need to prove that the partial sum sequence $\{A_n\}$ converges. Since we have no specific limit in mind, we will show that $\{A_n\}$ is a Cauchy sequence, and then invoke Theorem 2.20, page 116.

> We worked hard to prove the theorem; here comes the payoff.

Observe first that, because $\sum |a_n|$ converges, *its* partial sum sequence

$$C_1 = |a_1|, \ C_2 = |a_1| + |a_2|, \ C_3 = |a_1| + |a_2| + |a_3|, \ \ldots$$

is convergent, and therefore Cauchy. We will use this fact and the triangle inequality to show that $\{A_n\}$ is Cauchy, too.

Let $\epsilon > 0$ be given. Because $\{C_n\}$ is Cauchy we can choose N such that $|C_n - C_m| < \epsilon$ whenever $n > m > N$. The same N turns out to work for $\{A_n\}$, because if $n > m > N$, then

> Watch for the (finite) triangle inequality.

$$
\begin{aligned}
|A_n - A_m| &= |(a_1 + \cdots + a_m + a_{m+1} + \cdots + a_n) - (a_1 + \ldots a_m)| \\
&= |a_{m+1} + a_{m+2} + \cdots + a_n| \\
&\leq |a_{m+1}| + |a_{m+2}| + \cdots + |a_n| \\
&= C_n - C_m = |C_n - C_m| < \epsilon.
\end{aligned}
$$

Thus, $\{A_n\}$ is Cauchy and hence also convergent.

Invoking the triangle inequality again gives

$$|A_n| = |a_1 + a_2 + \cdots + a_n| \leq |a_1| + |a_2| + \cdots + |a_n| = C_n$$

for all n, and so

$$\left| \sum a_k \right| = \lim_{n \to \infty} |A_n| \leq \lim_{n \to \infty} C_n = \sum |a_k|.$$

This finishes the proof.

Conditional convergence. A series $\sum a_k$ for which $\sum |a_k|$ converges is called *absolutely convergent*. Theorem 2.27 says that every absolutely convergent series is also convergent in the ordinary sense. The converse, however, is false: A series $\sum a_k$ may converge while $\sum |a_k|$ diverges. A series with this property is called *conditionally convergent*.

EXAMPLE 6. The *alternating harmonic series*

> The signs alternate.

$$\sum_{k=1}^{\infty} (-1)^{k+1} \frac{1}{k} = 1 - \frac{1}{2} + \frac{1}{3} - \frac{1}{4} + \frac{1}{5} - \cdots$$

is conditionally convergent. Explain why. Can anything be said about limits?

SOLUTION. Taking absolute values of all terms gives the *ordinary* harmonic series

$$1 + \frac{1}{2} + \frac{1}{3} + \frac{1}{4} + \frac{1}{5} + \dots,$$

which we know to diverge. To see why the alternating version converges, consider the partial sums $\{S_n\}$:

$$1,\ 1 - \frac{1}{2},\ 1 - \frac{1}{2} + \frac{1}{3},\ 1 - \frac{1}{2} + \frac{1}{3} - \frac{1}{4},\ 1 - \frac{1}{2} + \frac{1}{3} - \frac{1}{4} + \frac{1}{5} \dots.$$

Here's a (rounded) numerical view of $S_1, \dots, S_{10}$:

$$1.00,\ 0.500,\ 0.833,\ 0.583,\ 0.783,\ 0.617,\ 0.760,\ 0.635,\ 0.746,\ 0.646, \dots.$$

Notice the pattern: Successive partial sums alternately increase and decrease, but with smaller and smaller gaps. In particular, for any N, every term S_n with $n > N$ lies *between* S_N and S_{N+1}. If $n > 1000$, for example, then

$$0.69265 \approx S_{1000} \leq S_n \leq S_{1001} \approx 0.69365.$$

The same holds for two terms. If $n > m > 1000$, then *both* S_n and S_m lie between 0.69265 and 0.69365, and so lie within 0.001 of each other. Using this idea we can prove that $\{S_n\}$ is Cauchy, and hence convergent. We leave details to the exercises.

So what *is* the limit? The striking but far-from-obvious exact answer turns out to be $\ln 2 \approx 0.69315$. A rigorous proof is beyond our scope, but numerical evidence shows we are in the ballpark.   ◊

Exercises

1. For each series following, find a formula or simple rule for the partial sums S_n. Then decide whether the series converges or diverges, and to what limit.

   (a) $\displaystyle\sum_{k=1}^{\infty} 0 = 0 + 0 + 0 + \dots$

   (b) $\displaystyle\sum_{k=1}^{\infty} 42 = 42 + 42 + 42 + \dots$

   (c) $\displaystyle\sum_{k=1}^{\infty} (-1)^k = -1 + 1 - 1 + \dots$

   (d) $\displaystyle\sum_{k=1}^{\infty} k = 1 + 2 + 3 + \dots$

(e) $\displaystyle\sum_{k=1}^{\infty} 0.99^{k-1}$

(f) $\displaystyle\sum_{k=1}^{\infty} \frac{1}{k^2 + k}$ (Hint: First guess a formula for S_n and prove it by induction.)

2. For a certain series $\sum a_n$, we know that $A_n = a_1 + a_2 + \cdots + a_n = 4 + 1/n$ for all $n \in \mathbb{N}$.

   (a) Does the series $\sum a_n$ converge? If so, to what limit? How do you know?

   (b) Does the *sequence* $\{a_n\}$ converge? If so, to what limit? How do you know?

   (c) Does the *series* $\sum A_n$ converge? If so, to what limit? How do you know?

   (d) Note that $a_1 = A_1 = 5$. Find a_2 and a_3.

   (e) Find a formula in terms of n for a_n.

3. This problem is about partial sums H_n of the harmonic series (see Example 1, page 121).

   (a) Oresme's (fourteenth-century) proof uses inequalities like the following:

   $$\frac{1}{3} + \frac{1}{4} > \frac{1}{2}; \quad \frac{1}{5} + \frac{1}{6} + \frac{1}{7} + \frac{1}{8} > \frac{1}{2}; \quad \frac{1}{9} + \cdots + \frac{1}{16} > \frac{1}{2}.$$

   Deduce that $H_{2^n} > \frac{n}{2}$, and so $H_n \to \infty$.

   (b) Consider the inequality

   $$H_{2n} = \left(1 + \frac{1}{2}\right) + \left(\frac{1}{3} + \frac{1}{4}\right) + \left(\frac{1}{5} + \frac{1}{6}\right) + \cdots + \left(\frac{1}{2n-1} + \frac{1}{2n}\right)$$

   $$\geq \frac{3}{2} + \frac{1}{2} + \frac{1}{3} + \cdots \frac{1}{n} = \frac{1}{2} + H_n.$$

   Use this to show (again) that $\{H_n\}$ diverges. (Hint: If $\{H_n\}$ were convergent, then the subsequence $\{H_{2n}\}$ would have to converge, too.)

   (c) Yet another proof that H_n converges follows from the inequality $H_n > \ln n$ for all positive integers n and the fact that $\ln n \to \infty$. Without a formal definition of $\ln n$ we can't prove this rigorously. Instead, draw a picture illustrating that $H_n > \ln n$; view $\ln n$ as the area under the graph $y = 1/x$ from $x = 1$ to $x = n$.

4. Decide whether each of the following series converges or diverges. The comparison test will be useful in some parts.

(a) $\displaystyle\sum_{k=1}^{\infty} \frac{1}{k + 2^k}$

(b) $\displaystyle\sum_{k=1}^{\infty} \frac{k}{2k^2 - 1}$

(c) $\displaystyle\sum_{k=1}^{\infty} \frac{k}{3^k}$ (Hint: Compare to a geometric series.)

(d) $\displaystyle\sum_{k=1}^{\infty} \frac{k^2 + 2}{3k^2 + 4}$

5. Decide whether each of the following series converges or diverges. Do any converge conditionally?

(a) $\displaystyle\sum_{k=1}^{\infty} \frac{(-1)^k}{3k^2 + 1}$

(b) $\displaystyle\sum_{k=1}^{\infty} \frac{k^2 - 1}{k^4 + 1}$

(c) $\displaystyle\sum_{k=1}^{\infty} \frac{\sin(k)}{3^k}$

(d) $\displaystyle\sum_{k=1}^{\infty} \frac{1}{2 + \sin(k)}$

6. A *telescoping series* is one for which the partial sums S_n "collapse" to some shorter, simpler form.

(a) For the series $\sum_{k=1}^{\infty}(\frac{1}{k} - \frac{1}{k+1})$, write out (by hand) S_2 and S_5. Do you see the "telescoping" behavior? Find a formula for S_n and deduce the limit. (Note: This series, rewritten as $\sum_{k=1}^{\infty} 1/(k^2 + k)$, appears in an earlier problem.)

(b) For the series $\sum_{k=1}^{\infty}(\frac{1}{k} - \frac{1}{k+2})$, find a formula for S_n and then find the limit. (Hints: Write out, say, S_{10} to see the pattern for S_n.)

(c) For the series $\sum_{k=1}^{\infty}(\frac{1}{k} - \frac{1}{k+10})$, find a formula for S_n and then find the limit. (Hints: Use ideas from the preceding parts. No proofs needed, but check your claim numerically, using technology.)

7. Prove the claim about sums and differences in Theorem 2.23, page 123. (Note that the claim amounts to saying that addition is commutative for convergent *infinite* sums; we already know that addition is commutative for *finite* sums.)

8. This problem ties up some loose ends concerning Proposition 2.25, page 125, which addresses convergence and divergence of $\sum_{k=1}^{\infty} a\, r^{k-1}$.

   (a) How does the series look if $r = 1$? What if $r = -1$? Prove (using partial sums) that the series diverges in both of these cases.

   (b) In the proof of Proposition 2.25 we used several reasonable-seeming facts about convergence and divergence of the sequence $\{r^n\}$, depending on the value of r. Prove them, as follows:

      (i) If r and L are any real numbers and $r^n \to L$, then $r^{n+1} \to L$, too.

      (ii) If r and L are any real numbers and $r^n \to L$, then $r^{n+1} = r\,r^n \to r\,L$.

      (iii) If $r \neq 0$ and $r^n \to L$, then $L = 0$. (Hint: Use the two preceding parts.)

      (iv) If $0 \leq r < 1$, then $r^n \to 0$. (Hint: Use a theorem about monotone sequences.)

      (v) If $-1 < r \leq 0$, then $r^n \to 0$. (Hint: Squeeze.)

      (vi) If $|r| \geq 1$, then $\{r^n\}$ diverges. (Hint: It is enough to explain why $r^n \to 0$ is impossible.)

9. This problem is about the alternating harmonic series. See Example 6, page 129, where we said that the partial sum sequence $\{S_n\}$ is Cauchy.

   (a) Claim: For any positive integer N, every partial sum S_n with $n \geq N$ is *between* (or equal to one of) S_N and S_{N+1}. Prove this for $N = 1000$. (A general proof is similar.)

   (b) Show that, for any $\epsilon > 0$, we can find N such that $|S_N - S_{N+1}| < \epsilon$.

   (c) Use the preceding parts to show that $\{S_n\}$ is Cauchy.

10. This problem explores implications of Theorem 2.23, page 123. Throughout, let $\sum c_k$ be a convergent series and let $\sum d_k$ be a divergent series.

   (a) Use Theorem 2.23 to show that $\sum d_k/10$ diverges.

   (b) Can $\sum c_k d_k$ converge? Can it diverge? Give examples or proofs as needed.

   (c) Let A be a constant. Can $\sum A c_k$ diverge? Give a proof or counterexample.

(d) Is it possible that $d_k \leq c_k$ for all k? If so, give an example. If not, why not?

11. Let $\sum_{k=1}^{\infty} a_k$ and $\sum_{k=1}^{\infty} b_k$ be two series, and suppose that $a_k = b_k$ whenever $k > 42$.

  (a) Show that $\sum a_k$ converges if and only if $\sum b_k$ converges.
  (b) Suppose that $\sum_{k=1}^{\infty} a_k = 100$ and that $b_k = a_k + k^2$ for $1 \leq k \leq 42$. Find $\sum_{k=1}^{\infty} b_k$.

12. Let $\sum a_k$ be a series with positive terms. Show that if $\sum a_k$ converges, then $\sum a_k^2$ converges, too. (Hint: Compare the series $\sum a_k$ and $\sum a_k^2$.)

13. In discussing geometric series we invoked the formula

$$a + ar + ar^2 + ar^3 + \cdots + ar^{n-1} = a\,\frac{1 - r^n}{1 - r}$$

for the partial sums of a geometric series with $r \neq 1$ and a any real number. Prove by induction that this formula holds for all positive integers n.

2.6   Series 102: Testing for Convergence and Estimating Limits

The preceding section was about definitions and basic ideas. Now we proceed to deeper investigation of the two main questions about any given series: Does it converge? If so, to what limit?

We start by expanding our catalog of "known" series.

Meet the family: p-series. A *p-series* has the form

$$\sum_{k=1}^{\infty} \frac{1}{k^p} = \frac{1}{1^p} + \frac{1}{2^p} + \frac{1}{3^p} + \frac{1}{4^p} + \frac{1}{5^p} + \cdots,$$

where p is any fixed number. With $p = 1$ we have the harmonic series, which we've shown to *diverge*. With $p = 2$ we get

$$\sum_{k=1}^{\infty} \frac{1}{k^2} = 1 + \frac{1}{2^2} + \frac{1}{3^2} + \frac{1}{4^2} + \frac{1}{5^2} + \cdots,$$

which *converges*, as we'll prove in two different ways. The general story on convergence and divergence is simple:

Proposition 2.28. *The p-series $\sum_{k=1}^{\infty} \frac{1}{k^p}$ converges if $p > 1$ and diverges if $p \leq 1$.*

Notes toward a proof. For $p \leq 1$, divergence follows by comparison to the harmonic series: For all integers $k > 1$, we have

$$k^p \leq k, \quad \text{and so} \quad \frac{1}{k^p} > \frac{1}{k}.$$

Because the harmonic series diverges, so must *every* p-series with $p \geq 1$.

Showing convergence for $p > 1$ takes a little more effort. We illustrate the argument for $p = 1.3$ and leave a general proof to the exercises.

As with any positive series, convergence occurs if the partial sum sequence $\{S_n\}$ is bounded above. "Our" proof of this uses some clever algebraic bookkeeping to show that the subsequence

It gets easier when we know about integrals.

It's probably centuries old.

$$S_1, \, S_3, \, S_7, \, S_{15}, \, S_{31}, \, \ldots, \, S_{2^n-1}, \, \ldots$$

is bounded above; this implies, in turn, that the entire sequence $\{S_n\}$ is bounded above. The idea is to group summands into blocks of length 1, 2, 4, 8, ..., and then estimate the contribution of each block. We illustrate this for S_{31}. The inequality holds because the first term in each block is largest, and the last equality comes from counting the terms in each block:

Do you see why?

$$S_{31} = 1 + \underbrace{\frac{1}{2^{1.3}} + \frac{1}{3^{1.3}}} + \underbrace{\frac{1}{4^{1.3}} + \cdots + \frac{1}{7^{1.3}}} + \underbrace{\frac{1}{8^{1.3}} + \cdots + \frac{1}{15^{1.3}}} + \underbrace{\frac{1}{16^{1.3}} + \cdots + \frac{1}{31^{1.3}}}$$

$$< 1 + \underbrace{\frac{1}{2^{1.3}} + \frac{1}{2^{1.3}}} + \underbrace{\frac{1}{4^{1.3}} + \cdots + \frac{1}{4^{1.3}}} + \underbrace{\frac{1}{8^{1.3}} + \cdots + \frac{1}{8^{1.3}}} + \underbrace{\frac{1}{16^{1.3}} + \cdots + \frac{1}{16^{1.3}}}$$

$$= 1 + \frac{2}{2^{1.3}} + \frac{4}{4^{1.3}} + \frac{8}{8^{1.3}} + \frac{16}{16^{1.3}} = 1 + \frac{1}{2^{0.3}} + \frac{1}{4^{0.3}} + \frac{1}{8^{0.3}} + \frac{1}{16^{0.3}}.$$

Here comes the punchline: The last quantity can be rewritten in *geometric* form:

$$S_{31} < 1 + \frac{1}{2^{0.3}} + \frac{1}{4^{0.3}} + \frac{1}{8^{0.3}} + \frac{1}{16^{0.3}}$$

$$= 1 + \frac{1}{2^{0.3}} + \left(\frac{1}{2^{0.3}}\right)^2 + \left(\frac{1}{2^{0.3}}\right)^3 + \left(\frac{1}{2^{0.3}}\right)^4$$

$$< \sum_{k=1}^{\infty} \left(\frac{1}{2^{0.3}}\right)^{k-1} \approx \sum_{k=1}^{\infty} 0.81225^{k-1} = \frac{1}{1 - 0.81225} \approx 5.326.$$

Now we are getting somewhere. We've shown, by painstaking calculation, that S_{31}, and (by the same argument) *all* partial sums S_{2^n-1}, are bounded above by the sum—about 5.326—of a convergent geometric series. This completes the proof for $p = 1.3$; other exponents $p > 1$ can be handled similarly.

Mathematica says $S_{1000} \approx 3.51$ and $S_{2000} \approx 3.59$; the numerical evidence is reassuring but only anecdotal.

Sharper Tests for Convergence

Now that we understand the convergence behavior for two important families—geometric and p-series—we can study convergence and divergence more effectively. First we need some more discriminating convergence tests.

Limit comparison test. The ordinary comparison test (Theorem 2.26, page 127) is simple and powerful, but can be hard to apply. The limit comparison test can simplify the work.

Theorem 2.29 (Limit comparison test). *Consider two series $\sum a_k$ and $\sum b_k$ with positive terms, and suppose that $\{a_k/b_k\}$ converges to a finite limit L.*

(a) *If $\sum b_k$ converges, then $\sum a_k$ converges, too.*

(b) *If $L \neq 0$, then either both $\sum a_k$ and $\sum b_k$ converge or both diverge.*

Before proving the theorem, let's try it out.

EXAMPLE 1.   We showed in Example 2, page 122, that the series $\sum_{k=1}^{\infty} b_k = \sum_{k=1}^{\infty} 1/(k^2 + k)$ converges, with sum 1. Use limit comparison to deduce that some other series converge.

SOLUTION.   We have already shown that the p-series $\sum a_k = \sum 1/k^2$ converges, but the proof was laborious. Limit comparison is easier. Here is the key limit:

$$\lim_{k \to \infty} \frac{a_k}{b_k} = \lim_{k \to \infty} \frac{1/k^2}{1/(k^2 + k)} = \lim_{k \to \infty} \frac{k^2 + k}{k^2} = \lim_{k \to \infty} 1 + \frac{1}{k} = 1,$$

which implies (again) that $\sum 1/k^2$ converges.

Similar arguments apply to uglier series. If, say,

$$\sum_{k=1}^{\infty} a_k = \sum_{k=1}^{\infty} \frac{k^2 + 7k - 3\sin k}{2k^4 - k},$$

then a look at powers of k in the numerator and denominator suggests limit comparison with $\sum 1/k^2$. The strategy turns out to work:

Note the algebra with sequence limits.

$$\lim_{k \to \infty} \frac{\frac{k^2+7k-3\sin k}{2k^4-k}}{\frac{1}{k^2}} = \lim_{k \to \infty} \frac{k^4 + 7k^3 - 3k^2 \sin k}{2k^4 - k}$$

$$= \lim_{k \to \infty} \frac{1 + 7/k - 3/k^2 \cdot \sin k}{2 - 1/k^3} = \frac{1}{2}.$$

Because the limit is finite, the ugly series converges.   ◇

We'll prove Theorem 2.29 using two auxiliary facts, each of interest in its own right.

Proposition 2.30. *Suppose $a_k = b_k$ for all $k > N$, where N is any constant. Then either both $\sum a_k$ and $\sum b_k$ converge (not necessarily to the same limits) or both diverge.*

Proof (sketch): The trick is to compare the partial sum sequences $\{A_n\}$ and $\{B_n\}$, which turn out to resemble each other. Here is the idea for $N = 42$; a general proof is almost identical. If $a_k = b_k$ for $k > 42$, and we know $n > 42$, then

$$
\begin{aligned}
A_n &= a_1 + \cdots + a_{42} + a_{43} + \cdots + a_n \\
&= a_1 + \cdots + a_{42} + b_{43} + \cdots + b_n \\
&= (a_1 + \cdots + a_{42}) - (b_1 + \cdots + b_{42}) + (b_1 + \cdots + b_{42}) + b_{43} + \cdots + b_n \\
&= A_{42} - B_{42} + B_n.
\end{aligned}
$$

Thus, $A_n = A_{42} - B_{42} + B_n$ for $n \geq 42$, which implies that the sequences $\{A_n\}$ and $\{B_n\}$ either both converge or both diverge. If, in fact, $B_n \to B$, then $A_n \to A_{42} - B_{42} + B$. $\qquad\square$

Proposition 2.31 (Bounded comparison test). *Consider two series $\sum a_k$ and $\sum b_k$ with positive terms, and suppose that the sequence $\{a_k/b_k\}$ is bounded above.*

(a) *If $\sum b_k$ converges, then $\sum a_k$ converges, too.*

(b) *If $\sum a_k$ diverges, then $\sum b_k$ diverges, too.*

Proof: The hypothesis means that, for some positive number M, we have $a_k < Mb_k$ for all k. Now if $\sum b_k$ converges, then $\sum Mb_k$ also converges. By the (ordinary) comparison test, $\sum a_k$ must also converge, as claimed in (a). Part (b) is the contrapositive of (a). $\qquad\square$

> The second series is a constant multiple of the first.

The limit comparison test is a special case of Proposition 2.31, so the proof is easy.

Proof (of Theorem 2.29): Because $\{a_k/b_k\}$ converges, it is also bounded above, and so Proposition 2.31 implies part (a).

Now suppose, as in (b), that $L \neq 0$. In this case $\{b_k/a_k\}$ converges to $1/L$, and we can apply (a) to conclude that if $\sum a_k$ converges, $\sum b_k$ must do so, too. This establishes the claim in (b) about convergence; the statement about divergence is equivalent. $\qquad\square$

> Sorting out the *a*'s and *b*'s is a little confusing ... but it all works out.

The ratio test. For a geometric series, the ratio of successive terms is a constant r, and convergence depends on the magnitude of r. The ratio test generalizes this principle to a large class of series that, although not geometric, behave "in the limit" something like geometric series. The theorem makes these ideas precise.

Theorem 2.32 (Ratio test). *Consider a series* $\sum a_k$ *with positive terms, and suppose that the sequence* $\{a_{k+1}/a_k\}$ *converges to a finite limit L.*

(a) *If* $L < 1$ *then* $\sum a_k$ *converges.*

(b) *If* $L > 1$ *then* $\sum a_k$ *diverges.*

(c) *If* $L = 1$ *the test is inconclusive;* $\sum a_k$ *may converge or diverge.*

Let's try it out.

EXAMPLE 2.   What does the ratio test say about

$$1 + \frac{1}{1} + \frac{1}{2} + \frac{1}{6} + \frac{1}{24} + \cdots = \sum_{k=0}^{\infty} \frac{1}{k!}?$$

What does it say about the harmonic series?

SOLUTION.   The ratio test says that the first series converges. The ratio in question collapses nicely:

$$\frac{a_{k+1}}{a_k} = \frac{1/(k+1)!}{1/k!} = \frac{k!}{(k+1)!} = \frac{1 \cdot 2 \cdot \cdots \cdot k}{1 \cdot 2 \cdot \cdots \cdot k \cdot (k+1)} = \frac{1}{k+1}.$$

The last quantity converges to zero as $k \to \infty$, so the ratio test guarantees convergence.

The limit-of-ratios calculation is just as easy for the harmonic series:

$$\frac{a_{k+1}}{a_k} = \frac{1/(k+1)}{1/k} = \frac{k}{k+1} \to 1 \quad \text{as } k \to \infty.$$

Because the limit is one, however, the ratio test says nothing. The wasted work is annoying, but only mildly so: we knew already from other arguments that the harmonic series diverges.   ◊

Proof (of the ratio test): We prove (a) and leave (b) and (c) to the exercises. Given $L < 1$, as in (a), choose any real number r with $L < r < 1$. Then the geometric series $1 + r + r^2 + \ldots$ converges, and we'll use "bounded comparison" (Proposition 2.31) to show that $\sum a_k$ converges, too.

Doing so requires a basic observation about limits and an algebraic trick. The observation is that, because $a_{k+1}/a_k \to L$ and $L < r$, there must be a number N such that

$$\frac{a_{k+1}}{a_k} < r \quad \text{whenever} \quad k > N.$$

To complete the proof, using Proposition 2.31, we'll show that the sequence $\{a_k/r^k\}$ is bounded. To do this it is enough to assume $k > N$. In this case,

The ratio test wasn't the right tool for this job.

Plenty of choices exist.

The remaining terms, with $k \leq N$, form a finite set.

we can write (here comes the algebraic trick)

$$\frac{a_k}{r^k} = \frac{a_k}{a_{k-1}} \frac{a_{k-1}}{a_{k-2}} \frac{a_{k-2}}{a_{k-3}} \cdots \frac{a_{N-1}}{a_N} \frac{a_N}{r^k}$$

$$< \underbrace{r \cdot r \cdot r \cdot \ldots r}_{k - N \text{ factors}} \cdot \frac{a_N}{r^k} = \frac{r^{k-N} a_N}{r^k} = \frac{a_N}{r^N}.$$

Now the last quantity is constant (independent of k), and so the sequence is bounded, as desired.   □

Not-necessarily-positive series. Most of our convergence tests require *positive* series, and so don't apply directly to series like

$$\sum_{k=1}^{\infty} \frac{\sin k}{2^k} = \frac{\sin 1}{1} + \frac{\sin 2}{4} + \frac{\sin 3}{8} + \frac{\sin 4}{16} + \cdots$$

and

$$\sum_{k=1}^{\infty} \frac{(-1)^{k-1}}{\sqrt{k}} = \frac{1}{\sqrt{1}} - \frac{1}{\sqrt{2}} + \frac{1}{\sqrt{3}} - \frac{1}{\sqrt{4}} + \frac{1}{\sqrt{5}} - \cdots,$$

which have both positive and negative terms. What can we do?

We've already seen one work-around: test $\sum |a_k|$, not $\sum a_k$, for convergence. Theorem 2.27, page 128, guarantees that if the former converges, so does the latter. Indeed, we already used this method (Example 5, page 127) to show that the first series converges.

This approach fails for the second series, however, because

$$\sum_{k=1}^{\infty} \left| \frac{(-1)^{k-1}}{\sqrt{k}} \right| = \frac{1}{\sqrt{1}} + \frac{1}{\sqrt{2}} + \frac{1}{\sqrt{3}} + \cdots,$$

which *diverges*. We need another strategy.

It's a *p*-series with *p* = 1/2.

The series $\sum (-1)^{k-1}/\sqrt{k}$ resembles the *alternating harmonic series* $\sum (-1)^{k-1}/k$, which we've shown to converge. (See Example 6, page 129.) The former series turns out to converge, too, and for essentially the same reasons. The following theorem applies to these and to other *alternating series*—ones for which successive terms alternate in sign. It is convenient to write such a series in the form $\sum_{k=1}^{\infty} (-1)^{k-1} a_k = a_1 - a_2 + a_3 - \ldots$, where $a_k \geq 0$ for all k.

A series that starts with $a_1 + a_2$ − ... needs a tiny change in notation, but no new ideas.

Theorem 2.33 (Alternating series test). *Let $\sum_{k=1}^{\infty} (-1)^{k-1} a_k$ be an alternating series such that*

(i) the sequence $\{a_k\}$ is positive and decreasing: $a_1 \geq a_2 \geq a_3 \geq \cdots > 0$;

(ii) $a_k \to 0$ as $k \to \infty$.

Then the series converges, and the limit S lies between any two successive partial sums S_n and S_{n+1}.

EXAMPLE 3.   Finish off the series $\sum_{k=1}^{\infty}(-1)^{k-1}/\sqrt{k}$.

SOLUTION.   Conditions (i) and (ii) of the theorem hold for our series:

$$\frac{1}{\sqrt{1}} > \frac{1}{\sqrt{2}} > \frac{1}{\sqrt{3}} > \cdots > \frac{1}{\sqrt{k}} \to 0,$$

as needed. Thus our series converges to some limit S, which lies *between* any two successive partial sums. For instance, we have

$$S_{100} \approx 0.55502 < S < 0.65453 \approx S_{101}$$

and

$$S_{1002} \approx 0.58911 < S < 0.62068 \approx S_{1003}. \qquad \diamond$$

Proof (of Theorem 2.33): The proof depends on a pleasing pattern among the partial sums:

$$S_2 \le S_4 \le S_6 \le S_8 \le \cdots \le S_7 \le S_5 \le S_3 \le S_1.$$

All of these inequalities hold because the terms alternate in *sign* but decrease in *size*. The reasons are straightforward but pretty, and best thought through for oneself.

Start small. See, e.g., why $S_5 \ge S_8$.

In particular, the odd-index subsequence $S_1, S_3, S_5, \ldots$ is *decreasing* and bounded below (by any even-index term, such as S_8), while the even-index subsequence $S_2, S_4, S_6, \ldots$ is *increasing* and bounded above. Hence both of these subsequences converge. They tend to the *same* limit, moreover, because

$$S_{2k} - S_{2k-1} = a_{2k},$$

and the right side tends to zero as n tends to infinity. This implies, finally, that $\{S_n\}$ itself converges, as desired, and that the limit, S, satisfies the main inequality:

$$S_2 \le S_4 \le S_6 \le S_8 \le \cdots \le S \le \cdots \le S_7 \le S_5 \le S_3 \le S_1. \qquad \square$$

Estimating Limits

We have now acquired and sharpened several tools for detecting whether a lot of series converge or diverge. That's nice, but what about the limits themselves? The bad news is that *exact* limits may be hard or impossible to find. (The family of geometric series is a rare but important exception.) The good news is that the same tools, cleverly employed, can also help us *estimate* limits with good precision.

EXAMPLE 4.   We used the ratio test in Example 2 to show that the series

$$\sum_{k=0}^{\infty} \frac{1}{k!} = 1 + \frac{1}{1} + \frac{1}{2} + \frac{1}{6} + \frac{1}{24} + \frac{1}{120} + \cdots$$

converges. What can we say about a limit?

SOLUTION.   Listing the first few (rounded) partial sums S_n,

  1, 2, 2.5, 2.66667, 2.70833, 2.71667, 2.71806, 2.71825, 2.71828, 2.71828,

suggests a limit somewhere near $S_{10} \approx 2.71828$. Could the limit be the number e? Can we be sure? The S_n are strictly increasing, after all, so what would prevent, say, $S_{123456789} > 3$? How much can the *rest* of our series

$$\frac{1}{11!} + \frac{1}{12!} + \frac{1}{13!} + \cdots = \sum_{k=11}^{\infty} \frac{1}{k!},$$

contribute, at most?

The comparison test idea gives a quick (if slightly crude) answer. Since

$$\frac{1}{11!} < \frac{1}{2^{10}}, \quad \frac{1}{12!} < \frac{1}{2^{11}}, \quad \frac{1}{13!} < \frac{1}{2^{12}}, \quad \cdots,$$

we can compare with a *geometric* series, for which we *do* know a limit:

$$\frac{1}{11!} + \frac{1}{12!} + \frac{1}{13!} + \cdots < \frac{1}{2^{10}} + \frac{1}{2^{11}} + \frac{1}{2^{12}} + \cdots = \frac{1}{2^9} \approx 0.002.$$

This tells us that $S_{10} \approx 2.71828$ underestimates the true limit by less than 0.002, so we can have confidence in (at least) the first three digits.   ◊

Bounding the tail: lessons from Example 4. For any convergent series $\sum a_k$ and any partial sum S_n we can write

$$\sum_{k=1}^{\infty} a_k = (a_1 + a_2 + \cdots + a_n) + (a_{n+1} + a_{n+2} + \ldots) = S_n + R_n,$$

where $R_n = \sum_{k=n+1}^{\infty} a_k$ is the nth *upper tail*, or *remainder*.

The idea illustrated in Example 4 is to *bound the tail*: If we can somehow show that $|R_n|$ is small, then the limit $S = S_n + R_n$ must be close to the *finite* sum S_n, which we can compute without any worries over convergence. Upper tails are often bounded by comparison to suitable geometric series, for which (atypically) we can find exact limits.

We should confess, finally, that the *exact* limit of the series in Example 4 is well known to be $e \approx 2.71828182846$. A rigorous proof requires methods we haven't developed; our point here is that reasonable accuracy is available using basic arguments.

EXAMPLE 5.   An *alternating* version of the series in Example 4 looks like this:

$$1 - \frac{1}{1} + \frac{1}{2} - \frac{1}{6} + \frac{1}{24} - \frac{1}{120} + \cdots = \sum_{k=0}^{\infty} \frac{(-1)^k}{k!}.$$

What can we say about a limit?

SOLUTION.   First, there *is* a limit.   We showed in Example 2 that our series converges *absolutely*, and so it also converges in the ordinary sense.   Second, our tail estimate from Example 4 can be recycled here:

Says Theorem 2.27, page 128.

The last inequality comes from comparison to the geometric series.

$$|R_{10}| = \left| -\frac{1}{11!} + \frac{1}{12!} - \frac{1}{13!} + \cdots \right| \le \frac{1}{11!} + \frac{1}{12!} + \frac{1}{13!} + \cdots < \frac{1}{2^9}.$$

For the present series we find $S_{10} \approx 0.367879$, and we conclude that the exact limit lies somewhere within about 0.002 of S_{10}.

But we can do better—with less work—using Theorem 2.33, which guarantees that the exact limit lies between *any* two successive partial sums. Calculating $S_9 \approx .3678792$ and $S_{10} \approx 0.3678795$ gives us about six decimal places of accuracy. Again, our estimate plays well with the *exact* limit, which can be shown by other methods to be $1/e \approx 0.367879441$.   ◊

Exercises

1. Use methods of this section to prove as efficiently as possible that each series converges or diverges. (Similar series appeared in exercises for the preceding section; now we know more techniques.)

   (a) $\displaystyle\sum_{k=1}^{\infty} \frac{1}{2^k - 1}$

   (b) $\displaystyle\sum_{k=1}^{\infty} \frac{k}{2k^2 - 1}$

   (c) $\displaystyle\sum_{k=1}^{\infty} \frac{k}{3^k}$

   (d) $\displaystyle\sum_{k=1}^{\infty} \frac{k^2 + 2}{3k^2 - 2}$

2. Use methods of this section, where possible, to decide whether each of the following series converges or diverges. Do any converge conditionally? (Similar series appeared in exercises for the preceding section. With new techniques we can handle some more efficiently.)

(a) $\displaystyle\sum_{k=1}^{\infty} \frac{(-1)^k}{3k^2 + 1}$

(b) $\displaystyle\sum_{k=1}^{\infty} \frac{k^2 + k}{k^4}$

(c) $\displaystyle\sum_{k=1}^{\infty} \frac{\sin(k)}{3^k - 1}$

(d) $\displaystyle\sum_{k=1}^{\infty} \frac{1}{2 + \sin(k)}$

3. Decide whether each of the following series converges or diverges; use the ratio test.

(a) $\displaystyle\sum_{k=1}^{\infty} \frac{2^k}{3k^2}$

(b) $\displaystyle\sum_{k=1}^{\infty} \frac{k^2}{1.3^k}$

(c) $\displaystyle\sum_{k=1}^{\infty} \frac{2^k + k}{3^k}$

4. Each series following involves a constant $a \geq 0$. Decide for each series following which values of a give a convergent series. (The ratio test will be useful.)

(a) $\displaystyle\sum_{k=1}^{\infty} a^k$

(b) $\displaystyle\sum_{k=1}^{\infty} \frac{a}{1.3^k}$

(c) $\displaystyle\sum_{k=1}^{\infty} \frac{a^k}{1.3^k}$

(d) $\displaystyle\sum_{k=1}^{\infty} a^k 2^k$

(e) $\displaystyle\sum_{k=1}^{\infty} \frac{1}{ak^2}$

5. For which values of x does each of the following series converge *absolutely*? (Series like this are called *power series*.)

(a) $\displaystyle\sum_{k=0}^{\infty} x^k$

(b) $\displaystyle\sum_{k=1}^{\infty} kx^{k-1}$

(c) $\displaystyle\sum_{k=1}^{\infty} \frac{x^k}{k}$

(d) $\displaystyle\sum_{k=0}^{\infty} k!x^k$

6. Prove or disprove that each of the following series converges.

(a) $\displaystyle\sum_{k=1}^{\infty} \frac{100^k}{k!}$

(b) $\displaystyle\sum_{k=1}^{\infty} \frac{\sin k}{1.0001^k}$

(c) $\displaystyle\sum_{k=1}^{\infty} \frac{k^2 + k + 1}{k^3 + k^2 + k + 1}$

(d) $\displaystyle\sum_{k=1}^{\infty} \frac{(-1)^k}{\ln\ln(k + 100)}$

7. In both parts, assume $a_k > 0$ for all k.

   (a) Give examples to show that if $\sum a_k$ diverges, then $\sum a_k^2$ may either converge or diverge.

   (b) Show that if $\sum a_k$ converges, then $\sum a_k^2$ converges, too. (Hint: Use limit comparison.)

8. In the situation and notation of Theorem 2.29, page 136, show that if $\sum b_k$ diverges and $a_k/b_k \to \infty$, then $\sum a_k$ diverges, too. (Hint: If $a_k/b_k \to \infty$, then $b_k/a_k \to 0$.)

9. (This generalizes Problem 11, page 134.) Complete and prove the following version of Proposition 2.30, page 137: Consider two series $\sum a_k$ and $\sum b_k$ such that $a_k = b_k$ for $k > N$. If $\sum b_k$ diverges, then $\sum a_k$ diverges, too. If $\sum b_k$ converges to B, then $\sum a_k$ converges to _____.

10. In Example 2, page 138, we used the ratio test to show that $\sum_{k=0}^{\infty} 1/k!$ converges. Prove the same thing using the comparison or limit comparison test.

11. Let $\{S_n\}$ be the partial sum sequence for the series $\sum_{k=1}^{\infty} \frac{1}{\sqrt{k}}$. Show that $S_n > \sqrt{n}$ for all n; conclude that the series diverges.

12. In our proof sketch for Proposition 2.28, page 134, we showed that $\sum 1/k^p$ converges if $p = 1.3$. Show in a similar way that $\sum 1/k^p$ converges for *every* $p > 1$. (Hint: Writing $p = 1 + s$ simplifies the algebra slightly; then s plays the role of 0.3 in the given proof.)

13. Consider the series $\sum_{k=1}^{\infty} 1/(2^{k-1} + 1)$, which converges to some limit S.

    (a) Explain why the series converges.

    (b) *Mathematica* approximates the tenth partial sum: $S_{10} \approx 1.26255$. Give an upper bound for R_{10}, the corresponding upper tail.

    (c) Use results of the previous part to give a small interval guaranteed to contain S.

14. Consider the series $\sum_{k=1}^{\infty} (-1)^{k-1}(1/(2^{k-1} + 1))$ (the alternating version of the series in the preceding problem).

    (a) Explain why this series converges to some limit S.

    (b) Find an interval of length less than 0.001 that is guaranteed to contain S.

    (c) By how much can the partial sum S_{100} differ from the limit S? Why?

15. For any p-series $\sum_{k=1}^{\infty} \frac{1}{k^p}$ with $p > 0$, it can be shown that all upper tails satisfy

$$R_n = \sum_{k=n+1}^{\infty} \frac{1}{k^p} < \int_n^{\infty} \frac{dx}{x^p}.$$

    (a) Draw a picture to illustrate this inequality. (Leaf through any standard calculus text, if necessary.)

    (b) Estimate $\sum_{k=1}^{\infty} \frac{1}{k^3}$ with error less than 0.001. (Technology will be helpful.)

16. Does each of the following series converge absolutely, converge conditionally, or diverge? Prove your answers. (It's OK to assume basic facts about geometric series, p-series, etc.)

    (a) $\sum_{k=1}^{\infty} \left(\frac{1}{k} - \frac{1}{k+3} \right)$

    (b) $\sum_{k=1}^{\infty} \frac{\cos k}{2^k + 2k + 2}$

(c) $\displaystyle\sum_{k=1}^{\infty}(-1)^k\frac{81}{347k}$

2.7   Lim sup and lim inf: A Guided Discovery

This brief section, designed for self-study, explores two useful generalizations of the "ordinary" limit of a sequence: the *limit superior* and the *limit inferior* (also known as the *upper limit* and the *lower limit*, respectively).

Results of this section are not needed in later sections of the book.

We will discuss unbounded sequences below.

Defining the lim sup. Given any bounded sequence $\{x_n\}$, we can define the new sequence $\{v_n\}$

$$v_1 = \sup\{x_1, x_2, x_3, \dots\}; \quad v_2 = \sup\{x_2, x_3, x_4, \dots\}; \quad \dots$$
$$v_n = \sup\{x_n, x_{n+1}, x_{n+2}, \dots\}.$$

(We'll call this the *sup-sequence* associated to $\{x_n\}$.)

Now we can give the definition:

$$\limsup x_n = \lim_{n\to\infty} v_n.$$

In brief: $\limsup x_n = \lim_{n\to\infty} \sup\{x_k \mid k \geq n\}$.

Basic problems.

1. For each of the following sequences $\{x_n\}$, find the corresponding sequence $\{v_n\}$; then find $\limsup x_n$.

   (a) $\{x_n\} = \{1, 1, 1, 1, \dots\}$.

   (b) $\{x_n\} = \left\{2, \dfrac{3}{2}, \dfrac{4}{3}, \dfrac{5}{4}, \dfrac{6}{5}, \dots\right\}$.

   (c) $\{x_n\} = \left\{2, \dfrac{3}{2}, -\dfrac{3}{2}, \dfrac{4}{3}, -\dfrac{4}{3}, \dots\right\}$.

   (d) $\{x_n\} = \left\{1, \dfrac{1}{2}, \dfrac{2}{3}, \dfrac{3}{4}, \dfrac{4}{5}, \dots\right\}$.

2. Why must all the v_i exist in the sup-sequence? Why must $\{v_n\}$ have a limit? (Cite appropriate theorems.)

3. How are $\lim x_n$ and $\limsup x_n$ related to each other for a bounded sequence $\{x_n\}$? Can one exist but not the other? Can both exist but be different?

The lim inf. The lim inf is closely analogous to the lim sup.

1. Carefully state an appropriate definition for the lim inf of a bounded sequence. Then find $\liminf x_n$ for each of the example sequences in the first
   problem.

2. Find a sequence $\{x_n\}$ with $\limsup x_n = 5$ and $\liminf x_n = -2$.

3. Prove that if $\limsup x_n = 5$ and $\liminf x_n = 5$, then $\lim x_n = 5$, too.

4. It seems reasonable that $\liminf x_n \leq \limsup x_n$ for any bounded sequence
   $\{x_n\}$. Prove this. What happens if the two are equal?

Algebra with the lim inf and the lim sup. The ordinary limit has nice algebraic
properties: for example, $\lim (x_n + y_n) = \lim x_n + \lim y_n$ (assuming that both
limits on the right side exist). Do lim inf and lim sup have similar properties?

1. Suppose $\limsup x_n = S$, $\liminf x_n = I$, and k is a constant. What can
   be said about $\limsup kx_n$ and $\liminf kx_n$? (Hints: Experiment with your
   example from above, and with several possible values of k.)

2. If $\limsup x_n = L$ and $\limsup y_n = M$, what can be said about $\limsup(x_n
   + y_n)$? (Show that $\limsup(x_n + y_n) \leq L + M$.)

Unbounded sequences. If $\{x_n\}$ is unbounded *above*, then the sup-sequence
$\{v_n\}$ doesn't exist, but it is natural to define $\limsup x_n = \infty$. Similarly, if $\{x_n\}$
is unbounded *below*, then we define $\liminf x_n = -\infty$.

1. Find the lim sup and the lim inf of the sequence $-1, -2, -3, -4, \ldots$. Find
   a sequence $\{x_n\}$ with $\limsup x_n = \infty$ and $\liminf x_n = -\infty$.

2. Show that *every* sequence (bounded or not) has a lim sup and a lim inf.

3. If $\limsup x_n = +\infty$, must $\lim x_n = +\infty$, too? If $\limsup x_n = -\infty$,
   must $\lim x_n = -\infty$, too?

Lim sups, lim infs, and subsequences. The lim sup and the lim inf of a sequence $\{x_n\}$ are closely connected to subsequences of $\{x_n\}$, as the following
proposition suggests.

Proposition 2.34. *Let $\{x_n\}$ be a sequence.*

(a) *If $\{x_n\}$ is unbounded above (so $\limsup x_n = \infty$), then some subsequence
    $\{x_{n_k}\}$ diverges to infinity.*

(b) *If $\{x_n\}$ is bounded above, with $\limsup x_n = L$, then some subsequence
    $\{x_{n_k}\}$ converges to L.*

Proof: Claim (a) is essentially Lemma 2.14, page 110.

To prove claim (b), let's construct a subsequence $\{x_{n_k}\}$ such that, for all k,

$$L - \frac{1}{k} \le x_{n_k} \le v_k = \sup\{x_k, x_{k+1}, \ldots,$$

where $\{v_k\}$ is the usual sup-sequence. Now the sequences at left and right both clearly converge to L, so our subsequence $\{x_{n_k}\}$ is "squeezed" to the same limit.

To construct the desired subsequence, observe first that $L - 1 < L \le v_1$. Thus $L - 1$ is *not* an upper bound for $\{x_1, x_2, x_3, \ldots\}$, so we can find x_{n_1} with $L - 1 < x_{n_1} \le v_1$, as desired. Similarly, $L - 1/2$ is *not* an upper bound for $\{x_{n_1+1}, x_{n_1+2}, x_{n_1+3}, \ldots\}$, so there is some x_{n_2} with $n_2 > n_1$ and $L - 1/2 < x_{n_2} \le v_{n_2} \le v_2$, as desired. Continuing this process produces the desired subsequence $\{x_{n_1}, x_{n_2}, x_{n_3}, \ldots\}$. □

Use these ideas in the following problems.

1. Show that if $\limsup x_n = 5$, then there is a subsequence $\{x_{n_k}\}$ that converges to 5. Could some other subsequence converge to 6? To 4? Explain your answers.

2. Show that if $\{x_n\}$ has a subsequence $\{x_{n_k}\}$ that converges to 5, then

$$\liminf x_n \le 5 \le \limsup x_n.$$

3. Show that if $\limsup x_n = \infty$, then there is a subsequence $\{x_{n_k}\}$ with $x_{n_k} \to \infty$.

More on the "sup-sequence." The following problems explore further the correspondence between the original sequence $\{x_n\}$ and the "sup-sequence" $\{v_n\}$. We can ask, for instance, about "inverting" this correspondence.

1. For each of the following sequences $\{v_n\}$, find (if possible) two *different* sequences $\{x_n\}$ that give the same sup-sequence $\{v_n\}$.

   (a) $\{v_n\} = \left\{2, 2, \dfrac{3}{2}, \dfrac{3}{2}, \dfrac{4}{3}, \dfrac{4}{3}, \ldots\right\}$

   (b) $\{v_n\} = \{2, 2, 2, 2, 2, 2, \ldots\}$

   (c) $\{v_n\} = \{2, 1.1, 1.01, 1.001, 1.0001, \ldots\}$

2. For which sequences $\{v_n\}$ can there be more than one associated sequence $\{x_n\}$? State and prove a claim.

CHAPTER 3

Limits and Continuity

3.1 Limits of Functions

This chapter is about continuous functions and their properties. As a foretaste, consider this reasonable-sounding claim about a function f:

> If $f(-1) = -2$ and $f(1) = 5$, then $f(a) = 0$ for some $a \in (-1, 1)$.

Is this true? If a graph passes through the points $(-1, -2)$ and $(1, 5)$, must it also cross the x-axis somewhere in between? Must there also be an input b with $f(b) = \pi$?

Both answers turn out to be "yes"—if the function f is *continuous* everywhere along the input interval $[-1, 1]$. In this case, answers follow from the famous *intermediate value theorem*, which we will carefully state and prove later in this chapter. To prove such results we'll draw on earlier work with completeness, suprema and infima, limits, the Bolzano–Weierstrass theorem, and more. Our work starts, as usual, with clear definitions of words like "limit" (as applied to functions) and "continuous."

The IV theorem says that the range of a continuous function has no "gaps"; details come later.

Defining Limits of Functions

Most of us emerge from elementary calculus courses with a rough and ready view of function limits. To say that $\lim_{x \to 3} f(x) = 5$, for example, means something like

> $f(x)$ approaches 5 as x approaches 3, or $f(x) \approx 5$ when $x \approx 3$.

These views are not incorrect, but they are too vague to be useful for building theory or proving theorems. What exactly does "approaches" mean? *How* close to 3 must x be for $x \approx 3$ to hold? We need a precise definition.

Definition 3.1 (Limit of a function). Let f be a function whose domain includes an open interval I containing a, except perhaps for $x = a$. Let L be a number. We write $\lim_{x \to a} f(x) = L$ if, for every $\epsilon > 0$, there exists $\delta > 0$ such that

$$|f(x) - L| < \epsilon \quad \text{whenever } x \in I \text{ and } 0 < |x - a| < \delta.$$

Observe, right away, two important subtleties about domains:

- *At $x = a$:* The value $f(a)$ itself *need not* be defined for the limit to exist at $x = a$. Indeed, the whole point of finding such a limit is often to do with a "missing" or mysterious function value. If $f(a)$ *is* defined, that's fine, but it's irrelevant to the definition, which avoids all mention of $f(a)$.

- *Near $x = a$:* For the key ϵ–δ condition to make good sense—let alone hold true—$f(x)$ must be defined for all x (other than a) that are within δ of a. We will sometimes summarize these conditions by requiring that the domain of f include some set of the form $I \setminus \{a\}$, where I is any open interval containing a.

EXAMPLE 1. Let $f(x) = (3x^2 - 3)/(x - 1)$. Use Definition 3.1 to show that $\lim_{x \to 1} f(x) = 6$.

SOLUTION. Note first that all is well with domains: $f(1)$ is undefined, but $f(x)$ makes good sense for *all* other x. Next, for $x \neq 1$ we have

$$f(x) = \frac{3x^2 - 3}{x - 1} = \frac{3(x - 1)(x + 1)}{x - 1} = 3x + 3,$$

and so

$$|f(x) - 6| = |3x + 3 - 6| = |3x - 3| = 3\,|x - 1|.$$

Thus,

$$3\,|x - 1| < \epsilon \quad \Longleftrightarrow \quad |x - 1| < \frac{\epsilon}{3}.$$

This shows that $\delta = \epsilon/3$ "works" for any given ϵ, and we're ready for the brief formal proof.

Let $\epsilon > 0$ be given, and set $\delta = \epsilon/3$. This δ works, because if $0 < |x - 1| < \delta$, then

$$|f(x) - 6| = \left| \frac{3x^2 - 3}{x - 1} - 6 \right| = 3\,|x - 1| < 3\delta = \epsilon,$$

where we used factoring in the second equality.

Note that the preceding paragraph stands on its own as a proof—the discussion beforehand is there "only" for motivation. $\Diamond$

EXAMPLE 2. Let a and k be any constants. It is no surprise that

$$\lim_{x \to a} x = a \quad \text{and} \quad \lim_{x \to a} k = k,$$

but how do these facts follow from Definition 3.1?

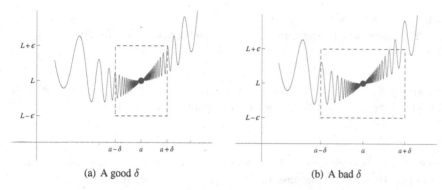

(a) A good δ (b) A bad δ

Figure 3.1. Picturing function limits.

SOLUTION.   With $f(x) = x$ for the first limit $f(x) = k$ for the second, verifying Definition 3.1 is an easy exercise.   ◇

Picturing limits. Function limits can be understood graphically. For any given ϵ and δ, we can look at the tiny rectangular "window" in the xy-plane, centered at (a, L), in which $L - \epsilon < y < L + \epsilon$ and $a - \delta < x < a + \delta$. This window has "half-height" ϵ and "half-width" δ; two possibilities are shown in Figure 3.1.

From this viewpoint, the ϵ–δ condition is satisfied if the graph of f stays *inside* this rectangle all the way from left to right, as in Figure 3.1(a). If the graph "escapes" at top or bottom, as in Figure 3.1(b), the chosen δ does *not* work for the given ϵ. The limit is L if, for any choice of ϵ (the half-height), there exists some good half-width δ.

EXAMPLE 3.   The function

$$f(x) = \begin{cases} 1 & \text{if } x \in \mathbb{Q} \\ 0 & \text{if } x \notin \mathbb{Q} \end{cases}$$

is called the *characteristic* function of the set $\mathbb{Q}$. Does $\lim_{x \to 0} f(x)$ exist?

It is useful in illustrating "bad" behavior; we'll see it again. Can you picture the graph?

SOLUTION.   The (unsurprising) answer is no. To prove this formally, suppose toward contradiction that L is a limit, and set $\epsilon = 0.001$. By our assumption, there must be some $\delta > 0$ with

$$|f(x) - L| < 0.001 \quad \text{whenever} \quad x \in (-\delta, \delta).$$

But (as we've shown) every interval $(-\delta, \delta)$ contains both rational and irrational numbers. If s is such a rational and t an irrational, then both

$$|f(s) - L| = |1 - L| < 0.001 \quad \text{and} \quad |f(t) - L| = |0 - L| < 0.001,$$

which is clearly absurd.   ◊

The next example illustrates what can—and cannot—be inferred from the existence of a limit.

EXAMPLE 4.   Let f be a function for which $\lim_{x \to a} f(x) > 0$. Must $f(x) > 0$ also hold for x *near* a? What can be said about another continuous function g if $\lim_{x \to a} g(x) = 0$?

SOLUTION.   Concerning f, the answer is yes—except *at* $x = a$, where we know nothing. To see why, suppose $\lim_{x \to a} f(x) = L > 0$, and set $\epsilon = L/2$. By definition of the limit, there is some $\delta > 0$ that "works" for this ϵ in the sense that

$$|f(x) - L| < \frac{L}{2} \quad \text{whenever} \quad 0 < |x - a| < \delta.$$

Rewriting these inequalities gives

$$f(x) > \frac{L}{2} > 0 \quad \text{whenever} \quad x \in (a - \delta, a + \delta) \quad \text{and} \quad x \neq a,$$

as desired.

Concerning g, the answer is "not much." Near $x = a$ the function g could be *positive* (if, say, $g(x) = (x - a)^2$) or *negative* (if, say, $g(x) = -(x - a)^2$). Or perhaps $g(x) = x - a$, which *changes sign* at $x = a$.   ◊

Basic Properties of Limits

Limits of functions have a lot in common with limits of sequences. This is hardly surprising—sequences *are* functions, after all—but it's a good thing, because most of our hard work with sequence limits will transfer readily to function limits. It is easy to prove, for instance, that function limits have the same nice algebraic properties that sequence limits enjoy. The following technical lemma helps clarify the connection.

Limits respect sums, products, quotients, etc.

Lemma 3.2. *Let $f(x)$ be defined on $I \setminus \{a\}$, where I is an open interval containing a, and let L be a number. The following are equivalent:*

(i) $\lim_{n \to \infty} f(x_n) = L$ *for every sequence* $\{x_n\} \subset I \setminus \{a\}$ *with* $x_n \to a$.

(ii) $\lim_{x \to a} f(x) = L$

Proof: To show that (i) implies (ii), let's assume that (ii) fails and construct a sequence $\{x_n\}$ that violates (i).

Let's prove the contrapositive, in other words.

For (ii) to fail means that there is some positive ϵ, say ϵ_0, for which *no* $\delta > 0$ works. In particular, $\delta = 1$ fails, so there must be some x_1 in $I \setminus \{a\}$ such that

$$0 < |x_1 - a| < 1 \quad \text{but} \quad |f(x_1) - L| \geq \epsilon_0.$$

Because $\delta = 1/2$ also fails, there is some x_2 with

$$0 < |x_2 - a| < 1/2 \quad \text{but} \quad |f(x_2) - L| \geq \epsilon_0.$$

Continuing in this way, we construct a sequence $\{x_n\}$ of points in $I \setminus \{a\}$ such that, for each n,

$$0 < |x_n - a| < \frac{1}{n} \quad \text{but} \quad |f(x_n) - L| \geq \epsilon_0.$$

In particular, $\{f(x_n)\}$ does *not* converge to L. On the other hand, $\{x_n\}$ converges to a; the squeeze principle assures this because

$$a - \frac{1}{n} < x_n < a + \frac{1}{n}$$

for all n. Thus our sequence $\{x_n\}$ violates (i), as desired, and we've shown that (i) $\implies$ (ii).

Showing (ii) $\implies$ (i) is easier; we'll leave it as an exercise. $\qquad\square$

Lemma 3.2 lets us exploit earlier work with sequences to find, without much effort, a lot of function limits that would be tedious or difficult to handle from scratch.

EXAMPLE 5. Let $f(x) = (3x^2 + 5x + 2)/(2x - 7)$. Use Lemma 3.2 to prove that $\lim_{x \to 5} f(x) = \frac{102}{3}$.

SOLUTION. Proving such a thing straight from the ϵ–δ definition could get ugly. With Lemma 3.2, it is easy.

To get started, let $\{x_n\}$ be any sequence with $x_n \to 5$. By Lemma 3.2, it is enough to show that

$$\lim_{n \to \infty} f(x_n) = \lim_{n \to \infty} \frac{3x_n^2 + 5x_n + 2}{2x_n - 7} = \frac{102}{3}.$$

This, in turn, follows from different parts of Theorem 2.5, page 96, which allows algebraic combinations like the following for limits of *sequences*:

No worries here about zero denominators.

$$\lim_{n \to \infty} \frac{3x_n^2 + 5x_n + 2}{2x_n - 7} = \frac{3(\lim x_n)^2 + 5(\lim x_n) + 2}{2(\lim x_n) - 7}$$

$$= \frac{3 \cdot 5^2 + 5 \cdot 5 + 2}{2 \cdot 5 - 7} = \frac{102}{3},$$

as claimed. $\Diamond$

New function limits from old. The following theorem does for functions what Theorem 2.5, page 96, does for sequences.

Theorem 3.3 (Algebra with limits). *Let $f(x)$ and $g(x)$ be defined for all inputs x in $I \setminus \{a\}$, where I is any open interval containing a, and suppose that*

$$\lim_{x \to a} f(x) = L \quad and \quad \lim_{x \to a} g(x) = M.$$

Then

$$\lim_{x \to a} (f(x) \pm g(x)) = \lim_{x \to a} f(x) \pm \lim_{x \to a} g(x) = L \pm M;$$
$$\lim_{x \to a} (f(x) \cdot g(x)) = \lim_{x \to a} f(x) \cdot \lim_{x \to a} g(x) = L \cdot M;$$
$$\lim_{x \to a} \frac{f(x)}{g(x)} = \frac{\lim_{x \to a} f(x)}{\lim_{x \to a} g(x)} \quad if \, M \neq 0.$$

Observe:

- *Existence:* Theorem 3.3 is more than a recipe for calculating new limits from old; it is also a guarantee that various limits (those on the left of each equation) actually *exist*. In the case of quotients, for existence, one might reasonably worry about small or vanishing denominators; Theorem 3.3 assures us that all is well if $M \neq 0$.

- *Something to combine:* The how-to-combine-limits rules in Theorem 3.3 aren't much good until we have some already-proved basic limits to combine. The good news is that a few very, very basic limits, such as

$$\lim_{x \to a} x = a \quad and \quad \lim_{x \to a} k = k$$

We addressed these in Example 2.

go a long way. Applying Theorem 3.3, repeatedly as necessary, lets us calculate limits like these without too much effort:

$$\lim_{x \to 5} \frac{3x^2 + 5x + 2}{2x - 7} = \frac{3 \cdot 5^2 + 5 \cdot 5 + 2}{2 \cdot 5 - 7} = 34;$$
$$\lim_{x \to 5} \frac{317x^2 - 5\frac{x+2}{x+3}}{42.07^5 - x^5} = \frac{317 \cdot 5^2 - 5\frac{5+2}{5+3}}{42.07^5 - 5^5} \approx 0.5117.$$

(We calculated the first limit, a bit differently, in Example 5.)

Proof: All parts of Theorem 3.3 follow easily when we combine the analogous results for sequences (Theorem 2.5, page 96) with Lemma 3.2, above. Concerning products, for instance, we consider any sequence $\{x_n\}$ in $I \setminus \{a\}$. By hypothesis, the new sequences $\{f(x_n)\}$ and $\{g(x_n)\}$ converge to L and M, respectively. By Theorem 2.5, the product *sequence* $\{f(x_n)g(x_n)\}$ converges to LM, and Lemma 3.2 assures us that $\lim_{x \to a} f(x)g(x) = LM$, too. The remaining parts are similar. $\square$

Squeezing functions. A squeeze principle works for function limits. We've already proved the sequence version (Theorem 2.6, page 99):

Proposition 3.4 (The squeeze principle). *Let $f(x)$, $g(x)$, and $h(x)$ be defined for all inputs x in $I \setminus \{a\}$, where I is an open interval containing a, and suppose that*

$$f(x) \leq g(x) \leq h(x) \quad \text{for all } x \in I \setminus \{a\}.$$

If $\lim_{x \to a} f(x) = L$ and $\lim_{x \to a} h(x) = L$, then $\lim_{x \to a} g(x) = L$, too.

The proof, like that of Theorem 3.3, amounts to combining the sequence version with Lemma 3.2. We'll leave that to the exercises.

The limit-squeezing *idea* is simple. The tricky bit in practice is to find helpful squeezing inequalities.

EXAMPLE 6. Squeeze something to find $\lim_{x \to 0} x \sin(1/x)$ and $\lim_{x \to 0} \frac{\sin x}{x}$.

Plotting these functions might be useful.

SOLUTION. The fact that $|x \sin(1/x)| \leq |x|$ for all $x \neq 0$ suggests a simple squeezing inequality:

$$-|x| \leq x \sin\left(\frac{1}{x}\right) \leq |x|.$$

Clearly, $\pm|x| \to 0$ as $x \to 0$, so the middle quantity tends to zero, too.

Finding a good squeezing inequality for the second limit takes more effort. Here is one possibility:

$$\cos x \leq \frac{\sin x}{x} \leq \frac{1}{\cos x} \quad \text{if } x \in (-1, 1) \quad \text{and} \quad x \neq 0.$$

This does the job: Since $\cos x \to 1$ as $x \to 0$, the left- and right-hand functions tend to 1, and hence so does the middle function. The squeezing inequality needs proof too, of course; we'll leave that to the exercises to avoid distraction.   ◊

Beyond Vanilla: More Limit Flavors

We'll find uses for several variations on the basic limit theme discussed above. Here are some samples of *limits at infinity* and *infinite limits*:

$$\lim_{x \to \infty} \frac{4x + 3}{2x + 1} = 2; \quad \lim_{x \to 0} \frac{1}{x^2} = \infty; \quad \lim_{x \to -\infty} (x^2 - x^3) = \infty.$$

Here are some *one-sided limits*:

$$\lim_{x \to 0+} \ln x = -\infty; \quad \lim_{x \to 0+} \frac{x}{1 + \sqrt{x}} = 0; \quad \lim_{x \to 3-} \frac{x - 3}{|3 - x|} = -1.$$

Following are several formal definitions; we leave some others, all in the same spirit, as exercises. As with Definition 3.1, each part here involves some technical assumption about domains, needed to ensure that the key inequalities make sense.

Familiar functions from calculus seldom cause trouble on this score.

Definition 3.5 (Variant limits).   Let f be a function and let a and L be real numbers.

- *Right-hand limit:* Let $f(x)$ be defined for all inputs x in some open interval $I = (a, b)$. We say $\lim_{x \to a+} f(x) = L$ if, for every $\epsilon > 0$, there exists $\delta > 0$ such that

$$|f(x) - L| < \epsilon \quad \text{whenever } x \in I \text{ and } a < x < a + \delta.$$

- *Left-hand infinite limit:* Let $f(x)$ be defined for all inputs x in some open interval $I = (b, a)$. We say $\lim_{x \to a-} f(x) = -\infty$ if, for every $M > 0$, there exists $\delta > 0$ such that

$$f(x) < -M \quad \text{whenever } x \in I \text{ and } a - \delta < x < a.$$

- *Limit at infinity:* Let $f(x)$ be defined for all inputs x in some open interval $I = (b, \infty)$. We say $\lim_{x \to \infty} f(x) = L$ if, for every $\epsilon > 0$, there exists $N > 0$ such that

$$|f(x) - L| < \epsilon \quad \text{whenever } x \in I \text{ and } x > N.$$

- *Infinite (two-sided) limit:* Let $f(x)$ be defined for all inputs x in an open interval I containing a, except perhaps at $x = a$. We say $\lim_{x \to a} f(x) = \infty$ if, for every $M > 0$, there exists $\delta > 0$ such that

$$f(x) > M \quad \text{whenever } x \in I \text{ and } 0 < |x - a| < \delta.$$

- *Infinite limit at infinity:* Let $f(x)$ be defined for all inputs x in some open interval $I = (b, \infty)$. We say $\lim_{x \to \infty} f(x) = \infty$ if, for every $M > 0$, there exists $N > 0$ such that

$$f(x) > M \quad \text{whenever} \quad x > N.$$

EXAMPLE 7.   Prove these limit claims:

(a) $\lim_{x \to \infty} \dfrac{1}{x} = 0;$ (b) $\lim_{x \to \infty} \dfrac{3x + 2}{x + 5} = 3;$ (c) $\lim_{x \to 0+} \dfrac{2x}{1 + \sqrt{x}} = 0.$

SOLUTION.   Note first that the functions in all three limits are fine as regards domains. The first is defined for $x > 0$, the second for $x > -5$, and the third for $x \geq 0$.

Limits (a) and (b) have close analogues for sequences—$\lim_{n \to \infty} \frac{1}{n} = 0$ and $\lim_{n \to \infty} (3n + 2)/(n + 5) = 3$—and the proofs for functions are almost identical to those for sequences. For (b), for instance, we proved the sequence version in

Example 2, page 85, and the function version is almost the same. First, we see that for any positive ϵ,

$$\left|\frac{3x+2}{x+5} - 3\right| = \left|\frac{3x+2-3(x+5)}{x+5}\right| = \frac{13}{x+5}.$$

(It is OK to drop the absolute value because $x + 5 > 0$ certainly holds for the large positive x in which we're interested.) Now

$$\frac{13}{x+5} < \epsilon \iff \frac{13}{\epsilon} - 5 < x,$$

and so for any given $\epsilon > 0$ we can set $M = 13/\epsilon$. This M works in the appropriate definition, because if $x > M$, then (skipping some details from just above)

$$\left|\frac{3x+2}{x+5} - 3\right| = \frac{13}{x+5} < \frac{13}{M} = \epsilon,$$

as desired. Limit (a) is left to you.

For (c), only *positive* x matter, and so

$$\left|\frac{2x}{1+\sqrt{x}} - 0\right| = \frac{2x}{1+\sqrt{x}} < 2x < \epsilon \iff x < \frac{\epsilon}{2}.$$

Thus, for given $\epsilon > 0$ the value $\delta = \epsilon/2$ works: if $0 < x < \delta$, then

$$\left|\frac{2x}{1+\sqrt{x}} - 0\right| =< 2x < 2\delta = \epsilon,$$

as the definition requires.   ◊

All in the limit family. All of these limit variants are close kin to each other. Limits at infinity, for instance, are one-sided in the sense that ∞ is approachable only from below, and $-\infty$ only from above. The following proposition makes some of this kinship explicit:

Similar results apply to limits that involve $-\infty$.

Proposition 3.6. *Let a and L be finite numbers, and f a function.*

(i) $\lim_{x\to a} f(x) = L$ *if and only if* $\lim_{x\to a+} f(x) = \lim_{x\to a-} f(x) = L$.

(ii) $\lim_{x\to\infty} f(x) = L$ *if and only if* $\lim_{x\to 0+} f\left(\frac{1}{x}\right) = L$.

(iii) $\lim_{x\to\infty} f(x) = \infty$ *if and only if* $\lim_{x\to\infty} \frac{1}{f(x)} = 0$.

Observe, especially, what the proposition says about *existence*: if the limits on either side of "if and only if" exist, then so must the limits on the other side. We will leave formal proofs to the exercises, but illustrate the idea of (i) with an example.

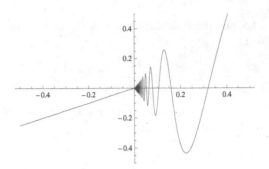

Figure 3.2. A piecewise-defined function.

EXAMPLE 8.   Figure 3.2 shows the function

$$g(x) = \begin{cases} x/2 & \text{if } x < 0, \\ 2x\sin(1/x) & \text{if } x > 0. \end{cases}$$

Show that $\lim_{x\to 0} g(x) = 0$.

SOLUTION.   Let $\epsilon > 0$ be given; we need a $\delta > 0$ that works for ϵ. Because of g's peculiar two-sided nature, we'll first find positive numbers δ_{left} and δ_{right} that work for ϵ on the left and on the right of zero, respectively.

Here "right" and "left" mean $x > 0$ and $x < 0$, respectively.

Matters are simplest on the left. There $g(x) = x/2$, and so

$$\left| g(x) - 0 \right| = \left| \frac{x}{2} \right| < \epsilon \quad \Longleftrightarrow \quad |x| < 2\epsilon,$$

which means that $\delta_{\text{left}} = 2\epsilon$ works (and also that $\lim_{x\to 0-} g(x) = 0$). On the right, we have $g(x) = 2x\sin(1/x)$, and so

$$\left| g(x) - 0 \right| = \left| 2x\sin(1/x) \right| \le |2x| < \epsilon \quad \Longleftrightarrow \quad |x| < \frac{\epsilon}{2},$$

which means that $\delta_{\text{right}} = \epsilon/2$ works (and that $\lim_{x\to 0+} g(x) = 0$).

Combining these results means that $|g(x) - 0| < \epsilon$ holds for all nonzero x in the asymmetric interval

$$(-\delta_{\text{left}}, \delta_{\text{right}}) = (-\epsilon/2, 2\epsilon).$$

Obviously, $|g(x) - 0| < \epsilon$ also holds for nonzero x in the (smaller) *symmetric* interval $(-\delta_{\text{left}}, \delta_{\text{left}})$. This means that $\delta = \delta_{\text{left}} = \epsilon/2$ works for ϵ in the desired limit. We're done.   ◇

More algebra. These newfangled limits—if *finite*—enjoy all of our by-now-familiar algebraic and squeezing properties. Infinite limits, on the other hand, need special care.

EXAMPLE 9.   Find and prove (or re-prove) these limits:

$$\lim_{x\to\infty} \frac{1}{x+\sqrt{x}+3}; \quad \lim_{x\to\infty} \frac{3x+2}{x+5}; \quad \lim_{x\to\infty} x^2 - 1000x; \quad \lim_{x\to\infty} \sqrt{x^2+2x} - x.$$

SOLUTION.   The first limit yields to squeezing. For all $x > 0$, we have

$$0 < \frac{1}{x+\sqrt{x}+3} < \frac{1}{x};$$

since $\lim_{x\to\infty} \frac{1}{x} = 0$, we must have $\lim_{x\to\infty} 1/(x+\sqrt{x}+3) = 0$, too.

We found the second limit from scratch in Example 7. With algebra we can reduce the limit to something simpler (also handled in Example 7):

$$\lim_{x\to\infty} \frac{3x+2}{x+5} = \lim_{x\to\infty} \frac{3+2/x}{1+5/x} = \frac{3+2\lim 1/x}{1+5\lim 1/x} = \frac{3+0}{1+0}.$$

The limit $\lim_{x\to\infty} x^2 - 1000x$ involves the difference of two quantities, each tending to infinity. Which tendency "wins"? It is easy to guess that x^2 overwhelms $1000x$ for large x, and so, presumably $\lim_{x\to\infty} x^2 - 1000x = \infty$. Algebra helps confirm this. For any given $M > 0$, we have

$$x^2 - 1000x = x(x - 1000) > M \quad \text{if } x > M \text{ and } x \geq 1001,$$

so we can use $N = \max\{M, 1001\}$ in the appropriate definition.

Check details for yourself.

The function $\sqrt{x^2+2x} - x$ is another difference of two quantities that diverge to infinity. Plotting the function or plugging in large values of x suggests that the limit is one. We can show this algebraically—with a little effort. The trick is to multiply and divide by the *conjugate* expression:

Try it.

$$\sqrt{x^2+2x} - x = \frac{\left(\sqrt{x^2+2x} - x\right)\left(\sqrt{x^2+2x} + x\right)}{\sqrt{x^2+2x} + x}$$

$$= \frac{2x}{\sqrt{x^2+2x} + x} = \frac{2}{\sqrt{1+2/x} + 1}.$$

(We divided numerator and denominator by x in the last equality.) This is progress: to show that $\lim_{x\to\infty} \sqrt{x^2+2x} - x = 1$ it remains only to show that $\sqrt{1+2/x} \to 1$ as $x \to \infty$. We'll leave this reasonable-seeming result to the exercises. $\diamond$

Exercises

1. We said in Example 2 that if a and k are any constants, then

$$\lim_{x \to a} x = a \quad \text{and} \quad \lim_{x \to a} k = k.$$

   Use Definition 3.1 to prove this.

2. Suppose $f(x) = 0$ if $x \neq 0$ and $f(0) = 42$. Show that $\lim_{x \to 0} f(x) = 0$. What's $\lim_{x \to 42} f(x)$?

3. Guess a value for each of the following limits; prove your answers using Definition 3.1, page 149.

   (a) $\lim_{x \to 1} 2x + 3$

   (b) $\lim_{x \to -1} \dfrac{x^2 - 1}{x + 1}$

4. Guess a value for each of the following limits; prove your answers using Definition 3.1, page 149.

   (a) $\lim_{x \to 1} \dfrac{x^2 + x - 2}{x - 1}$

   (b) $\lim_{x \to 0} x \dfrac{2 + \sin x}{3 - \cos x}$

5. For a function f and an input $x = a$ it may (or may not) happen that $\lim_{x \to a} f(x) = f(a)$. If $f(x) = 3x + 5$, for instance, it is easy to show that $\lim_{x \to 42} f(x) = 3 \cdot 42 + 5 = f(42)$.

   In each case following, decide whether this happens. If so, prove it; if not, say why not. (Theorem 3.3, page 154, may be useful.)

   (a) $f(x) = x^2$; $a = 42$.

   (b) $f(x) = x^2$; $a = $ any constant.

   (c) $f(x) = \dfrac{x^2 - 4}{x - 2}$; $a = 2$.

   (d) $f(x) = \dfrac{x^2 - 4}{x - 2}$; $a = 3$.

6. See directions in Problem 5. (At a domain endpoint, use the appropriate one-sided limit.)

   (a) $f(x) = \sqrt{x}$, $a = 0$.

(b) $f(x) = |x|$; $a = 0$.

(c) $f(x) = 1 + \pi x + ex^2 + \pi ex^3$; $a = $ any constant.

7. Prove Proposition 3.4. (One way is to use Theorem 2.6, page 99, and Lemma 3.2.)

8. Show that (ii) implies (i) in Lemma 3.2.

9. We defined several variations on the basic limit theme in Definition 3.5. Here we explore two more members of the family.

   (a) Give a precise definition for the expression $\lim_{x \to -\infty} f(x) = L$.

   (b) Give a precise definition for $\lim_{x \to 0^-} g(x) = L$.

   (c) Use the preceding parts to show that $\lim_{x \to -\infty} f(x) = L$ if and only if $\lim_{x \to 0^-} f(1/x) = L$.

10. This problem explores limit properties of the function $f(x) = \sqrt{x}$.

    (a) Let $a > 0$. Show $\lim_{x \to a} f(x) = f(a) = \sqrt{a}$. Use an ϵ–δ proof; the inequality

    $$\left| \sqrt{x} - \sqrt{a} \right| = \frac{|x - a|}{\sqrt{x} + \sqrt{a}} \leq \frac{|x - a|}{\sqrt{a}}$$

    may help.

    (b) Show $\lim_{x \to 0^+} f(x) = f(0) = 0$. Use an ϵ–δ proof.

    (c) We claimed in Example 9 that $\sqrt{1 + 2/x} \to 1$ (or, equivalently, that $\sqrt{1 + 2/x} - 1 \to 0$) as $x \to \infty$. Deduce this by showing that the squeezing inequality $0 < \sqrt{1 + 2/x} - 1 < 2/x$ holds for all $x > 0$.

11. Use a computer plotting utility in this problem, which explores the graphical interpretation of Definition 3.1, page 149. No proofs needed. Throughout, let $f(x) = x^2 - 6x$.

    (a) Assume (it is easy to show, but don't bother) that $\lim_{x \to 1} f(x) = -5$. For $\epsilon = 0.01$, find a positive δ that works, and illustrate your answer by sketching the graph of f in a well-chosen window of half-height ϵ and half-width δ.

    (b) It is a fact that $\lim_{x \to 100} f(x) = 9400$. For $\epsilon = 0.01$, find a positive δ that works, and illustrate your answer by sketching the graph of f in a well-chosen window of half-height ϵ and half-width δ.

    (c) It is true that $\lim_{x \to 10} f(x) = 40$. Set $\epsilon = 0.01$. Does $\delta = 0.001$ work in this case? Sketch the graph of f in an appropriate window to illustrate your answer.

12. Suppose that $f(x) = 0$ if $x \neq 0$ and $f(0) = 42$. Explain carefully why $\lim_{x \to a} f(x) = 0$ for all real numbers a.

13. Suppose that $g(x) = 0$ if $x \notin \mathbb{Z}$ and $g(n) = 42$ if $x \in \mathbb{Z}$. Explain carefully why $\lim_{x \to a} g(x) = 0$ for all real numbers a.

14. Suppose that $h(x) = 42$ if $x = 1, 1/2, 1/3, \ldots$ and $h(x) = 0$ otherwise. What can be said about $\lim_{x \to a} h(x)$? Why?

15. Let S be a finite set of real numbers. Suppose that $j(s) = 42$ if $s \in S$ and $j(x) = 0$ for all other real x. Show that $\lim_{x \to a} j(x) = 0$ for all real numbers a.

16. Explain why the function g in Example 8, page 158, satisfies the squeezing inequality $-|2x| \leq g(x) \leq |2x|$. Use this to give another proof that $\lim_{x \to 0} g(x) = 0$.

17. In Example 6, we used the squeezing inequality

$$\cos \theta \leq \frac{\sin \theta}{\theta} \leq \frac{1}{\cos \theta}.$$

(We said x, not θ.) Show as follows that this holds for nonzero θ in $(-\pi/2, \pi/2)$.

(a) Explain why it is good enough to show this for $\theta \in (0, \pi/2)$.

(b) Give a *geometric* proof of the (equivalent) inequality $\sin \theta \cos \theta \leq \theta \leq \tan \theta$. (Hint: Draw the angle θ into the first quadrant of the unit circle in the usual way. Then find right triangles whose *areas* represent the left- and right-hand quantities above. What area does θ represent?)

3.2   Continuous Functions

In everyday use the word "continuous" means something like "unbroken" or "without gaps." In elementary calculus, for instance, a continuous function may be thought of as one whose graph can be drawn without lifting the pencil. Such views can be useful, but they need sharpening for mathematical purposes. Pen-

Even sharp ones.

cils are not mathematical objects; to get started we need to describe continuity in mathematical language. Function limits are the key.

Definition 3.7 (Continuity of a function at a point). Let the function f be defined for all inputs x in an open interval I containing a. We say f is *continuous* at $x = a$ if, for every $\epsilon > 0$, there exists $\delta > 0$ such that

$$|f(x) - f(a)| < \epsilon \quad \text{whenever } x \in I \text{ and} \quad |x - a| < \delta.$$

Otherwise, f is *discontinuous* at $x = a$.

Observe:

- *In terms of limits:* All those ϵ's and δ's suggest a lurking limit. Sure enough, the definition boils down to this:

$$\lim_{x \to a} f(x) = f(a).$$

In words: The *value* $f(a)$ is also the *limit* of $f(x)$ as x approaches a.

- *No domain gap allowed:* For f to be continuous at a it is necessary—but not sufficient—that $\lim_{x \to a} f(x)$ exist. Two additional requirements apply: (i) the value $f(a)$ must exist; (ii) $f(a)$ must *be* the limit.

- *What if a is an endpoint?* Is the function $f(x) = \sqrt{x}$ continuous at $x = 0$? The answer will turn out to be yes—but note that the definition above doesn't say either way, because f is not defined on any *open* interval containing $x = 0$. Later we will treat the case where $x = a$ is an endpoint of a function's domain; right now we will stick to the basic version.

- *No surprises:* Continuity of f at a means, roughly speaking, that f "springs no surprises" at a: The value $f(a)$ can be inferred from values $f(x)$ for x *near a*. In terms of ϵ and δ, the condition is that $f(x)$ stays within ϵ of $f(a)$ as long as x stays within δ of a.

- *Continuity along sequences:* Continuous functions play well with sequences. If f is a function and $\{x_n\}$ a sequence, both defined on an interval I with $a \in I$, and f is continuous at $x = a$, then

$$x_n \to a \implies f(x_n) \to f(a).$$

This fact should sound reasonable in light of the "no surprises" property just discussed. A formal proof can be based on Lemma 3.2, page 152.

Let's see some examples and non-examples.

EXAMPLE 1. Where is the function $f(x) = \dfrac{x^2 - 4x - 5}{x^2 - 25}$ continuous? Can any discontinuities be "fixed"?

SOLUTION. The live question for any given a is whether

$$\lim_{x \to a} f(x) = \lim_{x \to a} \frac{x^2 - 4x - 5}{x^2 - 25} = \frac{a^2 - 4a - 5}{a^2 - 25} = f(a).$$

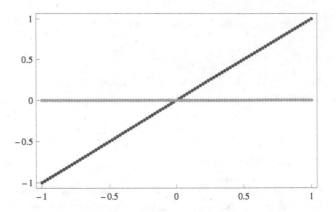

Figure 3.3. A function continuous at only one point.

Summarized in Theorem 3.3,
page 154.

Thanks to algebraic properties of limits the answer is yes—except perhaps at $a = \pm 5$, where the denominator vanishes. So f *is* continuous at all $a \neq \pm 5$.

What if $a = \pm 5$? The short answer is that f is discontinuous at both points, because both $f(5)$ and $f(-5)$ are undefined. On the other hand, factoring gives

$$f(x) = \frac{(x-5)(x+1)}{(x-5)(x+5)} = \frac{x+1}{x+5}$$

for $x \neq \pm 5$, from which it is easy to see (or prove, but we won't bother) that

$$\lim_{x \to 5} f(x) = \frac{6}{10}, \quad \text{but} \quad \lim_{x \to -5} f(x) \text{ does not exist.}$$

Plot *f* to see for yourself; a gap might appear at *x* = 5, depending on the technology.

This means, in turn, that we can *make* f continuous at $a = 5$ by defining $f(5) = 3/5$, but that the discontinuity at $a = -5$ is there to stay. A look at the graph of f suggests the same conclusion.   ◇

EXAMPLE 2.   Consider the functions

$$\operatorname{sign}(x) = \begin{cases} -1 & \text{if } x < 0, \\ 0 & \text{if } x = 0, \\ 1 & \text{if } x > 0, \end{cases} \quad \text{and} \quad g(x) = \begin{cases} x & \text{if } x \in \mathbb{Q}, \\ 0 & \text{if } x \notin \mathbb{Q}. \end{cases}$$

Figure 3.3 hints at the graph of g.

Sketch your own.

SOLUTION.   As its graph suggests, the sign function is *discontinuous* at $a = 0$, because $\lim_{x \to 0} \operatorname{sign}(x)$ does not exist. But the sign function is continuous at all $a \neq 0$, because (as the graph also suggests) $\lim_{x \to a} \operatorname{sign}(x) = \pm 1 = \operatorname{sign}(a)$. Formal proofs are straightforward; see the exercises.

The function g is stranger, and its graph is much weirder than Figure 3.3 suggests. For example, neither "line" contains any unbroken segment, but each "line" contains infinitely many points, so densely packed that between any two points lie infinitely many more.

Bizarre as g seems, it is not hard to show that g is continuous at $a = 0$ but discontinuous elsewhere. To show continuity at $a = 0$, we need to check that $\lim_{x \to 0} g(x) = g(0) = 0$. This follows from the squeezing inequality

$$-|x| \leq g(x) \leq |x|.$$

Since the left- and right-hand quantities clearly tend to zero, so must $g(x)$.

For $a \neq 0$, on the other hand, $\lim_{x \to a} g(x)$ does not exist. One way to prove this is to consider two sequences $\{r_n\}$ and $\{p_n\}$, each converging to a, with $r_n \in \mathbb{Q}$ and $p_n \notin \mathbb{Q}$ for all n. Then we have   *Such sequences do exist.*

$$g(r_n) = r_n \to a \quad \text{while} \quad g(p_n) = 0 \to 0;$$

thus, $\lim_{x \to a} g(x)$ does not exist, and so g is discontinuous. $\Diamond$

Continuity at domain endpoints: use one-sided limits. The domain of $f(x) = \sqrt{x}$ includes the point 0, but nothing to the *left* of 0, and so the basic definition can't apply. Still, it seems reasonable to call f continuous at 0, because $f(x) = \sqrt{x} \approx f(0) = 0$ for all *positive* x near 0. The following definition offers a natural way out of this mild impasse.

Definition 3.8 (Continuity at a left endpoint). Suppose that f is defined on an interval $I = [a, b)$, but not on any open interval containing a. We say f is *continuous at* $x = a$ if, for every $\epsilon > 0$, there exists a number $\delta > 0$ such that

$$|f(x) - f(a)| < \epsilon \quad \text{whenever } x \in I \text{ and } \quad a \leq x < a + \delta.$$

Equivalently, f is continuous at $x = a$ if $\lim_{x \to a^+} f(x) = f(a)$.

The idea for continuity at a *right* endpoint is essentially the same. In practice,   *We leave it to you.* endpoint continuity comes up less often than the standard version—after all, an interval has only two endpoints, but infinitely many points in between.

EXAMPLE 3.   Check that the function $f(x) = \sqrt{x}$ is continuous at $x = 0$.

SOLUTION.   We only need to show that $\lim_{x \to 0^+} \sqrt{x} = \sqrt{0} = 0$. Doing so is straightforward. For any $\epsilon > 0$, the choice $\delta = \epsilon^2$ works:

$$\text{if} \quad 0 < x \leq \delta, \quad \text{then} \quad 0 < \sqrt{x} \leq \sqrt{\delta} = \epsilon,$$

as desired. $\Diamond$

Continuity on a set. As Examples 1 and 2 illustrate, a function may be continuous at some points of its domain, and discontinuous elsewhere. Here is the pertinent definition:

Definition 3.9 (Continuity on a set). A function f is continuous on a set S if f is continuous at each point $a \in S$.

We showed in Example 2, for instance, that the strange function g is continuous on the set $\{0\}$, but not on any larger set. The sign function, by contrast, is continuous on the complementary set $\mathbb{R} \setminus \{0\}$, and the function f in Example 1 is continuous on $\mathbb{R} \setminus \{-5, 5\}$. The function $f(x) = \sqrt{x}$ in Example 3 is continuous on $[0, \infty)$. It is possible to concoct terribly-behaved functions, continuous on small or strange sets, but continuity is the rule for the familiar functions studied in calculus.

New Continuous Functions from Old

As we've already done with limits, sequences, and other objects, we can combine "old" functions in various ways to produce new ones. We will show, next, that combining *continuous* ingredients usually produces continuous results.

Elementary functions. The standard functions of calculus have formulas, such as

$$\exp(x^2 + 5x + \sin x), \quad \sin\left(\frac{\cos x}{x}\right), \quad \text{and} \quad \arctan\left(\ln\left(\frac{\cos x}{3^x + 1}\right)\right).$$

Such functions—even the nasty ones—are called *elementary*, because they're built up, using composition and algebra, from simpler basic "elements," like x^2, $\sin x$, and $\ln x$.

Elementary functions come in families. Here are several:

- *polynomials:* functions of the form $p(x) = a_0 + a_1 x + a_2 x^2 + \cdots + a_n x^n$, where n is any positive integer and the a_i are constants;

- *rational functions:* functions of the form $r(x) = p(x)/q(x)$, where p and q are polynomials;

- *exponential and trigonometric functions:* $\exp x$, $\sin x$, $\cos x$, $\tan x$, $\sec x$, $\csc x$, $\cot x$.

- *inverse functions:* logs, inverse trigonometric functions, roots, etc., such as $\ln x$, $\arctan x$, $\sqrt{x}$;

- *hybrids:* functions formed from those above by composition and algebraic combinations.

Are elementary functions continuous? The short answer is yes—but only at points a where their formulas make good sense. (If a happens to be a domain endpoint, then continuity is understood in the sense of Definition 3.8.) In other words, an elementary function is continuous throughout its *natural domain*. This domain condition is often clear at a glance. The middle function above, for instance, is obviously undefined at $x = 0$, so there's no hope of continuity there. The right-hand function, by contrast, has an uglier natural domain—both the arctangent and the log might cause trouble.

Showing continuity of elementary functions, even ugly ones, on their natural domains is easy once we know that the basic functions (x^2, $1/x$, $\sin x$, e^x, $\ln x$, etc.) are all continuous on *their* domains. We address this problem briefly below—but only briefly to keep focus on our main concerns.

Proposition 3.10. *Each of the functions*

$$c(x) = k \, (any\ constant), \quad i(x) = x, \quad \sin x, \quad \arctan x, \quad e^x, \quad \ln x$$

is continuous on its natural domain.

Continuity of c and i at any input a amounts to nothing more than that $\lim_{x \to a} c = c$ and $\lim_{x \to a} x = a$; both claims are very easy to show. Proving continuity of the remaining functions rigorously takes more effort—starting with clear *definitions* of these functions. We omit details here, but see the exercises for further discussion.

Continuous combinations. Now that we have some continuous functions to combine, let's combine some, both algebraically and with composition.

Proposition 3.11 (Algebraic combinations are continuous). *Suppose that functions f and g are continuous at a point a. Then the functions*

$$f \pm g, \quad fg, \quad and \quad \frac{f}{g}$$

are all continuous at a (for f/g we require $g(a) \neq 0$).

Proposition 3.12 (Composites are continuous). *Suppose g is continuous at a and f is continuous at $b = g(a)$. Then the composite function $f \circ g(x) = f(g(x))$ is continuous at $x = a$.*

Proof (s): Proving Proposition 3.11 boils down to verifying algebraic properties of *limits*. We've done that already; see Theorem 3.3 (page 154) and its proof.

We'll prove Proposition 3.12 in good ϵ–δ style. For simplicity we'll assume that f and g are defined on *open* intervals about b and a, respectively. To this end,

The proof involves two δ's, chosen one after the other.

In other words, a and b are not domain endpoints.

let $\epsilon > 0$ be given. Since f is continuous at b, there exists $\delta_1 > 0$ such that

$$|f(y) - f(b)| < \epsilon \quad \text{whenever} \quad |y - b| < \delta_1.$$

Similarly, since g is continuous at a and $\delta_1 > 0$ there exists a $\delta_2 > 0$ so that

$$|g(x) - b| < \delta_1 \quad \text{whenever} \quad |x - a| < \delta_2.$$

To finish the proof, note that this δ_2 works for the original $\epsilon > 0$:

$$|x - a| < \delta_2 \implies |g(x) - g(a)| < \delta_1$$
$$\implies |f(g(x)) - f(g(a))| < \epsilon,$$

We will explore another proof, using sequences, in the exercises.

as desired.   □

Now we can put the pieces together.

Proposition 3.13 (Elementary functions are continuous). *The elementary functions are continuous at all points of their natural domains.*

About proofs. The general point is that elementary functions are combinations, of the types treated in Propositions 3.11 and 3.12, of the continuous functions discussed in Proposition 3.10. For instance, writing the polynomial $p(x) = 3x^2 + \pi x - 7$ in the form

$$3x^2 + \pi x - 7 = 3 \cdot x \cdot x + \pi \cdot x - 7$$

shows explicitly that p is built by multiplication and addition from constant functions and the function $i(x)$, both of which are continuous according to Proposition 3.10.

Continuity of all six basic trigonometric functions on their domains follows from continuity of the sine function and such identities as

$$\cos x = \sin\left(x + \pi/2\right); \quad \tan x = \frac{\sin x}{\cos x}; \quad \sec x = \frac{1}{\cos x}.$$

In a similar way, the functions

$$r(x) = \frac{3x^2 + 5x - 7}{x^2 - 1} \quad \text{and} \quad g(x) = \arctan\left(\ln\left(\frac{\cos x}{3^x + 1}\right)\right)$$

are built up from functions shown (or claimed) to be continuous in Proposition 3.10. A full proof for all cases would require rigorous definitions and proofs of continuity for logarithmic, inverse trigonometric, and other such functions.

Exercises

1. It's no surprise that the identity function $f(x) = x$ is continuous at every domain point $x = a$. Show this carefully using Definition 3.7, page 162.

2. It's no surprise that the constant function $f(x) = 42$ is continuous at every domain point $x = a$. Show this carefully using Definition 3.7, page 162.

3. This problem explores the ϵ–δ definition of continuity for the function $f(x) = 345x + 678$.

   (a) Let $a = 3$ in the notation of Definition 3.7, page 162, and set $\epsilon = 0.001$. Find a value of δ, as large as possible, that works with this ϵ.

   (b) Let $a = 3$, as above, and fix any $\epsilon > 0$. Find a value of δ, as large as possible, that works with this ϵ. Conclude that f is continuous at $x = 3$.

   (c) Let a be any number and $\epsilon > 0$ any positive number. Find a value of δ that works in this situation. Conclude that f is continuous at all points $a \in \mathbb{R}$.

4. This problem explores roles of ϵ and δ in Definition 3.1, page 149.

   (a) If $u = 3$ and $f(x) = x^2$, then (assume this—it's true) $\lim_{x \to 3} f(x) = 9$. Let $\epsilon = 1$. Explain carefully why $\delta = 0.16$ works in the definition. (Hint: Look at the domain interval $(2.84, 3.16)$.)

   (b) Explain why $\delta = 0.17$ does *not* work in the situation of the preceding part.

   (c) If $a = 10$ and $f(x) = x^2$, then $\lim_{x \to 10} f(x) = 100$. Again let $\epsilon = 1$. Does $\delta = 0.16$ work now? If not, find a value of δ (as big as possible) that does work.

   (d) If $a = 10$ and $f(x) = x^2$, then (again) $\lim_{x \to 10} f(x) = 100$. Find a value of δ (the bigger the better) that works for $\epsilon = 0.1$.

   (e) If $a = 0$ and $f(x) = x^2$, then $\lim_{x \to 0} f(x) = 0$. Find a value of δ (as big as possible) that works for $\epsilon = 0.01$.

   (f) If $a = 1$ and $g(x) = x^{10}$, then $\lim_{x \to 1} g(x) = 1$. Find a value of δ (as big as possible) that works for $\epsilon = 1$.

5. Suppose in both parts following that f is continuous at $x = c$ and g is *not* continuous at $x = c$.

   (a) Show that $f + g$ is *not* continuous at $x = c$.

(b) Give examples to show that fg may or may not be continuous at $x = c$.

6. Here is the *negation* of the definition of continuity: f is *not* continuous at $x = a$ if $\exists \epsilon > 0$ such that $\forall \delta > 0$ $\exists x$ such that $|x - a| < \delta$, but $|f(x) - f(a)| \geq \epsilon$. Use this in each part below.

   (a) Show that $g(x) = \begin{cases} 1 + x & \text{if } x \leq 2 \\ 4 & \text{if } x > 2 \end{cases}$ is *not* continuous at $x = 2$.

   (b) Show that $f(x) = \begin{cases} x & \text{if } x \leq 2 \\ 5 - x & \text{if } x > 2 \end{cases}$ is *not* continuous at $x = 2$.

7. Let A, B, and c be any real numbers—positive, negative, or zero. Use Definition 3.7 to prove the unsurprising fact that $f(x) = Ax + B$ is continuous at $x = c$.

8. Use Definition 3.7 to show that $f(x) = |x|$ is continuous at every domain point $a \in \mathbb{R}$.

9. Suppose that a function f is continuous at $x = 42$. In each part following, use ϵ and δ to show that the new function (concocted from f) is also continuous at $x = 42$.

   (a) $g(x) = f(x)/345$
   (b) $h(x) = f(x) + 345$
   (c) $k(x) = |f(x)|$

10. Suppose $\{x_n\}$ is a sequence such that $x_n \to 0$. Explain using results of this section why $\cos(x_n) \sin(x_n)) \to 0$.

11. Suppose $\{x_n\}$ is a sequence such that $x_n \to 0$. Explain using results of this section why $\cos(\sin(x_n)) \to 1$.

12. We said in Example 2, page 164, that the sign function is discontinuous at $a = 0$ but continuous everywhere else. Prove this in two parts:

    (a) $\lim_{x \to 0} \text{sign}(x)$ does not exist;
    (b) $\lim_{x \to a} \text{sign}(x) = \text{sign}(a)$ if $a \neq 0$.

13. Suppose $f : \mathbb{R} \to \mathbb{R}$ is continuous at $x = 0$ and $f(1/n) > 0$ for all n. Show that $f(0) \geq 0$.

14. Suppose $f(3) = 5$ and f is continuous at 3. Show that there exists $\delta > 0$ such that $4.999 < f(x) < 5.001$ for $x \in (3 - \delta, 3 + \delta)$.

15. Suppose $f(3) > 5$ and f is continuous at 3. Show that there exists $\delta > 0$ such that $f(x) > 5$ for $x \in (3 - \delta, 3 + \delta)$.

16. Let f be the function whose graph consists of the two line segments joining $(0,0)$, $(1,1)$, and $(2,-1)$. Show that f is continuous at $x = 0$ and at $x = 1$. (In fact, f is continuous on all of $[0, 2]$.)

17. A function f has a *removable discontinuity* at a point $x = a$ if (i) f is discontinuous at a; but (ii) we can define (or redefine) $f(a)$ so that f becomes continuous at $x = a$. In each part, decide whether the function has a removable discontinuity at the given point. If so, explain how to remove it. If not, why not?

  (a) $f(x) = \dfrac{1}{x}$; $a = 0$.

  (b) $f(x) = \dfrac{x^2 - 4}{x + 2}$; $a = -2$.

  (c) $f(x) = x \sin \dfrac{1}{x}$; $a = 0$.

18. If $f : I \to R$ and $g : I \to R$ are any functions defined on I, then we can form new functions $\max\{f, g\}$ and $\min\{f, g\}$ in the "obvious" way: $\max\{f, g\}(x) = \max\{f(x), g(x)\}$ for all $x \in I$, and similarly for $\min\{f, g\}$.

  (a) Sketch graphs of $\max\{f, g\}$ and $\min\{f, g\}$ if $f(x) = \sin x$ and $g(x) = \cos x$ on $I = [-2\pi, 2\pi]$.

  (b) Show that
  $$\max\{f(x), g(x)\} = \frac{|f(x) - g(x)| + f(x) + g(x)}{2}.$$
  Find a similar "formula" for $\min\{f(x), g(x)\}$.

  (c) Show that if f and g are continuous on I, then so are $\max\{f, g\}$ and $\min\{f, g\}$. (Hint: Use (b) and the fact that $a(x) = |x|$ is continuous.) $f(x) = |x|$ is continuous at every domain point $a \in \mathbb{R}$.

19. Suppose that f is continuous on $\mathbb{R}$ and that $f(x) = 0$ if x is rational. Show that $f(x) = 0$ for *all* x.

20. Let I be an open interval containing zero, and $f : I \to \mathbb{R}$ any function that is bounded on I. Define a new function $g : I \to \mathbb{R}$ by $g(x) = x f(x)$.

  (a) Show that g is continuous at $x = 0$.

  (b) Suppose $a \neq 0$. Show that g is continuous at $x = a$ if and only if f is continuous at $x = a$.

3.3 Why Continuity Matters: Value Theorems

In the preceding section we defined continuity of functions, and investigated many examples and a few non-examples. In this section we explore some good-behavior properties of continuous functions. Most notable are two famous theorems on "intermediate" and "extreme" values of continuous functions.

Questions of values. Continuity is a form of good behavior, so it's natural to wonder what other pleasant properties continuous functions might have. Some such properties concern *output values* of a function f. We might reasonably hope, for instance, that if $f(1) = -7$ and $f(2) = 42$, then for any number v *between* -7 and 42 (the value v is "intermediate" between $f(1)$ and $f(2)$) there should be some input c between 1 and 2 with $f(c) = v$. The famous *intermediate value theorem* (Theorem 3.16 in this section) guarantees that this happy outcome does indeed occur, provided that f is continuous on the entire interval $[1, 2]$. In short, a continuous function f "assumes all intermediate values" between $f(1)$ and $f(2)$. Put graphically, the claim sounds reasonable: If the graph of f starts at height -7 and ends at height 42, then it must pass through all "intermediate" heights along the way. A rigorous proof is another matter.

Another value-related question concerns extremes: Must the function f above achieve smallest and largest values somewhere on the interval $[1, 2]$? If f happens to be increasing, there is no question: $f(1) = -7$ and $f(2) = 42$ are clearly the desired "extreme values." But arbitrary continuous functions can be much choppier; must *every* one achieve maximum and minimum values? The famous *extreme value theorem* (Theorem 3.17 in this section) says yes—again provided that f is continuous on the entire interval $[1, 2]$.

Properties of Continuous Functions

Our proofs of the big "value theorems" depend on some basic properties of continuous functions. These properties are of independent interest, offering new insight into what continuity means.

Sticky inequalities. Our first such property says, roughly, that if a continuous function satisfies a strict inequality *at* a point, then the same inequality holds ("persists") for inputs *near* that point.

We made up the fancy name.

Proposition 3.14 (The Principle of Persistent Inequalities (PoPI)). *Let f be continuous at c and defined on an interval I containing c; let K be any constant. If $f(c) < K$ then there is some $\delta > 0$ such that*

$$f(x) < K \quad \text{for all} \quad x \in (c - \delta, c + \delta).$$

(A similar PoPI holds if $f(c) > K$.)

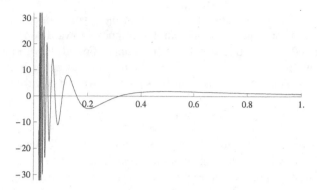

Figure 3.4. The function $h(x) = \sin(1/x)/x$; badly unbounded near $x = 0$.

Proof: For simplicity, we handle the case where c is not an endpoint of I. The case where c is an endpoint is similar.

Supposing, then, that $f(c) < K$, we set $\epsilon = K - f(c) > 0$, and choose $\delta > 0$ according to the definition of continuity of f at c. This δ satisfies our present claim, because if $|x - c| < \delta$, then

$$|f(x) - f(c)| < \epsilon = K - f(c).$$

In particular, we have

$$f(x) - f(c) < K - f(c) \implies f(x) < K,$$

as desired. $\square$

Boundedness. What, if anything, does continuity have to do with *boundedness* of a function f on an interval I? It is clear from easy examples that a continuous function f need not be bounded on I, even if I itself is bounded. For instance, the functions

$$f(x) = \frac{1}{x}, \quad g(x) = \frac{1}{x} + \frac{1}{x-1}, \quad \text{and} \quad h(x) = \frac{\sin(1/x)}{x}$$

are all continuous on $I = (0, 1)$, but f is unbounded above on I, g is unbounded above *and* below, and h behaves even worse. As Figure 3.4 suggests, h is unbounded both above and below on *every* interval $(0, \epsilon)$. But there's good news about continuous functions on *closed* intervals, and the proof involves an old friend.

Proposition 3.15. *Let f be continuous on the closed and bounded interval $[a, b]$. Then f is bounded on $[a, b]$; that is, there exist numbers m and M such that*

$$m \le f(x) \le M \quad \text{for all } x \in [a, b].$$

Proof: We will show that any function f with the given properties is bounded *above*; the proof for lower bounds is similar. Suppose, toward contradiction, that f is unbounded above. Because 1 is *not* an upper bound, there must be some $x_1 \in [a, b]$ with $f(x_1) > 1$. Because two is also not an upper bound, there is $x_2 \in [a, b]$ with $f(x_2) > 2$. Continuing this process, we can construct a sequence

$$x_1, x_2, x_3, \ldots \quad \text{with} \quad f(x_n) > n \quad \text{for all } n.$$

Below by *a* and above by *b*.

Now the sequence $\{x_n\}$ is bounded and so, by the Bolzano–Weierstrass theorem, has a subsequence $\{x_{n_k}\}$ that converges to some limit $x_0 \in [a, b]$.

Here comes our contradiction. Because f is continuous at x_0 and $x_{n_k} \to x_0$, we must have $f(x_{n_k}) \to f(x_0)$, too. This is impossible: For all k we have

$$f(x_{n_k}) > n_k \geq k.$$

This means that the sequence $\{f(x_{n_k})\}$ is unbounded, and hence divergent. $\square$

Proposition 3.15 applies, by the way, to the three functions f, g, and h discussed above—as long as we restrict domains to *closed* intervals, such as $[0.002, 0.9998]$, on which all three functions *are* continuous. (On this interval, $-500 < h(x) < 500$, for instance.)

Two Big Theorems

With Propositions 3.14 and 3.15 at the ready we can prove our two main theorems.

Theorem 3.16 (Intermediate value theorem (IVT)). *Let f be continuous on $[a, b]$, with $f(a) \neq f(b)$; let v be any number between $f(a)$ and $f(b)$ (i.e., $f(a) < v < f(b)$ or $f(b) < v < f(a)$). Then there exists c in (a, b) with $f(c) = v$.*

Proof: We discuss the case $f(a) < v < f(b)$; the argument for the case $f(b) < v < f(a)$ is similar.

$a \in S$, for instance.

By *b*, for instance.

Consider the set $S = \{x \in [a, b] \mid f(x) < v\}$. Because S is nonempty and bounded above, it has a least upper bound—say, c—somewhere in $[a, b]$. We'll show that c has all the properties claimed in the theorem.

Our first claim is that $f(c) \leq v$. This is trivial if $c \in S$. If $c \notin S$, then there is a sequence $\{s_n\}$ of members of S with $s_n \to c$. Because f is continuous at c we must have $f(s_n) \to f(c)$, and since $f(s_n) < v$ for all n it follows that $f(c) \leq v$. We see, too, that $c \neq b$, since we know $f(b) > v$.

That's Proposition 3.14, page 172.

To complete the proof, we show that $f(c) < v$ is impossible; the PoPI is the key. Indeed, if $f(c) < v$ holds, then the PoPI says that $f(x) < v$ must also hold for x in some small interval $(c - \delta, c + \delta)$. This is absurd—we chose c so that $f(x) \geq v$ for all $x > c$. Thus $f(c) = v$, as desired. $\square$

Theorem 3.17 (Extreme value theorem (EVT)). *If f is continuous on $[a, b]$, then f assumes both a maximum and a minimum value on $[a, b]$. That is, there exist $x_{\min}$ and $x_{\max}$ in $[a, b]$ such that*

$$f(x_{\min}) \le f(x) \le f(x_{\max}) \quad \text{for all } x \in [a, b].$$

Notice especially the important words "assumes," "maximum," and "minimum": The theorem guarantees that f *achieves*, not just *approaches*, biggest and smallest values on $[a, b]$.

Proof: Consider the output set

$$R = \text{range}(f) = \{f(x) \mid x \in [a, b]\}.$$

We defined "range" in Section 1.4.

Proposition 3.15 says that R is a bounded set of real numbers. Since R is also nonempty, the completeness axiom guarantees the existence of an infimum α and a supremum β. To finish the proof, we need only show that α and β are *members* of the range. With help—again—from Bolzano and Weierstrass we'll handle the case for β.

The case for α is almost identical.

Recall first that, since $\beta = \sup(R)$, there is a sequence $\{y_n\}$ contained in R with $y_n \to \beta$. (If $\beta \in R$, there is no harm done—we can just take $y_n = \beta$ for all n.) Now for each y_n there is at least one x_n in $[a, b]$ with $f(x_n) = y_n$. Choosing any one of these for each n produces a new sequence $\{x_n\}$ such that, for all n,

$$a \le x_n \le b \quad \text{and} \quad f(x_n) = y_n.$$

The sequence $\{x_n\}$ itself need not converge, but the Bolzano–Weierstrass theorem guarantees that some subsequence $\{x_{n_k}\}$ converges, to a limit we'll call $x_{\max}$; clearly, $x_{\max} \in [a, b]$. Since f is continuous at $x_{\max}$ and $x_{n_k} \to x_{\max}$, we must have

$$f(x_{\max}) = \lim_{k \to \infty} f(x_{n_k}) = \lim_{k \to \infty} y_{n_k} = \lim_{n \to \infty} y_n = \beta,$$

as claimed. $\square$

Bad values. Both the IVT and the EVT may fail, of course, if important hypotheses are violated. For example, the function $f(x) = 1/x$ is continuous on the *open* interval $I = (0, 1)$, but assumes neither a maximum nor a minimum value (for different reasons) on I. And the sign function, defined but *discontinuous* on the closed interval $[-10, 10]$, assumes *only one* intermediate value between $f(-10) = -1$ and $f(10) = 1$. (On the other hand, the sign function *does* assume maximum and minimum values.)

See Example 2, page 164.

Which one?

Using the Value Theorems

The intermediate and extreme value theorems (IVT and EVT) are essential and versatile tools for studying continuous functions. We illustrate with several brief examples.

Continuous functions map intervals to intervals. If $f : I \to \mathbb{R}$ is any function whose domain is an interval, then the *range* of f can be almost any subset of $\mathbb{R}$. For a constant function, for example, the range is just one point, while the sine function has range $[-1, 1]$, and the range of the sign function is the three-point set $\{-1, 0, 1\}$. For *continuous* functions, the picture is simpler.

Don't confuse the sine and sign functions.

Proposition 3.18. *If I is an interval and $f : I \to \mathbb{R}$ is continuous, then the range of f is either a single point (if f is constant) or an interval.*

The range might be an infinite interval, like $\mathbb{R}$.

Proof: Let J denote the range. Recall what it means to be an interval: if $u \in J$ and $w \in J$ (we might as well assume $u < w$) and v is any number such that $u < v < w$, then we need to show that $v \in J$, too. This follows immediately from the IVT. We know that $u = f(a)$ and $w = f(b)$ for some a and b in I, that f is continuous on $[a, b]$, and that v is "intermediate" between u and w, and so the IVT assures that $f(c) = v$ for some c between a and b, as we needed to show. $\square$

"Root" seems less ambiguous; we'll use it.

Odd-degree polynomials have real roots. A number r is a *root* (or a *zero*) of a function f if $f(r) = 0$. Thus, $p(x) = x^2 - 1$ has roots ± 1, $q(x) = x^2 + 1$ has *no* real roots, and

$$s(x) = x^5 - 15x^4 + 85x^3 - 225x^2 + 274x - 120$$
$$= (x - 1)(x - 2)(x - 3)(x - 4)(x - 5)$$

has roots $1, 2, 3, 4$, and 5. Deciding whether an arbitrary function has any roots, let alone finding them, can be challenging, but the IVT offers some encouragement for a large class of polynomials.

Proposition 3.19. *Every polynomial function*

$$p(x) = a_n x^n + a_{n-1} x^{n-1} + \cdots + a_1 x + a_0$$

of odd degree ($a_n \neq 0$ and n is odd) has at least one real root.

About the proof. Every polynomial function is continuous everywhere. If we can find any a and b with $p(a) < 0$ and $p(b) > 0$, then the IVT will guarantee that $p(r) = 0$ for some root r between a and b. Finding such a and b for an *odd*-degree polynomial $p(x)$ is indeed possible, because

$$\lim_{x \to \infty} p(x) = \pm\infty, \quad \text{while} \quad \lim_{x \to -\infty} p(x) = \mp\infty,$$

and so $p(x)$ must assume both positive and negative values. If, say, $p(x) = x^3 - 7x^2 + 5x + 3$, then $p(0) = 3$ while $p(-1) = -13$, so p must have a root in the interval $(-1, 0)$. We leave a more formal proof to the exercises.

Plotting p.

Fixed points. A number a is a *fixed point* of a function f if $f(a) = a$. (Equivalently, a is a root of the new function $f(x) - x$.) As with roots, the IVT can help assure that fixed points exist.

EXAMPLE 1.   Show that if $f : [0, 1] \to [0, 1]$ is continuous on $[0, 1]$, then f has a fixed point.

SOLUTION.   Note a crucial hypothesis about the codomain: $0 \le f(x) \le 1$ for all $x \in [0, 1]$. This means, geometrically, that the graph of f stays inside the square window $[0, 1] \times [0, 1]$ in the xy-plane. Our problem amounts to showing that this graph touches or crosses the line $y = x$ at least once. This may *seem* obvious, but we want proof, and the IVT will help.

See it in the notation?

Sketch for yourself.

The trick is to look at the new function $g(x) = f(x) - x$; note that g is also continuous on $[0, 1]$. Note also that

$$\text{(i)}\quad g(0) = f(0) \ge 0; \qquad \text{(ii)}\quad g(1) = f(1) - 1 \le 0.$$

If either $g(0) = 0$ or $g(1) = 0$, we're done already, so we may as well assume that both (i) and (ii) are *strict* inequalities. In this case, the IVT guarantees that $g(x) = 0$ holds for some $x \in (0, 1)$, as we aimed to show. $\diamond$

Maximum-minimum problems. Many hoary old problems of elementary calculus—enclose as much area as possible in a rectangular pigpen, one side up against the barn, with 100 yards of fence—depend implicitly on the EVT. The functions to be maximized or minimized are typically continuous and defined on closed intervals, and so the EVT guarantees that the desired maximum or minimum exists. Sometimes the functions and intervals in question are lightly disguised.

EXAMPLE 2.   Suppose a and b are nonnegative numbers with $a + b = 3$. Is there a maximum possible value of ab^2? Of b/a? If so, what are these values? How is the EVT involved?

SOLUTION.   If we write $b = 3 - a$, then it is natural to consider the functions

$$p(a) = ab^2 = a(3 - a)^2 \quad \text{and} \quad q(a) = \frac{b}{a} = \frac{3 - a}{a} = \frac{3}{a} - 1;$$

our question is whether these functions achieve maximum values for $a \in [0, 3]$. It is clear at a glance that $q(a) \to \infty$ as $a \to 0^+$, so q is *not* continuous on $[0, 3]$, and all EVT bets are off. The polynomial function p, by contrast, *is* continuous on $[0, 3]$, and so the EVT guarantees that a maximum exists—but gives no clue how to find it. With calculus methods (taboo for now, as we haven't gotten to derivatives) it is easy to show that the maximum value, four, is attained when $a = 1$. $\diamond$

At the formula or at a graph.

See the exercises for a non-calculus strategy.

Continuous inverses. Important calculus functions often come in "inverse pairs," like

$$f(x) = e^x \quad \text{and} \quad g(x) = \ln x;$$
$$f(x) = x^2 \quad \text{and} \quad g(x) = \sqrt{x};$$
$$f(x) = \tan x \quad \text{and} \quad g(x) = \arctan x.$$

With due care taken for domains, the inverse relationship means that $f(a) = b$ if and only if $g(b) = a$. For the last pair, for instance, we have $f(1.2) = \tan 1.2 \approx 2.57215$, while $g(2.57215) = \arctan(2.57215) \approx 1.2$.

It is natural to hope in this situation that if either of f and g is continuous, then so is its "partner." (Knowing this might be practically useful—sometimes it is easier to show directly that one rather than the other is continuous.) The following proposition addresses this situation.

Proposition 3.20. *Let $f : I \to J$ and $g : J \to I$ be inverse functions, where I and J are open intervals. If f is continuous on I, then g is continuous on J.*

Proof: Since f has an inverse, it must be one-to-one on I, and this leads to a useful claim:

> *If f is one-to-one and continuous on an interval I, then f is strictly monotone on I.*

Using the IVT.

The claim should seem plausible; we'll leave its proof as an exercise. Assuming the claim, we will handle the case where f is strictly increasing; the case for decreasing f is similar.

Let $b \in J$ be given. Then $g(b) = a$ for some $a \in I$; note that also $f(a) = b$. To see that g is continuous at b, let $\epsilon > 0$ be given. Since I is an open interval and $a \in I$, we may as well assume that the small *closed* interval $[a - \epsilon, a + \epsilon]$ is contained in I. Because f is increasing, we know that $f(a - \epsilon) < f(a) = b < f(a + \epsilon)$. To complete the proof, we now choose any $\delta > 0$ so that

If necessary, we can always take ϵ smaller.

$$f(a - \epsilon) < b - \delta < b + \delta < f(a + \epsilon).$$

To see that this δ works for the original ϵ, consider any y in J with

$$f(a - \epsilon) < b - \delta < y < b + \delta < f(a + \epsilon).$$

Applying the strictly increasing function g to all parts of this inequality, we get

$$(a - \epsilon) < g(y) < g(b + \delta) < (a + \epsilon),$$

which is another way of saying that $|g(y) - a| < \epsilon$. $\square$

Exercises

1. Assume in both parts of this problem that $f : [0, 2] \to \mathbb{R}$ is continuous on $[0, 2]$.

   (a) Describe (by graph or formula) such a function f where $f(0)$ is the minimum value and $f(1)$ is the maximum value of f on $[0, 2]$.

   (b) Suppose that f that attains a maximum value at $x = 1$. Show that f is *not* one-to-one.

2. Find examples of polynomials as prescribed. No formal proofs needed, but describe your answers clearly.

   (a) A polynomial of degree 5 with one root.

   (b) A polynomial of degree 4 with 3 roots.

   (c) A polynomial of degree 42 with no roots.

   (d) A polynomial of degree 42 with 41 roots.

3. This problem revisits Proposition 3.19, page 176. Throughout, let $p(x) = x^n + a_{n-1}x^{n-1} + \cdots + a_1 x + a_0$, where n is odd.

   (a) Explain why $\lim_{n \to \infty} p(x) = \infty$. (Similarly, $\lim_{n \to -\infty} p(x) = -\infty$.)

   (b) Use the definition of a limit at infinity to show that there exists $b \in \mathbb{R}$ with $p(b) > 0$. (For similar reasons, there is $a \in R$ with $p(a) < 0$.)

   (c) In practice, it is often easiest to find a and b as above by (educated!) guessing. (Plotting may also work, but that's too easy.) Use this approach and the IVT to show that $q(x) - x^5 - x^4 - x^3 - x^2 + x + 1$ has at least three roots. Can you also find the roots exactly, by algebra?

4. A function $f : \mathbb{R} \to \mathbb{R}$ achieves a maximum value at x_{max} if $f(x_{max}) \le f(x)$ for all $x \in \mathbb{R}$, and similarly for a minimum value at x_{min}.

   (a) What, if anything, does the EVT guarantee about maximum and minimum values of a continuous function $f : \mathbb{R} \to \mathbb{R}$ defined on all of $\mathbb{R}$?

   (b) *No* odd-degree polynomial achieves a maximum or a minimum value on $\mathbb{R}$. Explain why. (See an earlier problem.)

   (c) *Every* even-degree polynomial $q(x)$ achieves either a maximum or a minimum value on $\mathbb{R}$. Explain why. (Hint: This is just a little harder. Note that either (i) $q(x) \to \infty$ as $x \to \pm\infty$, or (ii) $q(x) \to -\infty$ as $x \to \pm\infty$. Show that q achieves a minimum in case (i) and a maximum in case (ii).)

(d) Can an even-degree polynomial $q(x)$ achieve *both* a maximum and a minimum value on $\mathbb{R}$? Explain.

5. This problem is about the functions $p(a) = a(3 - a)^2$ and $q(a) = \frac{3}{a} - 1$ discussed in Example 2, page 177.

   (a) Does p assume a minimum on $[0, 3]$? If so, what is it? What does the EVT say?

   (b) Does q assume a minimum on $(0, 3]$? If so, what is it? What does the EVT say?

   (c) We said, but didn't prove, in Example 2 that the maximum value of p on $[0, 3]$ is $p(1) = 4$. Prove it now, in two steps:

      (i) Factor $p(a) - 4$ (one factor is $a - 1$).

      (ii) Use your factorization to explain why $p(a) - 4 \leq 0$ when $a \in [0, 3]$.

6. Let P be an odd-degree polynomial. Show that $P(x) = \pi$ for at least one real input x. (Hint: See Proposition 3.19, page 176.)

7. (a) Find a continuous function $f : (0, 1) \to \mathbb{R}$ whose range is $[0, 1)$.

   (b) Find a continuous function $f : (0, 1) \to \mathbb{R}$ whose range is $[0, 1]$.

   (c) Is there a continuous function $f : [0, 1] \to \mathbb{R}$ whose range is $(0, 1]$)? Why or why not?

8. Does $f(x) = x^2 - 2x$ achieve maximum and/or minimum values somewhere in $(-1, 2)$? What about a supremum or infimum?

9. In each part following, either give an example or explain why none can exist.

   (a) A function $f : (0, 1) \to \mathbb{R}$, continuous on $(0, 1)$, that achieves a maximum but not a minimum on $(0, 1)$.

   (b) A function $g : [0, 1] \to \mathbb{R}$, continuous on $[0, 1]$, whose range contains π and e but not 3.

   (c) A polynomial function $p : \mathbb{R} \to \mathbb{R}$, with range $[-3, \infty)$.

   (d) A polynomial function $q : \mathbb{R} \to \mathbb{R}$, with range $(-3, \infty)$.

   (e) A function $r : \mathbb{R} \to \mathbb{R}$, continuous on $\mathbb{R}$, with range $(-3, \infty)$.

10. Suppose that $f : \mathbb{R} \to \mathbb{R}$ is continuous on all of $\mathbb{R}$, that $f(\pi) = 42$, and that $f(x) \leq 42$ for all x. It's clear (why?) that f is not onto. Show that f is also *not* one-to-one.

11. Let $f : [a, b] \to \mathbb{R}$ be continuous on $[a, b]$. Show that if f attains either its maximum or its minimum value at an interior point c, then f is *not* one-to-one.

12. Let I be an interval (open or closed) and suppose that $f : I \to \mathbb{R}$ is continuous and one-to-one. Show that f is strictly monotone.

13. Suppose that $f : \mathbb{R} \to \mathbb{R}$ is continuous and one-to-one, and that $f(1) > f(0)$. Show that $f(2) > f(1)$.

14. The Principle of Persistent Inequalities (Proposition 3.14, page 172), describes a nice property of continuous functions. Here we investigate two "one-sided" versions of the PoPI.

    A function f satisfies the PoPILTV (the "less-than" version of the PoPI) at $x = c$ if, for every number K with $f(c) < K$, there exists $\delta > 0$ such that $f(x) < K$ whenever $x \in (c - \delta, c + \delta)$. The PoPIGTV (the "greater-than" version) is the same, except that the condition $f(x) < K$ is replaced by $f(x) > K$.

    (a) Suppose f is continuous at $x = c$. The ordinary PoPI says that f satisfies the PoPILTV. Show that f also satisfies the PoPIGTV.

    (b) Suppose $g(x) = 1$ if $x > 0$ and $g(x) = 0$ otherwise. Show that g satisfies the PoPIGTV but not the PoPILTV at $x - 0$.

    (c) Suppose $h(x) = 1$ if $x = 0$ and $h(x) = 0$ otherwise. Which of the PoPIGTV and the PoPILTV does h satisfy at $x = 0$?

    (d) Show that a function $f : \mathbb{R} \to \mathbb{R}$ is continuous at $x = c$ if and only if f satisfies both the PoPIGTV and the PoPILTV at $x = c$.

15. Let $f : \mathbb{R} \to \mathbb{R}$ be a continuous function such that $f(x)$ is a rational number for every real number input x. Show that f must be a constant function.

16. Consider a function $f : \mathbb{R} \to \mathbb{R}$ and a domain point $c \in \mathbb{R}$. We say that f is *locally bounded* at $x = c$ if there is some $\delta > 0$ and some $M > 0$ such that $|f(x)| < M$ whenever $x \in (c - \delta, c + \delta)$. (In other words, f is bounded on some open interval centered at c.)

    (a) Show that $f(x) = x^2$ is locally bounded at $x = 42$. (Find specific values of δ and M; there are many possibilities.)

    (b) Show that if f is continuous at c, then f is locally bounded at c.

    (c) Find a function $f : \mathbb{R} \to \mathbb{R}$ that is locally bounded but not continuous at $x = 0$. Explain why your example works.

    (d) Find a function $f : \mathbb{R} \to \mathbb{R}$ that is *not* locally bounded at $x = 0$. Explain why your example works. (Be sure your example is *defined* for all inputs x.)

3.4 Uniform Continuity

Here's a definition with many familiar ingredients—but a new adjective:

Definition 3.21 (Uniform continuity). Let $I \subseteq \mathbb{R}$ be an interval and f a function defined on S. We say f is *uniformly continuous on* I if for each $\epsilon > 0$ there exists $\delta > 0$ such that

$$|f(x) - f(y)| < \epsilon \quad \text{whenever } x, y \in I \text{ and } \quad |x - y| < \delta.$$

What's new here? What's wrong with ordinary continuity, over which we've worked hard? We'll address these good questions right after some basic examples.

EXAMPLE 1. Both $f(x) = 3x + 5$ and $g(x) = x^2$ are continuous on $I = \mathbb{R}$. Are they also *uniformly* continuous on $\mathbb{R}$? Why or why not?

SOLUTION. The function f *is* uniformly continuous. For any inputs x and y, we have

$$|f(x) - f(y)| = |3x + 5 - 3y - 5| = 3\,|x - y|.$$

Thus, for any $\epsilon > 0$, the value $\delta = \epsilon/3$ does what Definition 3.21 asks:

$$|f(x) - f(y)| < \epsilon \quad \text{whenever } x, y \in \mathbb{R} \text{ and } \quad |x - y| < \delta = \frac{\epsilon}{3}.$$

The function g is *not* uniformly continuous. Showing this takes a little doing, but it is worthwhile to illustrate key ideas. We'll do it by setting $\epsilon = 1$ and showing that *no* positive δ works. Indeed, let any fixed $\delta > 0$ be given. First we choose an integer n with $1/n < \delta$, and set $x = n + 1/n$ and $y = n$. Then $|x - y| = 1/n < \delta$, but

$$|g(x) - g(y)| = |x^2 - y^2| = \left(n + \frac{1}{n}\right)^2 - n^2 = 2 + \frac{1}{n^2} > 2 > \epsilon.$$

We showed long ago that this is possible.

Thus, our attempted δ failed, and so would any other.   ◇

Uniform vs. ordinary continuity. As Example 1 suggests, uniform continuity is a *strictly stronger* property than ordinary continuity. A function that is *uniformly* continuous on I is also continuous on I in the ordinary sense, but the converse is false: A function (like $g(x) = x^2$ on $\mathbb{R}$) may be continuous but not uniformly continuous on I. Here are two ways to view the difference:

- *One δ-size fits all:* Ordinary continuity of f on I means that, having first specified $a \in I$ and $\epsilon > 0$, we can then choose $\delta > 0$ so that $|f(x) - f(a)| < \epsilon$ whenever $x \in I$ and $|x - a| < \delta$. In particular, δ may depend both on ϵ *and* on a.

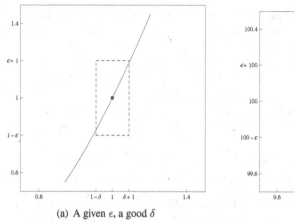 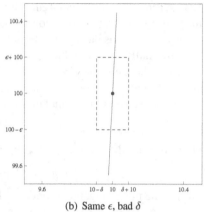

(a) A given ϵ, a good δ (b) Same ϵ, bad δ

Figure 3.5. Why $f(x) = x^2$ is not uniformly continuous on $\mathbb{R}$.

In uniform continuity, by contrast, δ may depend on ϵ but *not* on a. In this case, $|f(x) - f(y)| < \epsilon$ holds for *any* pair of inputs x and y in I. For given ϵ, in other words, a single choice of δ works "uniformly" across the interval I.

- *Window dimensions:* A function f is continuous at a if $\lim_{x \to a} f(x) = f(a)$. In Section 3.1 we described this condition graphically: For any given $\epsilon > 0$ we can choose $\delta > 0$ so that the graph of f stays inside the rectangular "window"

See Figure 3.1, page 151.

$$[a - \delta, a + \delta] \times [f(a) - \epsilon, f(a) + \epsilon]$$

all the way from left to right, never escaping out the top or bottom. (This window has dimensions $2\delta \times 2\epsilon$.) A function $f : I \to \mathbb{R}$ is continuous on I if, at every point of the graph, such a window can be found for every "half-height" ϵ, no matter how small.

To say that $f : I \to \mathbb{R}$ is *uniformly* continuous means, in this language, that for given $\epsilon > 0$ we can choose some $\delta > 0$ so that a $2\delta \times 2\epsilon$ window works at *every* point of the graph. This perspective suggests what went wrong with $g(x) = x^2$ in Example 1: as g rises more and more steeply over its domain, its graph tends more and more to escape from the top or bottom of a given window. Figure 3.5 gives some idea of the problem.

A function may be uniformly continuous on one set but not on another.

EXAMPLE 2. The function $f(x) = x^2$ is not uniformly continuous on $\mathbb{R}$. Is f uniformly continuous on $I = [-3, 42]$? Is f uniformly continuous on $(-2, 17)$?

SOLUTION. Yes, and yes—and the second yes follows immediately from the first. If *any* f is uniformly continuous on a set S, then f is automatically uniformly continuous on any smaller set $S' \subset S$.

Informally speaking, the first "yes" applies because f is "steeper" at $x = 42$ than anywhere else on $I = [-3, 42]$. Therefore, any δ that "works" for a given ϵ at $x = 42$ will also work elsewhere in I, where the graph is less steep.

Before starting the formal proof, note that for any x and y we have

$$|f(x) - f(y)| = |x^2 - y^2| = |x - y|\,|x + y|\,.$$

Now consider the last factor, $|x + y|$: For x and y in $[-3, 42]$ I we have

$$|x + y| \le |x| + |y| \le 42 + 42 = 84.$$

Putting the pieces together produces a key inequality: $|f(x) - f(y)| \le 84\,|x - y|$ for all x and y in I. This done, we're ready for our (very short!) formal proof.

Let $\epsilon > 0$ be given. Set $\delta = \epsilon/84$. This δ works, since if $x, y \in I$ and $|x - y| < \delta$, then

$$|f(x) - f(y)| = |x - y|\,|x + y| \le 84\,|x - y| < 84\delta = \epsilon,$$

as desired. $\Diamond$

This section's main theorem covers several of the preceding examples.

Theorem 3.22. *If f is continuous on a closed and bounded interval $[a, b]$, then f is uniformly continuous on $[a, b]$.*

Proof: Let $\epsilon > 0$ be given, and assume, toward contradiction, that *no* $\delta > 0$ works in the sense of Definition 3.21. Then $\delta = 1/n$ must fail for every positive integer n. This means that, for each n, there must exist x_n and y_n in $[a, b]$ such that

$$|x_n - y_n| < \frac{1}{n} \quad \text{but} \quad |f(x_n) - f(y_n)| \ge \epsilon.$$

The two sequences $\{x_n\}$ and $\{y_n\}$ so constructed are obviously bounded; applying the Bolzano–Weierstrass theorem to $\{x_n\}$ produces a subsequence $\{x_{n_k}\}$ that converges to some limit $x_0 \in [a, b]$. Now we extract the *same* subsequence $\{y_{n_k}\}$ from $\{y_n\}$; because

$$|x_{n_k} - y_{n_k}| < \frac{1}{n_k} \le \frac{1}{k}$$

for all k, we must have both $x_{n_k} \to x_0$ and $y_{n_k} \to x_0$.

Now a contradiction looms. Since f is continuous at x_0, we must have both

$$f(x_{n_k}) \to f(x_0) \quad \text{and} \quad f(y_{n_k}) \to f(x_0).$$

This is incompatible with our original arrangement that

$$|f(x_{n_k}) - f(y_{n_k})| > \epsilon$$

for all k, and so the proof is done. $\square$

Staying close. If uniform continuity seems forbiddingly technical or obscure, that's because we worked our way up to it gradually, via ordinary continuity at points and on sets. In another sense, however, uniform continuity of a function f on a set S is the cleanest and simplest version possible:

> $f(x)$ and $f(y)$ are within ϵ whenever x and y are within δ.

We will exploit this view later, when we study integrals. This basic idea—that f "preserves closeness"—translates naturally to Cauchy sequences, which are all about closeness.

Proposition 3.23. *If f is uniformly continuous on I and $\{x_n\} \subset I$ is a Cauchy sequence, then $\{f(x_n)\}$ is a Cauchy sequence, too.*

Proof: Let $\epsilon > 0$ be given. By hypothesis there is some $\delta > 0$ that works in the sense of uniform continuity of f on I. Because $\{x_n\}$ is a Cauchy sequence, we can choose N so that

Yes, we mean δ, not ϵ.

$$|x_n - x_m| < \delta \quad \text{whenever } n > m > N.$$

Because $x_n, x_m \in I$ for all n, our choice of δ implies that

$$|f(x_n) - f(x_m)| < \epsilon \quad \text{whenever } n > m > N,$$

as required.   □

EXAMPLE 3.   Let $f(x) = 1/x$. Use Proposition 3.23 to show that f is *not* uniformly continuous on $I = (0, \infty)$. Is f uniformly continuous on $[1, \infty)$?

SOLUTION.   We *could* use the definition to show directly that f is not uniformly continuous (see the exercises), but using Proposition 3.23 is shorter. We can simply observe that the sequence $\{1/n\}$ is contained in I and Cauchy, while the output sequence $\{f(x_n)\} = \{n\}$ is surely *not* Cauchy.

On the interval $[1, \infty)$, by contrast, f *is* uniformly continuous. To see this, note that for $x, y \in [1, \infty)$ we have

$$|f(x) - f(y)| = \left| \frac{1}{x} - \frac{1}{y} \right| = \left| \frac{y - x}{xy} \right| \le |y - x|.$$

This implies that, for any $\epsilon > 0$, the choice $\delta = \epsilon$ works in the definition of uniform continuity.   ◇

Staying bounded. Let $f : I \to \mathbb{R}$ be continuous on a bounded interval I. If $I = [a, b]$ is *closed*, then the EVT guarantees (among other things) that f is *bounded* on I. On an *open* interval $I = (a, b)$, by contrast, a continuous function may well be *unbounded*—as is $f(x) = 1/x$ on $(0, 1)$. But *uniformly* continuous functions behave better, as the following proposition shows. The proof involves ingredients that are familiar by now.

Proposition 3.24. *Let I be any bounded interval I, open or closed. If f is uniformly continuous on I, then f is bounded on I.*

Proof: Suppose toward contradiction that f is unbounded above. (A similar argument applies if f is unbounded below.) Then we can choose $x_1, x_2, x_3, \ldots$, all in I, such that

$$f(x_1) > 1, \quad f(x_2) > 2, \quad f(x_3) > 3, \quad \ldots.$$

By the Bolzano–Weierstrass theorem, some subsequence $\{x_{n_k}\}$ is convergent—and therefore Cauchy. Now Proposition 3.23 says that the image subsequence $\{f(x_{n_k})\}$ is also Cauchy—and therefore *bounded*. This is impossible, because we chose the x_n such that $f(x_{n_k}) > n_k > k$ for all k. $\square$

Exercises

1. Use Definition 3.21 (not theorems) in each part.

   (a) $f(x) = 5$ is uniformly continuous on $\mathbb{R}$.

   (b) $g(x) = 2x + 7$ is uniformly continuous on $\mathbb{R}$.

2. Use Definition 3.21 in each part.

   (a) Show that every constant function $f(x) = C$ is uniformly continuous on $\mathbb{R}$.

   (b) Show that every linear function $L(x) = Ax + B$ is uniformly continuous on $\mathbb{R}$. Does anything special happen if $A = 0$?

3. Show in two steps that $f(x) = 1/(x^2 + 1)$ is uniformly continuous on $\mathbb{R}$.

   (a) *Assume* in this part that, for all x and y, $|f(x) - f(y)| \leq |x - y|$. Show that, for any $\epsilon > 0$, the value $\delta = \epsilon$ works in Definition 3.21.

   (b) *Show* that, indeed, $|f(x) - f(y)| \leq |x - y|$ for all x and y. (This part involves "only" algebra—no ϵ or δ needed.)

4. Use Definition 3.21 (not theorems) in each part.

   (a) $h(x) = x^2 + 7$ is uniformly continuous on $[-100, 42]$.

(b) $l(x) = 1/x$ is not uniformly continuous on $(0, 1)$. Hint: Let $\epsilon = 1$ and show that no suitable δ can be chosen.

5. Theorem 3.22 implies that $f(x) = \dfrac{5}{x} + 3$ is uniformly continuous on the interval $[1/4, 10]$. Give an ϵ–δ proof of this fact.

6. Assume in both parts that $f : I \to \mathbb{R}$ is uniformly continuous on the interval I. Use Definition 3.21 in each part.

   (a) Show that the function $g(x) = -137f(x)$ is uniformly continuous on I.

   (b) Show that the function $h(x) = |f(x)|$ is uniformly continuous on I.

7. Throughout this problem, let $f(x) = x^2$ and $\epsilon = 1$. Since f is continuous at $c = 1$, we can find some $\delta > 0$ such that $|f(x) - f(1)| < 1$ when $|x - 1| < \delta$.

   (a) Show that $\delta = \sqrt{2} - 1 \approx 0.4142$ "works" in the situation described above.

   (b) Show that $\delta = \sqrt{101} - 10 \approx 0.0499$ works for $c = 10$. (As before, $f(x) = x^2$ and $\epsilon = 1$.)

   (c) In the spirit of the preceding parts, find values of δ that work for $c = 1$, $c = 10$, $c = 100$, and $c = 1000$. How do these values reflect the fact that f is *not* uniformly continuous on $\mathbb{R}$?

   (d) Now let $g(x) = 10x + 3$ and (again) $\epsilon = 1$. As in the preceding part, find values of δ that work for $c = 1$, $c = 10$, $c = 100$, and $c = 1000$. How do these values reflect the fact that f *is* uniformly continuous on $\mathbb{R}$?

   (e) The function f is uniformly continuous on the interval $I = [1, 1000]$. Find a value of δ that works for $\epsilon = 1$ in Definition 3.21.

8. Let I and J be any intervals, with $I \subseteq J$, and let f be a function. Show that if f is uniformly continuous on J, then f is uniformly continuous on I, too. (Note: The claim is pretty obvious; the point is to give a definition-based proof.)

9. Use Definition 3.21 to show in two steps that $f(x) = \sqrt{x}$ is uniformly continuous on $[0, \infty)$.

   (a) Show that if $0 \le y \le x$, then $\sqrt{x} - \sqrt{y} \le \sqrt{x - y}$.

   (b) Use the preceding part to finish the problem.

10. Consider the piecewise-linear function $f : \mathbb{R} \to \mathbb{R}$ with $f(x) = x$ for $x < 0$ and $f(x) = 42x$ for $x \geq 0$. Show that f is uniformly continuous on $\mathbb{R}$.

11. Let $f(x) = x^3$.

    (a) Use Definition 3.21 to show directly that f is uniformly continuous on $[-3, 42)$. (Hint: $a^3 - b^3 = (a - b)(a^2 + ab + b^2)$.)

    (b) Use Theorem 3.22 and the claim in Problem 8 to show (again) that f is uniformly continuous on $[-3, 42)$.

12. Show that if f is uniformly continuous on an interval I, then it is also continuous in the ordinary sense.

13. Assume in both parts that both f and g are uniformly continuous on an interval I.

    (a) Show that $f + g$ is uniformly continuous on I.

    (b) Give an example to show that fg need not be uniformly continuous on I (even though fg is continuous in the ordinary sense).

14. Assume in both parts that both f and g are uniformly continuous on an interval I.

    (a) Show that if $I = [a, b]$ is closed and bounded, then fg is uniformly continuous on I.

    (b) Show that if both f and g are bounded on I, then fg is uniformly continuous on I. (This is a little harder.)

15. A function $f : I \to \mathbb{R}$ is called *Lipschitz continuous* (after a nineteenth-century German mathematician) on I if, for some constant K, $|f(x) - f(y)| \leq K |x - y|$ holds for all x and y in I. Lipschitz and uniform continuity are both 'stronger" versions of ordinary continuity.

    (a) Show that if f is Lipschitz continuous on I, then it is also uniformly continuous on I.

    (b) Show that every linear function $L(x) = Ax + B$ is Lipschitz continuous on $\mathbb{R}$.

    (c) Suppose that f is Lipschitz continuous on an interval I. Show that the ratio $|f(x) - f(y)|/|x - y|$ is bounded for $x, y \in I$ and $x \neq y$.

    (d) We know (e.g., from Theorem 3.22) that $g(x) = \sqrt{x}$ is uniformly continuous on $[0, 1]$. Show that g is *not* Lipschitz continuous on $[0, 1]$. Hint: Let $x > 0$ and fix $y = 0$. Show that the ratio in the preceding part is not bounded in the desired sense.

16. In each part, either use Definition 3.21 or cite appropriate theorems or propositions.

    (a) $f(x) = x^3$ is not uniformly continuous on $\mathbb{R}$.

    (b) $g(x) = \sqrt{x}$ is uniformly continuous on $[1, \infty)$.

    (c) $h(x) = \cos(1/x)$ is not uniformly continuous on $(0, 1)$.

    (d) $k(x) = x \sin(1/x)$ is uniformly continuous on $(0, 1)$. (Hint: If we set $k(0) = 0$, then k is continuous on $[0, 1]$.)

17. Let $f : \mathbb{R} \to \mathbb{R}$ be a function. The condition

$$\forall \epsilon > 0 \quad \exists \delta > 0 \quad \text{such that} \quad |x - y| < \delta \implies |f(x) - f(y)| < \epsilon$$

defines uniform continuity of f on $\mathbb{R}$. What does each of the following conditions say about f on $\mathbb{R}$? Must such an f also be uniformly continuous on $\mathbb{R}$? If not, give an example.

    (a) $\forall \epsilon > 0$ and $\forall \delta > 0$, $|x - y| < \delta \implies |f(x) - f(y)| < \epsilon$.

    (b) $\exists \epsilon > 0$ such that $\forall \delta > 0$, $|x - y| < \delta \implies |f(x) - f(y)| < \epsilon$.

    (c) $\forall \epsilon > 0 \ \exists \delta > 1$ such that $|x - y| < \delta \implies |f(x) - f(y)| < \epsilon$.

    (d) $\forall \epsilon > 1 \ \exists \delta > 0$ such that $|x - y| < \delta \implies |f(x) - f(y)| < \epsilon$.

18. Claim: Suppose f is uniformly continuous on the half-open interval $(0, 1]$. Then we can "extend f" (i.e., define $f(0)$) to be uniformly continuous on the *closed* interval $[0, 1]$.

Prove this in two steps:

    (a) Set $f(0) = \lim_{n \to \infty} f(1/n)$. Explain why this makes sense—i.e., explain why the limit exists. (Hint: The sequence $1, 1/2, 1/3, \ldots$ is Cauchy; apply Proposition 3.23, page 185.)

    (b) (This is a little harder.) To show that f is uniformly continuous on all of $[0, 1]$, let $\epsilon > 0$. Set $\epsilon' = \epsilon/2$, and choose $\delta > 0$ that "works" for ϵ' in the definition of uniform continuity of f on $(0, 1]$. Show that this δ works for ϵ in the definition of uniform continuity of f on $[0, 1]$. (Hint: It is enough to show that $|f(x) - f(0)| < \epsilon$ when $|x - 0| < \delta$. To do this, choose any n_0 such that $0 < 1/n_0 < x$ and $|f(1/n_0) - f(0)| < \epsilon/2$. (Why is this possible?) Then use the triangle inequality.)

3.5   Topology of the Real Numbers

Topology is the mathematical study of objects and properties that stay unchanged after "continuous" transformations, like stretching and twisting, but may change after other operations, like cutting or puncturing. Topology might be said, more picturesquely, to be about pliable, rubber things; geometry, by contrast, is about rigid, steel things. A topologist, says an old joke, can't tell a coffee cup from a bagel, since either shape can be moulded from the other without cutting or puncturing.

These amusing metaphors only hint at the many mathematical faces of topology, some quite abstract. In real analysis, fortunately, topology focuses on concrete and familiar objects: individual real numbers (called *points*), various *sets* of real numbers, and relations between sets and points. Seen this way the subject is called *point-set topology*; here and in the next section we sample some highlights.

No coffee—but plenty of points and sets. Coffee cups and bagels aren't seen in real analysis. As rough analogues, however, consider two open intervals, say $I = (0, 1)$ and $J = (-13, 42)$. In obvious ways I can be stretched ("dilated") and slid ("translated"), without tearing or cutting, to coincide with J, and vice versa. In this sense I and J are topologically "the same". The closed interval $K = [0, 1]$, by contrast, is genuinely "different" from I and J: It contains left and right *endpoints* that can't be created or destroyed by stretching and sliding.

That's Theorem 3.17, page 175.

We've already seen that closed and open intervals like $[0, 1]$ and $(0, 1)$ have quite different properties in real analysis. The extreme value theorem, for instance, says that every continuous function assumes maximum and minimum values on $[0, 1]$. No such guarantee holds on the interval $(0, 1)$; a function continuous there need not even be bounded. In this sense the sets $[0, 1]$ and $(0, 1)$ are topologically different.

Consider $f(x) = 1/x$, for instance.

Individual points can also have differing topological properties with respect to sets. The points $x_1 = 0.5$ and $x_2 = 0.6$ are both *interior* to the set $[0, 1]$, and in this sense "the same". The point $x_3 = 1$, by contrast, is on the *boundary* of the set $[0, 1]$: *every* open interval, say $(0.999, 1.001)$ contains points both inside and outside $[0, 1]$.

Open and closed sets   Open and closed sets are the most important objects in topology. Among open sets, the most familiar are open *interval*, like $(0, 1)$. We've used open intervals freely throughout this book, and they're the basis for two more general definitions:

The word "basis" is used more formally in a theoretical development of topology.

Definition 3.25. A set $U \subseteq \mathbb{R}$ is called *open* if U is the union of any collection of open intervals. A set $A \subseteq \mathbb{R}$ is called *closed* if its complement $\mathbb{R} \setminus A$ is open.

. To begin, some comments and examples.

- $\mathbb{R}$ *and* $\emptyset$, *open and closed:* In these definitions the cases $U = \emptyset$, $A = \emptyset$, $U = \mathbb{R}$, and $A = \mathbb{R}$ are all allowable, and the union in the first definition is allowed to involve *any* collection—empty, finite, or infinite—of open intervals. Indeed, the set $U = \emptyset$ is both open—it's the union of no intervals at all—*and* closed—its complement is one giant interval, $\mathbb{R}$. Taking complements shows that $A = \mathbb{R}$ is also both closed and open.

  In one unlovely word, $\mathbb{R}$ and $\emptyset$ are "clopen".

- *Neighborhoods:* An interval $(x-\epsilon, x+\epsilon)$, with center x and positive radius ϵ, is called an ϵ-*neighborhood of* x. The notation $N_\epsilon(x)$ is convenient when we want to emphasize the center and radius of such an interval.

- *Elbow room:* Sets like $(0,1)$, $(0,1) \cup (2,3)$, and $(0,1) \cup (1,2) \cup (2,3) \cup \dots$ are written explicitly as unions of open intervals, and therefore obviously open by definition. Whether or not other sets are open may be less obvious, and the following alternative condition can be useful.

  A set $U \subseteq \mathbb{R}$ is open if and only if for each $x \in U$ there is some $\epsilon > 0$ such that $N_\epsilon(x) \subseteq U$.

  We can think of the interval $N_\epsilon(x)$ as "elbow room." For each element x, U holds not just x itself but also tiny left and right "ϵ-elbows" on either side of x. We see, for instance, that $S = (0,1]$ is not open: $1 \in S$ but no "right ϵ-elbow" of 1 fits into S. As another example, we see that the set $\mathbb{R} \setminus \mathbb{Z}$ is open: every non-integer, say $x = 2.93$, has *some* integer-free ϵ-neighborhood, say $(2.92, 2.94)$. We see, too, that the set $\mathbb{Z}$ is closed.

  Here $\epsilon = 0.01$.

- *Not open, not closed:* Many familiar subsets of $\mathbb{R}$ are neither open nor closed. The sets $\mathbb{Q}$ and $\mathbb{R} \setminus \mathbb{Q}$ are good examples. We've seen that *every* open interval contains both rationals and irrationals. Equivalently, neither $\mathbb{Q}$ nor $\mathbb{R} \setminus \mathbb{Q}$ contains *any* intervals; clearly, neither can be open.

Some examples will expand on these definitions and suggest new definitions, some of which we pursue in exercises.

EXAMPLE 1. Can or must a nonempty *finite* set $S = \{s_1, s_2, s_3, \dots, s_n\}$ of numbers be open or closed? What if $S = \{s_1, s_2, s_3, \dots\}$ is *countably infinite*?

SOLUTION. *Every* finite set is closed. To see why, suppose without loss of generality that $s_1 < s_2 < \cdots < s_n$. Now we can write

If not, we can reorder and rename the elements.

$$\mathbb{R} \setminus S = (-\infty, s_1) \cup (s_1, s_2) \cup (s_2, s_3) \cup \cdots \cup (s_n, \infty),$$

which shows that $\mathbb{R} \setminus S$ is open.

No nonempty finite or even countably infinite set is open: Open sets are unions of open *intervals*, and intervals are uncountable sets.   ◊

See Section 1.6 for more on countability.

We see in the next example that a countably infinite set may or may not be closed.

EXAMPLE 2.   Let

$$S = \left\{ 1, \frac{1}{2}, \frac{1}{3}, \frac{1}{4}, \dots \right\} \quad \text{and} \quad T = \left\{ 0, 1, \frac{1}{2}, \frac{1}{3}, \frac{1}{4}, \dots \right\}.$$

Is either S or T open or closed?

In the preceding example we gave another reason why S and T can't be open.

SOLUTION.   Neither S nor T includes any ϵ-neighborhoods at all, so neither S nor T is open.

What about complements? Notice first that we can write $\mathbb{R} \setminus T$ explicitly as an (infinite) union of intervals:

$$\mathbb{R} \setminus T = (-\infty, 0) \cup (1, \infty) \cup (1/2, 1) \cup (1/3, 1/2) \cup \dots.$$

Thus $\mathbb{R} \setminus T$ is open and so T is closed. The set $\mathbb{R} \setminus S$ differs from $\mathbb{R} \setminus T$ in only one element, 0. But that makes all the difference: *No* neighborhood $N_\epsilon(0)$ misses all points of S, and so $\mathbb{R} \setminus S$ is *not* open. We conclude that S is neither open nor closed, and that T is closed but not open.   ◇

EXAMPLE 3.   Let U_1 and U_2 be open sets, and A_1 and A_2 closed sets. What can be said about unions and intersections?

Do you see why? De Morgan can help.

SOLUTION.   We'll see that both $U_1 \cup U_2$ and $U_1 \cap U_2$ are open; it follows that both $A_1 \cup A_2$ and $A_1 \cap A_2$ are closed.

To see why $U_1 \cup U_2$ is open recall that U_1 and U_2 are themselves unions of intervals. Thus $U_1 \cup U_2$ is itself a union—the union of all the open intervals that make up either U_1 or U_2.

To see why $U_1 \cap U_2$ is open we use ϵ-neighborhoods. Suppose $x \in U_1 \cap U_2$. Since $x \in U_1$ there is some $\epsilon_1 > 0$ with $x \in N_{\epsilon_1}(x) \subseteq U_1$. Similarly, there is $\epsilon_2 > 0$ with $x \in N_{\epsilon_2}(x) \subseteq U_2$. Then, for $\epsilon = \min\{\epsilon_1, \epsilon_2\}$ we have $N_\epsilon(x) \subseteq U_1 \cap U_2$.   ◇

Example 3 points to more general results:

Proposition 3.26. *Unions and intersections of open and closed sets behave topologically as follows:*

(a) *The union of* any *collection of open sets is open. The intersection of any* finite *collection of open sets is open.*

(b) *The intersection of* any *collection of closed sets is closed. The union of any* finite *collection of closed sets is closed.*

Proof: The first assertion in (a) can be proved exactly as for unions of two open sets. We proved the second part in the preceding example. Part (a) implies (b) via De Morgan's laws.   □

Topology and sequences. Topology words and ideas are useful in studying sequences.

EXAMPLE 4.   Let $\{a_n\}$ be any sequence that converges to a, and U any open set with $a \in U$. Then $a_n \in U$ for all but finitely many n.

SOLUTION.   By hypothesis, there is some $\epsilon > 0$ with $N_\epsilon(a) \subseteq U$. Since $a_n \to a$, we can choose N such that $a_n \in N_\epsilon(a)$ for all $n > N$. So U contains all the a_n with at most finitely many exceptions: $a_1, a_2, \ldots, a_N$.   ◇

The preceding example suggests a more general property of closed sets.

> We've already seen that this holds for closed intervals.

Fact 3.27. *Suppose the set $A \subseteq \mathbb{R}$ is closed, and that $\{a_n\}$ is a sequence with $a_n \in A$ for all n. If $a_n \to a$, then $a \in A$.*

Proof: If not, then $a \in \mathbb{R} \setminus A$, an open set. Hence, for some $\epsilon > 0$ we have $a \in N_\epsilon(a) \subseteq \mathbb{R} \setminus A$. This means that $N_\epsilon(a)$ contains *none* of the a_n, which contradicts the claim in Example 4.   □

Limits and limit points.   A point a is defined to be a *limit point* of a set A if every open neighborhood $N_\epsilon(a)$ contains at least one element of A other than a.

EXAMPLE 5.   Specific examples illustrate some of the possibilities.

(a) Let $A = (0, 1]$. *All* points of A are limit points. The point $a = 0$ is also a limit point of A, even though $0 \notin A$. No other point in $\mathbb{R}$ is a limit point. For the point $a = 1.001$, for example, the neighborhood $N_{0.0001}(a)$ misses the set A entirely.

(b) Let $S = \left\{ 1, \dfrac{1}{2}, \dfrac{1}{3}, \dfrac{1}{4}, \ldots \right\}$, as in Example 2. *No* point of S is a limit point: *every* point of S has a (perhaps tiny) ϵ-neighborhood that contains no other point of S. The point 0, although not in S, *is* a limit point of S, since every set $N_\epsilon(0)$ contains infinitely points of S.

(c) For the set $\mathbb{Q}$ of rational numbers, *every* real number is a limit point.

Some of these points are pursued further in exercises.   ◇

As the parts of Example 5 suggest, limit points of a set A, whether or not in A, can be approximated closely by *other* points from within A. The analogy with limits of sequences is no accident:

Proposition 3.28. *A point a is limit point of a set A if and only if $a = \lim\{a_n\}$ for some sequence $\{a_n\}$ with $a_n \in A$ and $a_n \neq a$ for all n.*

Proofs are left as exercises.

Topology, functions, and continuity. Topological language is also convenient in describing behavior of functions.

EXAMPLE 6. Suppose the function $f : \mathbb{R} \to \mathbb{R}$ is continuous everywhere, and b is any constant. Show that the set $U = \{x \mid f(x) < b\}$ is open.

SOLUTION. If $U = \emptyset$ we're done already. If $U \neq \emptyset$ we can invoke the Principle of Persistent Inequalities (PoPI). Suppose, then, that $c \in U$, which means $f(c) < b$. By the PoPI, there exists $\delta > 0$ such that $f(x) < b$ for $x \in (c - \delta, c + \delta)$. But this is another way of saying that $N_\delta(c) \subseteq U$, so indeed U is open. $\Diamond$

> The PoPI is Proposition 3.14, page 172.

Inverse images and continuity. Continuity is fundamental in topology, just as in real analysis. Topological language offers a convenient and general description based on "inverse images": For any function $f : A \to B$ and any set $U \subseteq B$, the set

$$f^{-1}(U) = \{a \in A \mid f(a) \in U\}$$

> The notation f^{-1} has nothing to do with a reciprocal.

is the *inverse image* of U.

EXAMPLE 7. Suppose that $f : \mathbb{R} \to \mathbb{R}$ is continuous everywhere, and let $U = (a, b)$. Explain why $f^{-1}(U)$ is open.

SOLUTION. We did half the work just above. Unpacking the notation a bit gives

$$f^{-1}(U) = f^{-1}((a, b)) = \{x \mid f(x) < b\} \cap \{x \mid f(x) > a\}.$$

Of the last two sets on the right, the first is open by Example 6; the second is open by essentially the same argument. As the intersection of two open sets, $f^{-1}(U)$ is open, too. $\Diamond$

EXAMPLE 8. Define $f : \mathbb{R} \to \mathbb{R}$ by $f(x) = x^2$.

(a) If $U = (4, 9)$ then

$$f^{-1}(U) = \{x \in \mathbb{R} \mid 4 < x^2 < 9\} = (-3, -2) \cup (2, 3),$$

a union of two open intervals.

(b) If $U = (-42, -3)$ then $f^{-1}(U) = \emptyset$, since $f(x) \geq 0$ for all x.

(c) If $U = (-42, 3)$ then $f^{-1}(U) = f^{-1}([0, 3)) = \left(-\sqrt{3}, \sqrt{3}\right)$

(d) If $U = \{1, 2, 3\}$, then $f^{-1}(U) = \left\{\pm 1, \pm\sqrt{2}, \pm\sqrt{3}\right\}$.

$$\diamond$$

The pattern suggested above—inverse images of open sets are open, and similarly for closed sets—is no accident. Proposition 3.29 gives a brief but general description of continuity in topological language.

Proposition 3.29. *A function $f : \mathbb{R} \to \mathbb{R}$ is continuous on $\mathbb{R}$ if and only if $f^{-1}(U)$ is open whenever $U \subseteq \mathbb{R}$ is open.*

Proof: For the "if" part, fix $a \in \mathbb{R}$. Indeed, for any $\epsilon > 0$, we can let U be the open interval $(f(a) - \epsilon, f(a) + \epsilon)$. Then $a \in f^{-1}(U)$—an open set by our hypothesis—and so there is $\delta > 0$ such that $(a - \delta, a + \delta) \subseteq f^{-1}(U)$. Unraveling these notations shows that this is just another way of saying that $|f(x) - f(a)| < \epsilon$ when $|x - a| < \delta$. This is precisely the ϵ–δ condition for continuity of f at $x = a$, so the "if" part is proved.

Proving the "only if" part is similar. Here's a sketch: Suppose $U \subseteq \mathbb{R}$ is open. If $f^{-1}(U) = \emptyset$, we're done already. If not, and $a \in f^{-1}(U)$, then $f(a) \in U$, and since U is open we can choose $\epsilon > 0$ with $N_\epsilon(f(a)) \subseteq U$. By continuity of f at $x = a$ we can choose $\delta > 0$ with $N_\delta(a) \subseteq f^{-1}(U)$. Hence $f^{-1}(U)$ is open as claimed. □

Open intervals are good enough. The condition in Proposition 3.29 can be relaxed a little: f is continuous if and only if $f^{-1}(I)$ is open for every open *interval* $I = (a, b)$. This is because, roughly speaking, inverse images "play well" with unions: the inverse image of a union is the corresponding union of inverse images.

$\mathbb{R}$ is a "metric space". A *metric space* is a set X equipped with a reasonable notion of *distance*, known more formally as a *metric*. In real analysis the absolute value plays this role: For any numbers x and a, $|x - a|$ is the *distance* from x to a, and the inequality $|x - a| < \epsilon$ means that x is "within distance ϵ" of a.

The following abstract definition makes more general sense.

Definition 3.30. Let X be any set and $d : X \times X \to \mathbb{R}$ a function. We say d is a *metric* on X if the following conditions hold for all x, y, and z in X:

(i) $d(x, y) \geq 0$, and $d(x, y) = 0$ only when $x = y$ (aka, *positivity*)

(ii) $d(x, y) = d(y, x)$ (aka, *symmetry*)

(iii) $d(x, z) \leq d(x, y) + d(y, z)$ (aka, *triangle inequality*)

All parts of the definition are easy to check in our most familiar case, $X = \mathbb{R}$ and $d(x, y) = |x - y|$; the metric triangle inequality, in particular, is a long-familiar property of the absolute value.

Proposition 3.29 can be used to define continuity.

Further details are left as an exercise.

See the exercises for more on this.

Open balls, open neighborhoods. If d is a metric on a set X, $a \in X$ and $\epsilon > 0$, it makes sense to describe the set

$$B_\epsilon(a) = \{x \in X \mid d(x, a) < \epsilon\}$$

as the *open ball of radius ϵ about $x = a$.* When $d(x, y) = |x - y|$, this "ball" is really just an *interval* of radius ϵ, the set we've called an ϵ-neighborhood, and denoted $N_\epsilon(a)$. If X is a higher-dimensional set, like $\mathbb{R}^2$ or $\mathbb{R}^3$, with appropriate metrics, a "ball" might resemble a disk or a child's notion of a ball.

Building a metric space. Given any set X and any metric d, we can can create a useful topology on X—called a *metric space* topology—by defining open *sets* to be unions of open *balls*. Indeed, we did exactly that earlier in this section: real open intervals are actually open *balls*, and so our earlier definition is in fact the metric definition. Metric spaces are in some sense "nice," and this happy state is sometimes summarized by saying that $\mathbb{R}$ with its "usual" topology is is a metric space. We explore this idea further in exercises.

Exercises

1. We said in Example 3 that because the union of two open sets is open, "it follows" that the intersection of two closed sets is closed. Use De Morgan's laws to give details.

2. An open set $U \subseteq \mathbb{R}$ is defined as the union of *any* collection of open *intervals.*

   (a) Suppose U_1, U_2, U_3, ... are all open sets. Explain why the infinite union $U_1 \cup U_2 \cup U_3 \cup \ldots$ is open.

   (b) Suppose A_1, A_2, A_3, ... are all closed sets. Use De Morgan's laws (they hold for infinite unions and intersections, too) to explain why the infinite intersection $A_1 \cap A_2 \cap A_3 \cap \ldots$ is closed.

3. As in Problem 2, let U_i and A_i denote open and closed sets, respectively.

   (a) Give an example to show that $U_1 \cap U_2 \cap U_3 \cap \ldots$ can be the *closed* interval $[-1, 1]$

   (b) Give an example to show that $A_1 \cup A_2 \cup A_3 \cap \ldots$ can be the *open* interval $(-1, 1)$.

4. We said in Example 3 that because the union of two open sets is open, "it follows" that the intersection of two closed sets is closed. Use De Morgan's laws to give details.

5. We said in this section that a set $U \subseteq \mathbb{R}$ is open if and only if for each $p \in U$ there is some $\epsilon > 0$ such that $N_\epsilon(p) \subseteq U$.

    (a) Let $U = (0, \infty)$; we know U is open. For $p = 0.0042$ find a specific value of ϵ for which $N_\epsilon(p) \subseteq U$.

    (b) Let p be *any* element of $(0, \infty)$. Find a value of ϵ (depending on p, of course) for which $N_\epsilon(p) \subseteq U$.

    (c) Prove the statement above for *any* open set $U \subseteq \mathbb{R}$. (Hint: For the "only if" part note that if $p \in U$ then $p \in (a, b) \subseteq U$ for some open interval (a, b).)

6. Let $S \subseteq \mathbb{R}$ be any set. A point $p \in \mathbb{R}$ is called a *boundary point* of S if every ϵ-neighborhood $N_\epsilon(p)$ intersects both S and $\mathbb{R} \setminus S$.

    (a) Show that $S = \mathbb{R}$ has no boundary points.

    (b) What are the boundary points, if any, of $S = [0, 1]$ and $S = \{0, 1\}$? Are any boundary points of S contained in S?

    (c) Find the boundary points, if any, of $S = (0, 1)$? Are any boundary points contained in S?

    (d) Show that 1 is both a limit point and a boundary point of $(0, 1)$.

    (e) Show that a finite set F has no boundary points.

7. Here's the converse of the claim in Example 4, page 193: Let $\{a_n\}$ be a sequence, and a a number. Suppose that for any open set U with $a \in U$, $a_n \in U$ for all but finitely many n. Then $\{a_n\}$ converges to a.

8.   (a) Problem 6 illustrates that an open set U may *have* boundary points. Show that U *contains* no boundary points.

    (b) Suppose p is a boundary point of a closed set $A \subseteq \mathbb{R}$. Show that $p \in A$.

9. Let $S \subseteq \mathbb{R}$ be any set. A point $p \in S$ is an *interior point* of S if $N_\epsilon(x_0) \subset S$ for some $\epsilon > 0$.

    (a) Which points are interior to $[0, 1]$?

    (b) Show that every point in an open set U is interior to U.

    (c) Show that a finite set F has no interior points.

    (d) Every point $p \in \mathbb{R}$ is interior to $\mathbb{R}$, since $\mathbb{R}$ is open. Which points of $\mathbb{Q}$ are interior to $\mathbb{Q}$?

10. This problem resembles Example 8, page 194. Define $f : \mathbb{R} \to \mathbb{R}$ by $f(x) = 2x + 3$.

(a) Find $f^{-1}(\mathbb{R})$. Is the answer an open set?

(b) Find $f^{-1}((0, \infty))$. Is the answer an open set?

(c) Find $f^{-1}((-3, 42))$. Is the answer an open set?

(d) Let (a, b) be any open interval. Find $f^{-1}((a, b))$. Conclude that (as expected) f is continuous.

11. In the notation of Proposition 3.29, page 195, show the following.

(a) For any set $S \subseteq \mathbb{R}$, $f^{-1}(\mathbb{R} \setminus S) = \mathbb{R} \setminus f^{-1}(S)$.

(b) $f : \mathbb{R} \to \mathbb{R}$ is continuous on $\mathbb{R}$ if and only if $f^{-1}(A)$ is *closed* whenever $A \subseteq \mathbb{R}$ is *closed*.

12. Repeat Problem 10, but use the (continuous) function $f(x) = |x|$. Hint: For (d), use cases, depending on whether $0 \in (a, b)$.

13. A set $D \subseteq \mathbb{R}$ is *dense* in $\mathbb{R}$ if D has nonempty intersection with every open interval $(a, b) \subset \mathbb{R}$.

(a) Show that $\mathbb{Q}$ and $\mathbb{R} \setminus \mathbb{Q}$ are both dense in $\mathbb{R}$.

(b) Is the set $S = \left\{ \dfrac{n}{1000} \mid n \in \mathbb{Z} \right\}$ dense in $\mathbb{R}$? Why or why not?

14. Clean up any needed details to prove the "only if" part of Proposition 3.29, page 195.

15. Consider a function $f : \mathbb{R} \to \mathbb{R}$ and sets $A, B \subseteq \mathbb{R}$.

(a) Show that $f^{-1}(A \cap B) = f^{-1}(A) \cap f^{-1}(B)$.

(b) Show that $f^{-1}(A \cup B) = f^{-1}(A) \cup f^{-1}(B)$.

16. The "sign function" f :

$$\mathbb{R} \to \mathbb{R} \text{ given by } f(x) = \begin{cases} -1 & \text{if } x < 0, \\ 0 & \text{if } x = 0, \\ 1 & \text{if } x > 0, \end{cases}$$ is a poster child for discontinuity. (See Example 2, page 164.)

(a) Let $U = (0, 1)$. Find $f^{-1}(U)$. Is $f^{-1}(U)$ an open set?

(b) Find an open set $U \subseteq \mathbb{R}$ for which $f^{-1}(U)$ is not open.

(c) Find a closed set $A \subseteq \mathbb{R}$ for which $f^{-1}(A)$ is not closed.

17. We said in this section that the "usual" topology on $\mathbb{R}$ can be thought of as given by the metric $d(x, y) = |x - y|$. This means, in particular, that every open set $U \subseteq \mathbb{R}$ is a union of one or more open balls $B_\epsilon(a)$. For example, $U = (3, 7) = B_2(5)$. In that spirit, write each set following as a union (perhaps infinite) of open balls.

    (a) $(0, 1)$

    (b) $\mathbb{R}$

    (c) $\mathbb{R} \setminus \mathbb{Z}$

    (d) $(0, \infty)$

18. In the spirit of Definition 3.30, page 195, define for any real numbers x and
y, $d(x, y) = \begin{cases} 1 & \text{if } x \neq y \\ 0 & \text{if } x = y \end{cases}$.

    (a) Show that d is a metric on $\mathbb{R}$. (The resulting topology is called the *discrete topology*.)

    (b) Let $p \in \mathbb{R}$. What are the balls $B_1(p)$ and $B_2(p)$?

    (c) What are the open sets in this metric topology?

19. Define, for any real numbers x and y, $d(x, y) = \min\{|x - y|, 1\}$.

    (a) Show that d is a metric on $\mathbb{R}$.

    (b) Let $p \in \mathbb{R}$. What are the balls $B_1(p)$ and $B_2(p)$?

    (c) What are the open sets in this metric topology?

3.6 Compactness

Compactness is a "nice" topological property enjoyed by certain subsets of $\mathbb{R}$—e.g., finite sets, closed and bounded intervals, and convergent sequences—but not others. We've seen, for instance, that a function defined and continuous on a closed and bounded interval $[a, b]$ assumes maximum and minimum values, while a function defined and continuous on an *open* interval need not even be bounded.

    Formal definitions follow soon; here's a preview.

Almost finite. Compact sets need not be finite. We'll see, for instance, that the unit interval $[0, 1]$—an uncountably infinite set—is compact. On the other hand, $[0, 1]$ is both bounded and closed, and these properties turn out to be almost as good as finiteness for some purposes.

Finite sets are compact, though.

EXAMPLE 1. Following are some properties enjoyed both by *finite* sets $F \subset \mathbb{R}$ and *compact* sets $S \subset \mathbb{R}$.

Proofs are given below or are left to exercises.

(a) If F_1 and F_2 are finite, then so are $F_1 \cup F_2$ and $F_1 \cap F_2$. If S_1 and S_2 are compact, then so are $S_1 \cup S_2$ and $S_1 \cap S_2$.

(b) If F is finite and nonempty, then F contains maximum and minimum elements. If S is compact and nonempty, then S contains maximum and minimum elements.

We've discussed this before.
See, e.g., Problem 3, page 50.

(c) Every finite set F is bounded. Every compact set S is bounded.

(d) Every finite set F is closed. Every compact set S is closed.

(e) Let $f : \mathbb{R} \to \mathbb{R}$ be a continuous function. If $F \subset \mathbb{R}$ is finite and nonempty, then f achieves maximum and minimum values on F. If $S \subset \mathbb{R}$ is compact and nonempty, then f achieves maximum and minimum values on S.

$\Diamond$

Definitions. The "open cover" definition we'll soon give for compactness looks awkward at first glance. Indeed, other possible definitions of compactness (we'll see one soon) turn out to be equivalent for subsets of $\mathbb{R}$. But the open cover definition makes sense in any topological space, and turns out to be useful in and beyond real analysis.

First, an auxiliary definition and some basic examples:

Definition 3.31. Let $S \subseteq \mathbb{R}$ be any set of real numbers. An *open cover* of S is any collection $\mathcal{U}$ of open sets that "covers" S—i.e., for each $s \in S$ there is some $U \in \mathcal{U}$ with $s \in U$. A *subcover* is any subcollection $\mathcal{U}' \subseteq \mathcal{U}$ that still covers all of S; $\mathcal{U}'$ is "proper" if $\mathcal{U}' \subsetneq \mathcal{U}$.

EXAMPLE 2.   Let $S = (0, 2)$. The open sets

$$U_1 = (1, 2), \quad U_2 = (1/2, 2), \quad \ldots, \quad U_n = (1/n, 2), \quad \ldots$$

form an (infinite) open cover of S; we can write $\mathcal{U} = \{U_1, U_2, \ldots\}$. We don't really need all the sets in $\mathcal{U}$ to cover S. For instance, the proper but still infinite subcover

$$\mathcal{U}' = \{U_1, U_3, U_5, \ldots\}$$

does the job. $\Diamond$

EXAMPLE 3.   Let S be *any* set of real numbers, and fix $\epsilon > 0$. The collection $\mathcal{U} = \{N_\epsilon(s) \mid s \in S\}$ completely covers S.

Each point of S gets its own
ϵ-neighborhood.

Whether any proper subcover exists depends on S, and perhaps on ϵ. If, say, $S = \mathbb{Z}$ and $\epsilon = 0.1$, then *none* of the $N_\epsilon(s)$ can be omitted. If $S = \mathbb{R}$, on the other hand, the proper (but still infinite) subcover $\mathcal{U}' = \{N_\epsilon(q) \mid q \in \mathbb{Q}\}$ works.

It has about 10,000 members.

If $S = [0, 1000]$ and $\epsilon = 0.1$, then we can find a *finite* subcover, like this one:

$$\mathcal{U}' = \{N_{0.1}(x) \mid x = 0, 0.1, 0.2, \ldots, 999.9, 1000\}.$$

The same idea works for *any* bounded set $S \subset \mathbb{R}$ and *any* fixed $\epsilon > 0$. ◇
We're ready for that awkward definition, and some basic examples.

Definition 3.32. A set $S \subset \mathbb{R}$ is *compact* if every open cover of S has a *finite* subcover.

EXAMPLE 4. The empty set $\emptyset$ is compact. So is every *finite* set $F \subset \mathbb{R}$. The full set $\mathbb{R}$ is not compact. ◇

Proofs are left as exercises.

EXAMPLE 5. The open interval $(0, 2)$ is *not* compact. Why?

SOLUTION. The nested open intervals

$$\mathcal{U} = \{(1, 2), (1/2, 2), (1/3, 2), \ldots, (1/n, 2), \ldots\}$$

from Example 2 cover $(0, 1)$, but no *finite* subcollection suffices: If we stop with $(1/N, 2)$, then the set $(0, 1/N]$ is left uncovered. ◇

EXAMPLE 6. Let $\{a_n\}$ be any convergent sequence of real numbers, with limit a. Then the set $S = \{a, a_1, a_2, a_3, \ldots\}$ is compact. The set $S' = \{a_1, a_2, a_3, \ldots\}$ need not be compact.

SOLUTION. Let $\mathcal{U}$ be any open cover of S. Since $a \in S$ there is some open set U_a in $\mathcal{U}$ with $a \in U_a$. As we showed in Example 4, page 193, U_a contains all but (at most) finitely many of the a_n. We can cover these outliers, if any, with finitely many members of $\mathcal{U}$, say $U_1, U_2, \ldots U_N$. Thus $\mathcal{U}' = \{U_a, U_1, U_2, \ldots, U_N\}$ works as a finite subcover. We leave the claim about S' as an exercise. ◇

Properties of compact sets. Compact subsets of $\mathbb{R}$ have various nice properties, some analogous to those of finite sets; see Example 1 above. Here are some proofs.

Proposition 3.33. *If $S \subset \mathbb{R}$ is compact then S is (i) closed; and (ii) bounded.*

Proof: To prove (i) we check that $\mathbb{R} \setminus S$ is open: For any $p \in \mathbb{R} \setminus S$ we'll find a neighborhood of p that's disjoint from S. To wit, for each $\epsilon > 0$ we define $U_\epsilon = \{x \in \mathbb{R} \mid |x - p| > \epsilon\}$. The sets U_ϵ form an open cover $\mathcal{U}$ of S. By compactness, some *finite* collection $U_{\epsilon_1}, U_{\epsilon_2}, U_{\epsilon_3}, \ldots, U_{\epsilon_N}$ covers S. If we take ϵ_0 to be the *smallest* of these ϵ's, then we see that $S \subseteq U_{\epsilon_0}$. Thus, the open interval $(p - \epsilon_0, p + \epsilon_0)$ is disjoint from S, as desired.

There is a smallest one since our set of ϵ's is finite.

Proving (ii) is easier. The nested collection

$$\mathcal{U} = \{(-1, 1), (-2, 2), (-3, 3), \ldots\}$$

is an open cover of $\mathbb{R}$, and hence also of S. Because S is compact some *finite* subcollection $\{(-1,1),\ (-2,2),\ (-3,3),\ \ldots,\ (-N,N)\}$ covers S. This means that $-N < s < N$ for all $s \in S$, and so S is bounded.   □

A big theorem. Theorem 3.35 below is a famous result, dating from around 1900. It fully describes the compact subsets of $\mathbb{R}$ in down-to-earth language. We need just one more technical fact.

More general versions of Theorem 3.35 hold in some topological spaces other than $\mathbb{R}$.

Lemma 3.34. *Let $S \subset \mathbb{R}$ be compact and A a closed subset of S. Then A is compact.*

Proof: Let $\mathcal{U}$ be any open cover of A. Tossing one more open set, $\mathbb{R} \setminus A$, into $\mathcal{U}$ produces an open cover $\mathcal{V}$ of S. By compactness of S, some finite subcover $\mathcal{V}'$ covers S. Ejecting $\mathbb{R} \setminus A$ from $\mathcal{V}'$ gives the desired finite subcover of $\mathcal{U}$.   □

Theorem 3.35. *(Heine–Borel) A subset S of $\mathbb{R}$ is compact if and only if S is closed and bounded.*

Proof: The "only if" part is Proposition 3.33, above.

This part is clever and tricky; tread carefully.

And see the exercises.

It has to do with completeness; see Theorem 1.31, page 76.

Now for the "if" part. Although S itself need not be an interval, boundedness of S implies that $S \subseteq [a,b]$ for some closed interval $[a,b]$. By Lemma 3.34, it's good enough to show that every closed *interval* $[a,b]$ is compact. We do this for the closed interval $[0,1]$; the case for general $[a,b]$ follows easily. To do this we'll use the the *nested intervals property* of the real numbers.

Suppose then, toward contradiction, that some open cover $\mathcal{U}$ of $I_1 = [0,1]$ has *no* finite subcover. Then one or both of $[0,1/2]$ or $[1/2,1]$ has no finite subcover. Call such an interval I_2. Similarly, either the left half or the right half (or both) of I_2 has no finite subcover; let I_3 be such an interval. Continuing this process gives a nested sequence of closed intervals $I_1 \supset I_2 \supset I_3 \supset \ldots$, with $\text{length}(I_n) = 1/2^n$.

The nested interval property says that the intersection $I_1 \cap I_2 \cap I_3 \cap \ldots$ is a single point, say p. Since $p \in [0,1]$ we have $p \in U_0$ for some $U_0 \in \mathcal{U}$. Since U_0 is open, we know $p \in N_\epsilon(p) \subseteq U_0$ for some $\epsilon > 0$.

Now for any N large enough so $1/2^N < \epsilon$, we have $I_N \subseteq U_0$. Here comes our contradiction: I_N, which supposedly lacks *any* finite subcover from $\mathcal{U}$, is actually covered by just *one* set, U_0.   □

Compactness revisited: the sequence view. A close cousin of compactness as defined above is as follows:

Definition 3.36. A set $A \subseteq \mathbb{R}$ is called *sequentially compact* if every infinite sequence $\{a_n\}$ contained in A has a convergent subsequence whose limit is also in A.

The Bolzano–Weierstrass theorem asserts that every closed and bounded *interval* $A = [a, b]$ is sequentially compact. In fact, A need not be an interval:

Theorem 2.16, page 111

Proposition 3.37. *Every closed and bounded set $A \subset \mathbb{R}$ is sequentially compact.*

Proof: Let A be closed and bounded, and $\{a_n\}$ a sequence with $a_n \in A$ for all n. The sequence $\{a_n\}$ is bounded, so Bolzano–Weierstrass says there's a convergent subsequence $\{a_{n_k}\}$; call its limit a. Since A is *closed*, Fact 3.27, page 193, guarantees that $a \in A$, as desired.   □

The converse holds, too:

Proofs are in exercises.

Proposition 3.38. *A sequentially compact set $A \subset \mathbb{R}$ is closed and bounded.*

Now we have all the pieces for the big-picture result:

Theorem 3.39. *A set $A \subset \mathbb{R}$ is compact if and only if A is sequentially compact.*

EXAMPLE 7.   Use sequential compactness to show that if A and B are compact subsets of $\mathbb{R}$, then $A \cap B$ is compact, too.

SOLUTION.   Let $\{c_n\}$ be any sequence contained in $A \cap B$. Compactness of A means that some subsequence $\{d_n\}$ of $\{c_n\}$ converges to a point $d \in A$. It remains to show that $d \in B$. Compactness of B implies that some subsequence of $\{d_n\}$ converges to some point $e \in B$. But $\{d_n\}$ is a convergent sequence, so all subsequences converge to the same limit, d. Thus $d = e$ and we're done.   ◊

Using Compactness

Compact sets have, as claimed earlier, several pleasant properties that "play well" with continuous functions. Following are three (now) simple but important examples.

Proposition 3.40. *A nonempty compact set $K \subset \mathbb{R}$ has maximum and minimum elements.*

Proof: Since K is bounded, there exist $\alpha = \inf(K)$ and $\beta = \sup(K)$. By Lemma 2.9, page 100, there are sequences $\{a_n\}$ and $\{b_n\}$ in K with limits α and β, respectively. Since K is closed, Fact 3.27, page 193, guarantees that both $\alpha \in K$ and $\beta \in K$, as desired.   □

Proposition 3.41. *Let $f : \mathbb{R} \to \mathbb{R}$ be a continuous function, and $K \subset \mathbb{R}$ a compact set. The image set $f(K) = \{f(x) \mid x \in K\}$ is also compact.*

Another proof is outlined in
exercises.

Proof: Let's show that $f(K)$ is sequentially compact. To this end, let $\{y_n\}$ be any sequence contained in $f(K)$; we seek a convergent subsequence with limit in $f(K)$.

Since $\{y_n\} \subseteq f(K)$, there are $x_1, x_2, x_3, \ldots$, all in K, with $f(x_n) = y_n$ for all n. By compactness of K the sequence $\{x_n\}$ has a convergent subsequence $\{x_{n_k}\}$, with limit $x_0 \in K$. Setting $y_{n_k} = f(x_{n_k})$ for all k, and $y_0 = f(x_0)$ produces the sought-after subsequence of $\{y_n\}$. Continuity of f guarantees that

See page 163.

$$x_{n_k} \to x_0 \quad \Longrightarrow \quad f(x_{n_k}) \to f(x_0),$$

as desired.   □

Proposition 3.42. *Let $f : \mathbb{R} \to \mathbb{R}$ be continuous, and $K \subset \mathbb{R}$ a nonempty compact set. Then f achieves maximum and minimum values on K.*

The extreme value theorem
says the same thing for
compact intervals.

Proof: There is almost nothing to do. By Proposition 3.40, $f(K)$ includes maximum and minimum values.   □

EXAMPLE 8.   Suppose $K \subseteq \mathbb{R}$ is compact and $p \notin K$. Some point $x_0 \in K$ is "closest" to p.

SOLUTION.   The function $f(x) = |x - p|$ is everywhere continuous; note that $f(x) > 0$ for all $x \in K$. Proposition 3.42 says that f achieves a minimum $f(x_0) > 0$, as desired.   ◇

Note that x_0 is not necessarily
unique.

Exercises

1. Use Definition 3.31 to prove the claims in Example 4:

    (a) Every finite set $F = \{x_1, x_2, \ldots, x_n\} \subset \mathbb{R}$ is compact.

    (b) $\mathbb{R}$ is not compact.

    (c) $\emptyset$ is compact.

2. Use Definition 3.31 to show that $\mathbb{Z}$ is not compact.

3. Consider the set S' in Example 6.

    (a) Show that if $\{a_n\}$ is given by $a_n = 1/n$, then S' is *not* compact. (Find a suitable open cover.)

    (b) Suppose $\{a_n\}$ is the sequence $1, 0, 1/2, 0, 1/3, 0, \ldots$. Is the set $\{a_n\}$ compact? Why or why not?

    (c) Give an example of a *divergent* sequence $\{a_n\}$ for which the set $\{a_n\}$ is compact.

4. Give details in each part to prove Proposition 3.38: A sequentially compact set $S \subseteq \mathbb{R}$ is (a) closed and (b) bounded.

   (a) It's enough to show $\mathbb{R} \setminus S$ is open. Let $p \in \mathbb{R} \setminus S$. We're done if $N_\epsilon(p) \subseteq \mathbb{R} \setminus S$ for some $\epsilon > 0$. If not, then (why?) there's a sequence $\{s_n\}$ in S with

   $$|s_1 - p| < 1, \quad |s_2 - p| < \frac{1}{2}, \quad |s_3 - p| < \frac{1}{3}, \quad \ldots$$

   Then $s_n \to p$, which is a contradiction. (Why?)

   (b) Toward contradiction, suppose S is unbounded. Then there exists (why?) a sequence $\{s_n\} \subseteq S$ with $|s_1| > 1$, $|s_2| > 2$, $|s_3| > 3$, etc. The sequence $\{s_n\}$ has no convergent subsequence (why?).

5. We said in this section that if $[0, 1]$ is compact, then so is every closed interval $[a, b]$. Prove this using Proposition 3.31.

6. This problem is about the situation of Example 8, page 204. Use the notation there.

   (a) What's the closest point x_0 if $K = [0, 1]$ and $p = \pi$?
   Give an example of a compact set K in which there is more than one closest point to $p = \pi$. Could there be three "closest points"?

   (b) *Is there a closest point to $p = \pi$ in (the noncompact set) $[0, 1)$?*

   (c) *Is there a closest point to $p = \pi$ in $\mathbb{Z}$? In $\mathbb{Q}$?*

7. A junior-grade version of Proposition 3.40 says that if $f : \mathbb{R} \to \mathbb{R}$ is continuous and $K \subset \mathbb{R}$ is compact, then f is bounded on K—i.e., the set $\{f(x) \mid x \in K\}$ is bounded.

   Prove this by considering the open cover $\mathcal{U}$ consisting of sets $U_i = \{x \mid f(x) < |i|\}$ is bounded.

   Why are the U_i open? Why do they cover K? How does compactness help?

8. Explain why the extreme value theorem (see page 175) follows from Proposition 3.40 in this section.

9. There is *no* continuous function $f : \mathbb{R} \to \mathbb{R}$ such that $f([0, 1]) = (0, 1)$. Which result in this section says so?

10. Is there a continuous function $f : \mathbb{R} \to \mathbb{R}$ such that $f((0, 1)) = [0, 1]$? If so, give an example. If not, why not?

CHAPTER 4

Derivatives

4.1 Defining the Derivative

Calculus at last. We've worked hard to develop important calculus ingredients: functions, limits, continuity, etc. At last we are ready to put them together; we'll treat derivatives in this chapter and integrals in the next. Many derivative and integral formulas and calculations are, of course, already familiar from elementary calculus. The new twist is to derive such "known" facts rigorously, and to put them in theoretical context.

First comes the key definition:

Definition 4.1 (Derivative at a point). Let f be a function defined on an open interval I, and let a be a point in I. We say f is *differentiable* at a, with derivative $f'(a)$, if the limit

$$f'(a) = \lim_{x \to a} \frac{f(x) - f(a)}{x - a}$$

exists.

Let's test it on some familiar well- and ill-behaved functions.

EXAMPLE 1. Find, if possible, derivatives at $a = 3$ for the functions

$$c(x) = 5, \quad \ell(x) = 2x + 7, \quad q(x) = x^2, \quad \text{and} \quad a(x) = |x - 3|.$$

What general facts do the calculations suggest?

SOLUTION. The limit calculation for $c(x)$ is easy:

$$c'(3) = \lim_{x \to 3} \frac{c(x) - c(3)}{x - 3} = \lim_{x \to 3} \frac{5 - 5}{x - 3} = 0.$$

Thus, $c'(3) = 0$. Since the value $x = 3$ played no important role in the calculation, we'd guess (correctly) that $c'(a) = 0$ for *all* a.

Finding $\ell'(3)$ is easy, too:

$$\ell'(3) = \lim_{x \to 3} \frac{\ell(x) - \ell(3)}{x - 3} = \lim_{x \to 3} \frac{2x + 7 - 13}{x - 3} = \lim_{x \to 3} \frac{2x - 6}{x - 3} = 2.$$

Thus, $\ell'(3) = 2$, and (almost) the same calculation shows that $\ell'(a) = 2$ for *all* inputs a.

See the exercises.

Watch for a factoring trick.

For q, we need another limit calculation:

$$q'(3) = \lim_{x \to 3} \frac{q(x) - q(3)}{x - 3} = \lim_{x \to 3} \frac{x^2 - 9}{x - 3} = \lim_{x \to 3} (x + 3) = 6$$

Thus, $q'(3) = 6$. This time the calculation looks a bit different away from $x = 3$, so we defer (just briefly) finding $q'(x)$ for other inputs.

The derivative $a'(3)$, if it exists, is the value of

$$\lim_{x \to 3} \frac{a(x) - a(3)}{x - 3} = \lim_{x \to 3} \frac{|x - 3|}{x - 3} = \lim_{x \to 3} \begin{cases} 1 & \text{if } x > 3, \\ -1 & \text{if } x < 3. \end{cases}$$

Left- and right-hand limits do exist at $x = 3$, but they are unequal.

From the last form, we see that $a'(3)$ does not exist.   ◇

Sketch these functions to see the idea; no technology needed.

The results of Example 1 won't surprise any calculus veteran. Let's see what happens with some stranger functions, one of them an old friend.

EXAMPLE 2.   Are the functions

$$f(x) = \begin{cases} 1 & \text{if } x \in \mathbb{Q}, \\ 0 & \text{if } x \notin \mathbb{Q}, \end{cases} \quad \text{and} \quad g(x) = \begin{cases} x^2 & \text{if } x \geq 0, \\ 0 & \text{if } x < 0 \end{cases}$$

differentiable at $x = 0$?

SOLUTION.   The short answers are no and yes. The limits in question are

$$\lim_{x \to 0} \frac{f(x)}{x} \quad \text{and} \quad \lim_{x \to 0} \frac{g(x)}{x}.$$

Look what happens along a rational sequence tending to zero, and see the exercises.

The first limit clearly fails to exist, so f is not differentiable at $x = 0$. For the second limit, we just note that

$$\frac{g(x)}{x} = \begin{cases} x & \text{if } x \geq 0, \\ 0 & \text{if } x < 0, \end{cases}$$

and so

$$g'(x) = \lim_{x \to 0} \frac{g(x)}{x} = 0.$$ ◇

Different Views of the Derivative

The derivative is *defined* formally as a limit, but it has several familiar and useful interpretations. We review several briefly.

Average and instantaneous rates. The expression

$$\frac{f(x) - f(a)}{x - a},$$

known as a *difference quotient*, can be thought of as the *average* rate of change of $f(x)$ with respect to x over the input interval between a and x. The derivative is the *limit* of such rates as x approaches a, and so can be seen as the *instantaneous* rate of change of f when $x = a$.

We won't dwell now on the connection between derivatives and rates. But we should acknowledge that the idea is among the best and most useful ever had—by anybody. The power of calculus to describe and predict real-world change, from falling apples to celestial mechanics, is hard to overstate. As Isaac Newton and others discovered, calculus is the right "language" in which to express physical laws and to deduce their consequences.

Slopes. In elementary calculus, $f'(a)$ is sometimes described simply as the "slope of the graph" of f at $x = a$. To be a bit more precise, note that the difference quotient

$$\frac{f(x) - f(a)}{x - a}$$

is the slope of a certain *secant line*, which passes through the points $(x, f(x))$ and $(a, f(a))$ on the graph of f. If the derivative exists, then the limiting value $f'(a)$ can also be thought of as a slope, this time of the *tangent line* to the graph at $x = a$. Observe, too, that slope is itself a rate (of ascent, in this case), and hence a special case of the rate-of-change interpretation discussed above.

Linear approximation. Thinking of derivatives as slopes is simple and helpful, especially for well-behaved functions. But subtleties can arise. What is really meant, for example, by the slope of a *curve* at a point? What if the graph of f is not a smooth curve at all, but ragged or broken? To skirt such problems, we can consider *linear* functions, for which no such troubles arise.

And deservedly popular in elementary calculus courses.

Definition 4.2 (Linear approximation). Let f be a function and a a domain point for which $f'(a)$ exists. The *linear approximation to f at a* (also known as the *first-order approximation*) is the linear function ℓ_a defined by

$$\ell_a(x) = f(a) + f'(a)(x - a).$$

Observe some basic properties of the linear approximation:

- *Simplest examples:* If $f(x) = x^2$ and $a = 2$, then

$$\ell_2(x) = f(2) + f'(2)(x - 2) = 4 + 4(x - 2) = -4 + 4x.$$

  If $f(x) = x^2$ and $a = 0$, then

$$\ell_0(x) = f(0) + f'(0)(x - 0) = 0 + 0(x - 0) = 0.$$

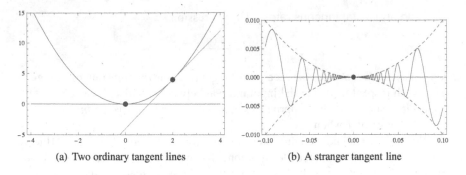

(a) Two ordinary tangent lines   (b) A stranger tangent line

Figure 4.1. Two views of linear approximation.

- *Shared values—up to a point:* The functions f and ℓ_a "agree to first order" at $x = a$:
  $$f(a) = \ell_a(x) \quad \text{and} \quad f'(a) = \ell'_a(a).$$
  (The second fact holds because ℓ_a, being linear, has *constant* derivative.) Away from $x = a$, anything can happen. In the case of $f(x) = x^2$ and $l_3(x) = 9+6(x-3)$, for example, we have $f(-42) = 1764$ and $f'(-42) = -84$, while $\ell_3(-42) = -261$ and $\ell'_3(-42) = 6$.

- *Graphical views:* The close connections between f and ℓ_a are reflected in their respective graphs. Indeed, the graph of ℓ_a is in a natural sense the line that "best fits" the f-graph at $x = a$. In elementary calculus parlance, this line is called the *tangent line* at $x = a$. The word "tangent" suggests "touching," and in many cases this line does indeed "touch but not cross" the graph of f at the point $(a, f(a))$, as illustrated in Figure 4.1(a). But in other cases the graphs of f and ℓ_a are seriously intertwined, as Figure 4.1(b) and the following example illustrate. With such possibilities in mind we will usually refer to "linear approximation" rather than "tangent lines."

EXAMPLE 3.   Show that the function $f : \mathbb{R} \to \mathbb{R}$, shown in Figure 4.1(b) and defined by
$$f(x) = \begin{cases} x^2 \sin(1/x) & \text{if } x \neq 0, \\ 0 & \text{if } x = 0, \end{cases}$$
is differentiable at $a = 0$. What is the linear approximation there?

SOLUTION.   The limit in question is
$$f'(0) = \lim_{x \to 0} \frac{f(x) - f(0)}{x - 0} = \lim_{x \to 0} \frac{x^2 \sin(1/x)}{x - 0} = \lim_{x \to 0} x \sin(1/x),$$

which we've already shown (by squeezing; see Example 6, page 155) to be zero. Thus $f'(0) = 0$, and so the corresponding linear approximation—also visible in Figure 4.1(b)—is just

$$\ell_0(x) = f(0) + f'(0)(x - 0) = 0.$$

The calculation was easy, but the situation is still undeniably strange: The graph of f crosses its tangent line *infinitely often* in *every* open interval $(-\delta, \delta)$ around $a = 0$. In the interval $(-.001, .001)$, for instance, we have $f\left(1/(k\pi)\right) = 0$ for all integers k with $|k\pi| > 1000$. $\Diamond$

Magnification. A useful geometric alternative to the derivative-as-slope view involves magnification factors. If $f'(a) = m$ (for magnification), then for x near a we have

$$\frac{f(x) - f(a)}{x - a} \approx m, \quad \text{or} \quad f(x) - f(a) \approx m(x - a),$$

which implies, in turn, that

$$|f(x) - f(a)| \approx |m|\, |x - a|.$$

Interpreting absolute values as distances, we see that the distance between outputs $f(x)$ and $f(a)$ is about $|m|$ times the distance between inputs x and a.

If $f(x) = x^2$, for instance, we have $f'(3) = 6$, and f maps the small input interval $(2.95, 3.05)$ to the output interval $(.7025, 9.3025)$, which is six times as long.

Derivatives as Functions

For a given function f the derivative $f'(a)$ depends on a. If, say, $f(x) = x^2$, then, as we have seen, $f'(a) = 2a$ holds for every input a, and we might simply write $f'(x) = 2x$. It is natural, in other words, to think of f' as a function in its own right, *derived* in a special way from f.

To avoid "symbol-creep" it is convenient to use the same input symbol, often x, for both f and f'. Thus we might write something like

Hence the term *derivative function*.

$$f'(x) = \lim_{t \to x} \frac{f(t) - f(x)}{t - x}$$

when we want to think of both f and f' as functions of x. The difference with earlier versions of the derivative limit is *entirely* notational—no new mathematics is involved.

EXAMPLE 4. If $f(x) = \sqrt{x}$, what's $f'(x)$? How are the domains of f and f' related to each other?

See the factoring trick?

SOLUTION. Let's calculate. For $x > 0$, we have

$$f'(x) = \lim_{t \to x} \frac{f(t) - f(x)}{t - x} = \lim_{t \to x} \frac{\sqrt{t} - \sqrt{x}}{t - x} = \lim_{t \to x} \frac{\sqrt{t} - \sqrt{x}}{(\sqrt{t} + \sqrt{x})(\sqrt{t} - \sqrt{x})}$$

$$= \lim_{t \to x} \frac{1}{\sqrt{t} + \sqrt{x}} = \frac{1}{\sqrt{x} + \sqrt{x}} = \frac{1}{2\sqrt{x}},$$

as we learned back in elementary calculus.

No derivative $f'(0)$ exists in this case. One problem is that f is not defined on any open interval containing 0, as the definition requires. Thus f' has domain $(0, \infty)$—a *proper subset* of $[0, \infty)$, the domain of f. ◇

Increase, decrease, and derivatives. As elementary calculus experience suggests, properties of the derivative function f' reflect corresponding properties of f. Where f' is positive, for instance, we expect f to be increasing, and where f' takes *large* positive values, we expect f to be rapidly increasing. These expectations will turn out to be well-founded, but giving precise proofs is a little trickier than one might expect. We'll return to such matters later in this chapter.

Higher Derivatives

The function f' may be differentiable in its own right, to produce a new function f'', the *second derivative* of f. Repeating the process produces *higher-order derivatives* f''', $f^{(4)}$, $f^{(5)}$, and so on. If $f(x) = x^2$, for example, then we have

$$f'(x) = 2x, \quad f''(x) = 2, \quad f'''(x) = 0 = f^{(4)}(x) = f^{(5)}(x) = \ldots.$$

Higher derivatives, better approximation. The *linear* (or *first-order*) approximation $\ell_a(x) = f(a) + f'(a)(x - a)$ "matches" f at $x = a$ in the sense that

$$f(a) = \ell_a(a) \quad \text{and} \quad f'(a) = \ell'_a(a).$$

Using second derivatives we can go a step further. The *quadratic* (aka *second-order*) approximation to f at $x = a$ is a quadratic function that fits f even more closely than does ℓ_a:

$$f(a) = q_a(a) \quad \text{and} \quad f'(a) = q'_a(a) \quad \text{and} \quad f''(a) = q''_a(a).$$

If, say, $f(x) = \cos x$ and $a = 0$, then we have (as you know from calculus)

$$f(0) = \cos 0 = 1; \quad f'(0) = -\sin 0 = 0; \quad f''(0) = -\cos 0 = -1.$$

Thus, $\ell_0(x) = 1 + 0x = 1$, and basic calculus shows that $q_0(x) = 1 - x^2/2$ "agrees" with f in the desired sense:

$$q(0) = 1; \quad q'_0(0) = 1; \quad f''(0) = -1.$$

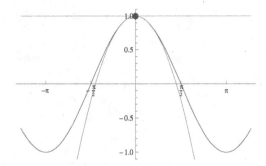

Figure 4.2. Linear and quadratic approximations to the cosine function.

(We will give a general formula for q_a in the next section.) Given the extra matching derivative, we expect q_0 to approximate f even better near $x = 0$ than does ℓ_0. Figure 4.2 doesn't disappoint.

Properties of Differentiable Functions

Any function f for which $f'(a)$ exists must be in some sense "nice" at $x = a$. The next theorem describes the most basic such form of good behavior.

Differentiability and continuity. A function continuous at a point need not be differentiable there. The function $h(x) = |x|$, for instance, is continuous but not differentiable at $a = 0$. (At all *other* inputs h is both continuous and differentiable.) But the converse assertion is true:

We showed something very similar in Example 1.

Theorem 4.3 (Differentiability implies continuity). *If f is differentiable at a, then f is also continuous at a.*

About the proof. To illustrate the idea, suppose $f(3) = 42$ and $f'(3) = 7$. To show continuity of f at $x = 3$, we need to show that $\lim_{x \to 3} f(x) = 42$, or, equivalently, that $\lim_{x \to 3} (f(x) - 42) = 0$. The calculation involves a little trick:

$$
\lim_{x \to 3} (f(x) - 42) = \lim_{x \to 3} \frac{f(x) - 42}{x - 3}(x - 3) = \lim_{x \to 3} \frac{f(x) - 42}{x - 3} \cdot \lim_{x \to 3} (x - 3)
$$
$$
= 7 \cdot 0 = 0,
$$

as desired. The general proof is similar, and left as an exercise.

Digging deeper. The following lemma delves a little further into good-behavior implications of differentiability.

Lemma 4.4. *Let f be a function defined in an open interval containing $x = a$, and suppose $f'(a)$ exists.*

(i) If $f'(a) \neq 0$, then $f(x) \neq f(a)$ for all $x \neq a$ in some interval $(a - \delta, a + \delta)$.

(ii) If $f'(a) = 0$, then—for every $\epsilon > 0$, no matter how small—there exists $\delta > 0$ such that
$$|f(x) - f(a)| \leq \epsilon |x - a|$$
whenever $|x - a| < \delta$.

(iii) If f has a local minimum or local maximum at $x = a$, then $f'(a) = 0$.

(iv) If $f'(a) > 0$, then there exists $\delta > 0$ such that (i) $f(x) < f(a)$ when $a - \delta < x < a$; and (ii) $f(x) > f(a)$ when $a < x < a + \delta$.

Proof: All parts follow from closer looks at the defining limit:
$$f'(a) = \lim_{x \to a} \frac{f(x) - f(a)}{x - a}.$$

To prove (i), suppose $f'(a) \neq 0$ and set $\epsilon = |f'(a)| > 0$. Because the preceding limit exists, we can choose $\delta > 0$ so that
$$\left| \frac{f(x) - f(a)}{x - a} - f'(a) \right| < \epsilon = |f'(a)|$$

See for yourself.

whenever $|x - a| < \delta$. In particular, we must have $f(x) \neq f(a)$ for all such x; otherwise the inequality above fails.

Claim (ii) essentially just restates the existence of the key limit. Because $f'(a) = 0$, we have
$$0 = \lim_{x \to a} \frac{f(x) - f(a)}{x - a} = \lim_{x \to a} \left| \frac{f(x) - f(a)}{x - a} \right|.$$

By definition of the limit, for given $\epsilon > 0$ there exists $\delta > 0$ such that, whenever $|x - a| < \delta$, we have
$$\left| \frac{f(x) - f(a)}{x - a} \right| < \epsilon, \quad \text{or, equivalently,} \quad |f(x) - f(a)| < \epsilon |x - a|,$$

as claimed.

We'll sketch a proof of (iii) assuming that f has a local *minimum* at $x = a$; the proof for a local maximum is similar. In this case there is an open interval $I = (a - \delta, a + \delta)$ with $f(a) \leq f(x)$ for $x \in I$. In particular,

Polishing the proof is an exercise.

$$\frac{f(x) - f(a)}{x - a} \geq 0 \quad \text{if } x > a \qquad \text{and} \qquad \frac{f(x) - f(a)}{x - a} \leq 0 \quad \text{if } x < a.$$

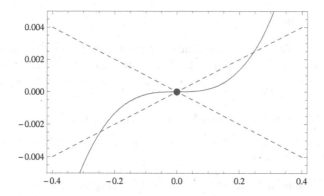

Figure 4.3. The function $f(x) = x - \sin x$: very flat near $x = 0$.

These inequalities imply, respectively, that

$$\lim_{x \to a^+} \frac{f(x) - f(a)}{x - a} \geq 0 \quad \text{and} \quad \lim_{x \to a^-} \frac{f(x) - f(a)}{x - a} \leq 0,$$

from which the only possible conclusion is that $f'(a) = 0$. □

EXAMPLE 5. If $f(x) = \sin x$ and $g(x) = x - \sin x$, then $f'(0) = 1$ and $g'(0) = 0$. What does Lemma 4.4 say about f and g at $a = 0$?

> We'll prove these things in the next section.

SOLUTION. Claim (i), applied to f, says that $\sin x \neq \sin 0 = 0$ for x in some interval $(-\delta, \delta)$. This is no surprise—even the big interval $(-\pi, \pi)$ works. Claim (ii), applied to g, says something more interesting. If, say, $\epsilon = 0.01$, then (ii) says that for x in some interval $(-\delta, \delta)$,

$$|x - \sin x| \leq 0.01 \, |x| \, .$$

Figure 4.3 illustrates graphically what this means—especially about how closely x approximates $\sin x$ near $x = 0$. The function g shows, finally, that the converse to claim (c) is *false*: Even though $g'(0) = 0$, g attains neither a maximum nor a minimum at 0. ◊

Exercises

1. Suppose $f'(x)$ exists for all x. Every calculus student knows that if $g(x) = f(x) + 3$, then $g'(a) = f'(a)$ for all a. Show this using Definition 4.1.

2. Consider a linear function $L(x) = Ax + B$. Show using Definition 4.1 that $L'(x)$ exists for all x, and that L' is a constant function.

3. Consider the functions $f(x) = x^2$ and $g(x) = \sqrt{x}$.

   (a) Carefully draw—by hand—the graphs of f and g in the "window" $0 \leq x \leq 2, 0 \leq y \leq 2$. Then draw the tangent lines to both f and g at the point $(1, 1)$. Note that all four graphs pass through the point $(1, 1)$.

   (b) Use Definition 4.1 to show that $f'(1) = 2$ and $g'(1) = 1/2$.

   (c) How does the symmetry of your picture in (a) reflect the values you found in (b)?

4. Use Definition 4.1 to find each of the following derivatives.

   (a) $f'(1)$ if $f(x) = 42$.

   (b) $g'(1)$ if $g(x) = x^3$.

   (c) $h'(1)$ if $h(x) = x^{42}$. (Hint: The difference quotient reduces to a nice form.)

   (d) $k'(1)$ if $k(x) = 1/(x + 3)$.

   (e) $m'(1)$ if $m(x) = 42 + x^3 + x^{42} + 1/(x + 3)$. (Hint: Use a fact about limits of sums.)

5. Consider the function $f(x) = 1/(x^2 + 5)$. Compare answers here to what you recall from elementary calculus.

   (a) Use Definition 4.1 to find $f'(0)$.

   (b) Use Definition 4.1 to find $f'(c)$ for any c.

6. (a) Use Definition 4.1 to show that if $f(x) = 1/x$ and $c \neq 0$, then $f'(c) = -1/c^2$.

   (b) Prove by induction: $(1/x^n)' = -n/x^{n+1}$ for all integers $n \geq 1$. (Note: It is OK to use the product rule; we'll prove it in the next section.)

7. Assume in this problem that $f : \mathbb{R} \to \mathbb{R}$ is differentiable at $x = 0$.

   (a) Show that if f is strictly increasing, then $f'(0) \geq 0$. (Hint: Look at the *sign* of the difference quotient.)

   (b) Suppose that $|f(x)| \leq |x|$ for all x. What can be said about $f'(0)$?

8. Assume in all parts that $f : \mathbb{R} \to \mathbb{R}$ is differentiable at $x = 0$.

   (a) Show that if $f(x) \geq f(0)$ for $x > 0$, then $f'(0) \geq 0$. (Hint: Consider $\lim_{x \to 0^+} (f(x) - f(0))/(x - 0)$.)

(b) Show that if $f(x) \geq f(0)$ for $x < 0$, then $f'(0) \leq 0$. (Hint: Consider $\lim_{x \to 0^-} (f(x) - f(0))/(x - 0)$.)

(c) What do the preceding parts say about the derivative at a minimum point of a function? (We'll further explore this famous connection soon.)

9. Consider continuous functions $f : \mathbb{R} \to \mathbb{R}$ and $g : \mathbb{R} \to \mathbb{R}$, with $f(0) = 0 = f'(0)$. Show that fg is differentiable at $x = 0$, and that $(fg)'(0) = 0$. (Use Definition 4.1; the product rule does not apply.)

10. Consider continuous functions $f : \mathbb{R} \to \mathbb{R}$ and $g : \mathbb{R} \to \mathbb{R}$ such that (i) $f(0) = 0$; (ii) $f'(0) = 2$; and (iii) $g(0) = 3$. Show that fg is differentiable at $x = 0$; find $(fg)'(0)$.

11. Suppose $f(x)$ is differentiable at $x = 2$, with $f'(2) = 3$. Let $g(x) = 4f(x) + 5$. Show that g is also differentiable at $x = 2$, with $g'(2) = 12$. Interpret the result geometrically.

12. Suppose $f(x)$ is differentiable at $x = 2$, with $f'(2) = 3$. Let $g(x) = f(x + 4) + 5$. We claim that $g(x)$ is differentiable at $x = -2$, and that $g'(-2) = 3$.

   (a) Give an informal geometric argument for the claim. (Hint: How are graphs of f and g related?)

   (b) Calculate $g'(-2)$ using Definition 4.1. (Hint: In the difference quotient for $g'(-2)$, write $x = y - 4$; note that $x \to -2$ is equivalent to $y \to 2$.)

13. Consider the function defined by

$$f(x) = \begin{cases} -x^2 & \text{if } x < 0, \\ x^2 & \text{if } x \geq 0. \end{cases}$$

   (a) Show that f is *continuous* at $x = 0$. (Hint: Use one-sided limits.)

   (b) Is f *differentiable* at $x = 0$? If so, find $f'(0)$.

14. Consider the function defined by

$$f(x) = \begin{cases} x^2 & \text{if } x < 1, \\ 3x - 2 & \text{if } x \geq 1. \end{cases}$$

   (a) Show that f is continuous but not differentiable at $x = 1$.

   (b) Make f differentiable at $x = 1$ by replacing $3x - 2$ with another linear expression.

15. Let $f : \mathbb{R} \to \mathbb{R}$ be any function such that $|f(x)| \leq x^2$ for all x. Show that $f'(0) = 0$.

16. Throughout this problem let $f : (-1, 1) \to \mathbb{R}$ be any bounded (not necessarily continuous) function.

   (a) Define a new function $g : (-1, 1) \to \mathbb{R}$ by $g(x) = xf(x)$. Show that g is continuous at $x = 0$.

   (b) Give an example to show that g need not be differentiable at $x = 0$. (Hint: Think absolutely.)

   (c) Define a new function $h : (-1, 1) \to \mathbb{R}$ by $h(x) = x^2 f(x)$. Show that h is differentiable at $x = 0$.

17. Let $f(x) = \begin{cases} x \sin(1/x) & \text{if } x \neq 0 \\ 0 & \text{if } x = 0 \end{cases}$. Show that f is continuous but not differentiable at $x = 0$.

18. Use Definition 4.1 to show that $g(x) = \dfrac{x^2 \cos x}{3 + \cos x}$ is differentiable at $x = 0$, with $g'(0) = 0$. (We don't know the quotient rule yet.)

19. Assume in this problem that both f and g are continuous at $x = 0$.

   (a) Show that if $f(x) > 42$ for all $x \neq 0$, then $f(0) \geq 42$.

   (b) Suppose that $g(0) = 42$, and let $h(x) = xg(x)$. Show that $h'(0)$ exists and find its value.

20. Expand the following proof sketch for Claim (iv) of Lemma 2. Set $\epsilon = f(a) > 0$. Since $\lim\limits_{x \to a} \dfrac{f(x) - f(a)}{x - a}$ exists, we can choose $\delta > 0$ so that $\left| \dfrac{f(x) - f(a)}{x - a} - f(a) \right| < \epsilon = f(a)$ when $0 < |x - a| < \delta$. This δ does what's needed.

21. If $f(x) = \sin(100x)$, then (just assume this) $f'(x) = 100 \cos(x)$.

   (a) What does Claim (iv) of Lemma 2, page 164, say about this f when $a = 0$?

   (b) Find a value of δ that works in this case.

4.2 Calculating Derivatives

Familiar Rules

Calculus veterans know well how to apply derivative rules (product, quotient, chain, etc.) in calculations like this one:

Or even worse.

$$\left(x^2 \sin(3x) + 4\ln x \right)' = 3x^2 \cos(3x) + 2x \sin(3x) + \frac{4}{x}.$$

Our interest here is less in applying the rules—you've suffered enough—than in stating them precisely and proving them rigorously. After all, derivatives *are* limits, so algebraic properties of limits—like those in Theorem 3.3—will be key.

Algebraic combinations. We have seen several ways to combine "old" functions f and g to form new ones. The combinations $f + g$, $f - g$, $f \cdot g$, and f/g are constructed *algebraically*, while $f \circ g$, $g \circ f$, and $f \circ g \circ f \circ g$ involve *composition*. Propositions 3.11 and 3.12 say that if f and g are *continuous* at the domain points in question, then so are all the combinations just mentioned.

See page 167.

We want to show now that similar properties hold for *derivatives*: if f and g are differentiable, and due care is taken with domains, then combinations are differentiable, too. We also expect that derivatives of combinations will be built, somehow, from derivatives of f and g.

We do need to avoid division by zero in combinations like f/g.

Theorem 4.5 (Derivatives of algebraic combinations). *Suppose that f and g are functions differentiable at a; let C and D be constants. Then the functions*

$$f \pm g, \quad Cf + Dg, \quad fg, \quad and \quad \frac{f}{g}$$

are all differentiable at a (for f/g we need $g(a) \neq 0$), with derivatives as follows:

- Linear combinations: $(Cf + Dg)'(a) = Cf'(a) + Dg'(a)$.

- Products: $(fg)'(a) = f'(a)g(a) + f(a)g'(a)$.

- Quotients: $\left(\frac{f}{g}\right)'(a) = \frac{g(a)f'(a) - f(a)g'(a)}{g(a)^2}$.

Notes on the theorem. Proofs follow or are left to the exercises. First, some informal observations:

- *Useful formulas—and more:* The theorem not only justifies some well-loved techniques from elementary calculus but also guarantees that, under the given hypotheses, all indicated derivatives *exist*. This existence follows, as we will see, from corresponding properties of *limits*.

- *Sums and constant multiples:* Both the *sum rule* and the *constant multiple rule* for derivatives are (easy) special cases of the result for linear combinations. Setting $C = D = 1$, for instance, gives one of these old favorites.

- *Linear combinations and linear transformations:* The rule for linear combinations can be phrased succinctly in the language of linear algebra: Differentiation is a linear transformation from one vector space to another. (Vectors, in this case, are functions.)

- *Approximate thinking:* One view of the theorem concerns linear approximation. By hypothesis, both f and g have linear approximations $\ell_{f,a}$ and $\ell_{g,a}$ at $x = a$:

$$f(x) \approx \ell_{f,a}(x) = f(a) + f'(a)(x - a);$$
$$g(x) \approx \ell_{g,a}(x) = g(a) + g'(a)(x - a).$$

The theorem tells, in effect, how to combine $\ell_{f,a}$ and $\ell_{g,a}$ to create new linear approximations to functions like $f + g$, fg, and f/g. For example, the linear combination rule says that for x near a we have

Following minor
re-arrangement.

$$3f(x) + 2g(x) \approx 3\ell_{f,a}(x) + 2\ell_{g,a}(x)$$
$$= (3f'(a) + 2g'(a))(x - a) + 3f(a) + 2g(a),$$

which should seem reasonable. Still more succinctly, we have $\ell_{3f+2g,a} = 3\ell_{f,a} + 2\ell_{g,a}$. The situation for products and quotients is a little more complicated; see the exercises.

Proof: All parts follow from manipulating the limits that define the derivatives in question. We treat one part in detail and leave the rest to the exercises.

For products, the theorem claims that

$$\lim_{x \to a} \frac{f(x)g(x) - f(a)g(a)}{x - a} = f(a)g'(a) + f'(a)g(a);$$

implicit, of course, is the claim that the limit exists. To see why, we manipulate the difference quotient limit, starting with a clever trick:

add and subtract the same thing

$$\lim_{x \to a} \frac{f(x)g(x) - f(a)g(a)}{x - a} = \lim_{x \to a} \frac{f(x)g(x) - f(a)g(x) + f(a)g(x) - f(a)g(a)}{x - a}$$

algebra with limits

$$= \lim_{x \to a} g(x) \frac{f(x) - f(a)}{x - a} + \lim_{x \to a} f(a) \frac{g(x) - g(a)}{x - a}$$

more limit algebra

$$= \lim_{x \to a} g(x) \lim_{x \to a} \frac{f(x) - f(a)}{x - a} + \lim_{x \to a} f(a) \lim_{x \to a} \frac{g(x) - g(a)}{x - a}$$

$$= g(a) f'(a) + f(a)g'(a).$$

Note especially the last step, in which we evaluated *four* limits, of which only one (the third) is completely trivial. The second and fourth limits *define* $f'(a)$ and $g'(a)$, which we've assumed to exist, and the first, $\lim_{x \to a} g(x) = g(a)$, holds because g is *continuous* at $x = a$.   □

Why? See Theorem 4.3, page 213.

EXAMPLE 1.   Finding derivatives of polynomials and rational functions, like

$$p(x) = 3x^7 - 5x + 2 \quad \text{and} \quad q(x) = \frac{3x^7 - 5x + 2}{x^3 - x},$$

is easy for calculus veterans. How is Theorem 4.5 involved?

SOLUTION.   Theorem 4.5 guarantees, first, that the derivative functions $p'(x)$ and $q'(x)$ exist. To prove that $p(x)$—or *any* polynomial function—is differentiable for all x, we only need to observe that $f(x) = x$ is differentiable, with $f'(x) = 1$ for all x. Now Theorem 4.5 implies that power functions like

Or prove ... it's easy.

$$x \cdot x = x^2, \quad x^2 \cdot x = x^3, \quad \ldots, \quad x^{41} \cdot x = x^{42}, \quad \ldots$$

are all differentiable in their own right, and that the familiar derivative formulas hold for *all* positive integer powers:

$$\left(x^2\right)' = 2x, \quad \left(x^3\right)' = 3x^2, \quad \ldots \quad \left(x^{42}\right)' = 42x^{41}, \quad \ldots.$$

(In the exercises, we will discuss two proofs, one by induction.)

We notice, too, that all polynomials are linear combinations of power functions, and so the theorem guarantees that all polynomials have derivatives. The function $q(x)$—like *every* rational function—is a quotient of differentiable functions, and so Theorem 4.5 guarantees differentiability except at roots of the denominator. For q there are three such roots; we will explore this function in more detail in the exercises.   ◇

EXAMPLE 2.   What does Theorem 4.5 say about derivatives of $f(x) = (x^2+1)^{15}$ and $g(x) = \sin x / e^x$?

SOLUTION.   The product rule in Theorem 4.5 can help with $f(x)$ if—but only if—we're willing to multiply out the 15th power. That is tedious for a human but no big deal for, say, *Mathematica*:

$$(x^2+1)^{15} = x^{30}+15x^{28}+105x^{26}+455x^{24}+1365x^{22}+3003x^{20}+\cdots+15x^2+1.$$

The result is easy, if laborious, to differentiate term by term. A wiser plan involves the chain rule; see below.

Differentiating g requires the quotient rule, of course, but first we need derivatives of the numerator and the denominator. Although familiar from elementary calculus, the formulas

$$(\sin x)' = \cos x \quad \text{and} \quad (e^x)' = e^x$$

are far from obvious, and we'll defend them below. Assuming them for the moment, the rest is easy. Since numerator and denominator are differentiable, and the denominator never vanishes, the quotient function is differentiable, with derivative

$$\left(\frac{\sin x}{e^x} \right)' = \frac{e^x \cos x - e^x \sin x}{e^x \cdot e^x} = \frac{\cos x - \sin x}{e^x}. \qquad \Diamond$$

Composition and the chain rule. Theorem 4.5 says nothing about functions like $h(x) = \sin(x^2)$ and $k(x) = \sin(\sin(\sin(e^x)))$, built by *composition* of differentiable functions. Theorem 4.6 assures us that such composites are indeed differentiable, and describes how the respective derivatives are combined.

As always, due care for domains is necessary.

Theorem 4.6 (The chain rule). *Suppose that g is differentiable at a and f is differentiable at $b = g(a)$. Then $f \circ g$ is differentiable at a, and*

$$(f \circ g)' (a) = f'(g(a)) \cdot g'(a).$$

EXAMPLE 3. The functions $f(x) = \sin x$ and $g(x) = x^2$ are differentiable for all x. Use the chain rule to differentiate

$$f \circ g(x) = \sin(x^2) \quad \text{and} \quad f \circ f \circ g(x) = \sin \left(\sin \left(x^2 \right) \right).$$

We could substitute x for a, but why bother?

SOLUTION. The chain rule applies directly to $f \circ g$:

$$(f \circ g)' (a) = f'(g(a)) \cdot g'(a) = \cos(a^2) \cdot 2a.$$

For $f \circ f \circ g$ we have to dig deeper:

$$\begin{aligned}
(f \circ f \circ g)' (a) &= f' (f \circ g(a)) \cdot (f \circ g)' (a) \\
&= f' (f \circ g(a)) \cdot f'(g(a)) \cdot g'(a);
\end{aligned}$$

we used the earlier calculation in the last step. Substituting the formulas for f and g gives the final answer:

$$(f \circ f \circ g)' (a) = \cos \left(\sin a^2 \right) \cdot \cos(a^2) \cdot 2a;$$

the three-fold product represents *two* uses of the chain rule. $\qquad \Diamond$

Here, before the proof, are some comments on the chain rule and why it is plausible.

- *Why multiply?* The chain rule says that the derivative of a *composition* is a certain *product* of derivatives. Thinking of derivatives as magnification factors suggests why multiplication is the right thing to do. Suppose, say, that $g'(a) = 2$ and $f'(b) = 3$. Then "lens" g and "lens" f magnify distances by factors of 2 and 3, respectively, and so we expect *six-fold* magnification on inputs sent first through g and then through f.

  We discussed the magnification view in the preceding section.

  Microscopes use the same principle.

- *It works for linear functions:* We can just calculate, without fancy proofs, that the chain rule holds for *linear* functions f and g. If $f(x) = Ax + B$ and $g(x) = Cx + D$, then $f'(x) = A$ and $g'(x) = C$ for all x, and

$$f \circ g(x) = Ag(x) + B = A(Cx + D) = ACx + (AD + B),$$

  and so

$$(f \circ g)'(x) = AC,$$

  as expected.

- *It works for nonlinear functions, too:* Differentiable functions are closely approximated by linear functions, and so it is reasonable to hope that differentiable functions might satisfy the same chain rule. This turns out to be true; the formal proof explains why.

Notes on a proof. We're given that both limits

$$\lim_{x \to a} \frac{g(x) - g(a)}{x - a} = g'(a) \quad \text{and} \quad \lim_{y \to b} \frac{f(y) - f(b)}{y - b} = f'(b)$$

exist, where $b = g(a)$, and we need to prove the related claim

$$\lim_{x \to b} \frac{f(g(x)) - f(g(a))}{x - a} = f'(b) \cdot g'(a) = f'(g(a)) \cdot g'(a).$$

First we apply some clever algebra:

$$\frac{f(g(x)) - f(g(a))}{x - a} = \frac{f(g(x)) - f(g(a))}{g(x) - g(a)} \cdot \frac{g(x) - g(a)}{x - a}$$
$$= \frac{f(y) - f(b)}{y - b} \cdot \frac{g(x) - g(a)}{x - a},$$

where $y = g(x)$ and $b = g(a)$. This looks promising. If $x \to a$, then $y \to b$ (because g is continuous at a), and so

$$\lim_{x \to a} \frac{f(g(x)) - f(g(a))}{x - a} = \lim_{y \to b} \frac{f(y) - f(b)}{y - b} \cdot \lim_{x \to a} \frac{g(x) - g(a)}{x - a}$$
$$= f'(b) \cdot g'(a),$$

just as we were hoping to show.

This argument works just fine—and shows that the chain rule holds—for *most* functions f and g. But one technical flaw needs attention: The key factorization

Look at the first denominator on the right.

$$\lim_{x \to a} \frac{f(g(x)) - f(g(a))}{x - a} = \lim_{x \to a} \frac{f(g(x)) - f(g(a))}{g(x) - g(a)} \cdot \lim_{x \to a} \frac{g(x) - g(a)}{x - a}$$

makes good sense if, but only if, $g(x) \neq g(a)$ for all $x \neq a$ in some interval containing a. For most functions g, this is the case, but there are exceptions. If g is *constant*, for instance, then our factorization is nonsense for *all* inputs x. But all is not lost: if g is constant, then $f \circ g$ is *also* constant, and so, clearly, both

$$g'(a) = 0 \quad \text{and} \quad (f \circ g)'(a) = 0,$$

and the chain rule still holds!

We're ready for the formal proof. As often happens, a special case takes some extra effort.

Proof (of the chain rule): We treat two cases separately.

Case 1: $g'(a) \neq 0$. In this case, $g(x) \neq g(a)$ for all $x \neq a$ in some interval containing a. We can therefore argue as above to obtain the chain rule.

See Lemma 4.4, page 214.

Case 2: $g'(a) = 0$. In this special case, we'll show that $(f \circ g)'(a) = 0$—again, as the chain rule claims. For notational simplicity, let's assume that $a = f(a) = 0$. Our claim then is that, for any $\epsilon > 0$, we can find $\delta > 0$ with

It makes no difference mathematically.

$$\left| \frac{f(g(x))}{x} \right| \leq \epsilon \quad \text{or, equivalently,} \quad |f(g(x))| \leq \epsilon |x|$$

whenever $x \in (-\delta, \delta)$.

To see this, let $\epsilon > 0$ be given; we can assume WLOG that $\epsilon < 1$. Note first that, because the limit

Challenging values of ϵ are near zero.

$$f'(0) = \lim_{x \to 0} \frac{f(x)}{x}$$

exists, the function $\frac{f(x)}{x}$ is bounded for x in some open interval $(-\delta_1, \delta_1)$. In other words, for some $M > 0$ we have

$$\left| \frac{f(x)}{x} \right| \leq M, \quad \text{or, equivalently,} \quad |f(x)| \leq M |x|$$

whenever $x \in (-\delta_1, \delta_1)$. We'll assume, again WLOG, that $M \geq 1$.

Next, because $g'(0) = 0$, we can choose $\delta_2 > 0$ so that

$$\left| \frac{g(x)}{x} \right| \leq \frac{\epsilon}{M} \quad \text{or, equivalently,} \quad |g(x)| \leq \frac{\epsilon}{M} |x|$$

whenever $x \in (-\delta_2, \delta_2)$. Finally, let δ be the smallest of δ_1, δ_2, and 1. This δ "works," because if $|x| < \delta$, then

$$|g(x)| < \frac{\epsilon}{M} |x| \leq \epsilon |x| < \epsilon\delta < \delta \leq \delta_2,$$

and so

$$|f(g(x))| < M |g(x)| < M \frac{\epsilon}{M} |x| = \epsilon |x|.$$

This proves the chain rule. $\qquad\square$

Derivatives of Elementary Functions

In Section 3.2, we discussed the "elementary functions" of calculus, built by combining basic function "elements": power functions, trigonometric functions, exponential functions, and their inverses. Proposition 3.10, page 167, says that every such elementary function is *continuous* throughout its natural domain. A similar claim holds for differentiability:

We didn't prove every detail rigorously.

Proposition 4.7 (Elementary functions are differentiable). *Let f be an elementary function defined on an open interval I containing a. Then f is differentiable at a.*

We've already done most of the work. Theorems 4.5 and 4.6 guarantee that a built-up function like

$$f(x) = \sin\left(\frac{\cos x + e^x}{\ln(2 + x^2)}\right)$$

is indeed differentiable (except perhaps at endpoints of its domain), provided that its basic building blocks—including the sine, cosine, exponential, and log functions—are all themselves differentiable functions. Proving this rigorously for every basic elementary function takes some work. That isn't our top priority, so we give only some brief samples here.

More details are in the exercises.

EXAMPLE 4. (Trigonometric derivatives). All six standard trigonometric derivative formulas follow from two basic limits:

$$\lim_{h \to 0} \frac{\sin h}{h} = 1 \quad \text{and} \quad \lim_{h \to 0} \frac{\cos h - 1}{h} = 0.$$

We proved the first in Example 6, page 155. The second follows from the first.

How? Why?

SOLUTION. All the remaining derivative formulas follow from the one formula $(\sin x)' = \cos x$ and from algebraic properties of sines and cosines. For example, we have

$$(\cos x)' = \left(\sin\left(x + \frac{\pi}{2}\right)\right)' = \cos\left(x + \frac{\pi}{2}\right) = -\sin x;$$

notice the chain rule implicit in the second equality, and two formulas relating sines and cosines. Here is another, this time thanks to the quotient rule:

$$(\tan x)' = \left(\frac{\sin x}{\cos x}\right)' = \frac{\cos^2 x + \sin^2 x}{\cos^2 x} = \sec^2 x.$$

To prove that $(\sin x)' = \cos x$, we'll use the well-known *addition formula for sines*:

$$\sin(x + h) = \sin(x)\cos(h) + \cos(x)\sin(h).$$

This will come in handy as we wrestle with the difference quotient:

$$\begin{aligned}
(\sin(x))' &= \lim_{h \to 0} \frac{\sin(x + h) - \sin(x)}{h} \\
&= \lim_{h \to 0} \frac{\sin(x)\cos(h) + \cos(x)\sin(h) - \sin(x)}{h} \\
&= \sin(x) \lim_{h \to 0} \frac{\cos(h) - 1}{h} + \cos(x) \lim_{h \to 0} \frac{\sin(h)}{h} \\
&= \sin(x) \cdot 0 + \cos(x) \cdot 1 = \cos(x),
\end{aligned}$$

as desired. $\Diamond$

EXAMPLE 5. (Exponential derivatives). The exponential derivative $(e^x)' = e^x$ follows from the single limit

$$\lim_{h \to 0} \frac{e^h - 1}{h} = 1.$$

How? Why?

SOLUTION. Assuming the limit above, we have

$$(e^x)' = \lim_{h \to 0} \frac{e^{x+h} - e^x}{h} = \lim_{h \to 0} \frac{e^x e^h - e^x}{h} = e^x \lim_{h \to 0} \frac{e^h - 1}{h} = e^x,$$

as expected. $\Diamond$

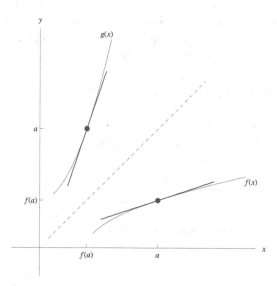

Figure 4.4. Derivatives of inverse functions f and g.

Derivatives of inverse functions. One loose end remains untied concerning differentiability of elementary functions: handling *inverses* of differentiable functions. (Logarithmic and exponential functions are inverses, for instance, as are $f(x) = x^3$ and $g(x) = \sqrt[3]{x}$.) It turns out that inverses of differentiable functions are indeed differentiable—if the usual care is taken with domains of definition and to avoid division by zero. Here is the key result:

Proposition 4.8 (Derivatives of inverse functions). *Let* $f : I \to J$ *and* $g : J \to I$ *be inverse functions, where* I *and* J *are open intervals. Suppose* f *is differentiable at* $a \in I$ *and* $f'(a) \neq 0$. *Then* g *is differentiable at* $f(a)$, *and*

$$g'(f(a)) = \frac{1}{f'(a)}.$$

Figure 4.4 illustrates the idea.

Notes on a proof. If we assume for the moment that g is differentiable at $f(a)$, the rest is easy, thanks to the chain rule. By definition of inverse functions, we have $g(f(x)) = x$ for all $x \in I$. Differentiating both sides gives

$$g(f(x)) = x \implies g'(f(x)) \cdot f'(x) = 1 \implies g'(f(a)) = \frac{1}{f'(a)}$$

whenever $f'(a) \neq 0$.

The harder part turns out to be showing what we assumed above: that $g'(f(a))$ exists. Figure 4.4 makes that assumption look reasonable, but proving it rigorously takes us a bit off our main course. We omit the detour.

EXAMPLE 6. (Differentiating the natural log). Use Proposition 4.8 to differentiate $\ln x$.

SOLUTION. If $f(x) = e^x$ and $g(x) = \ln x$, then we know f and g are inverses, and $f'(x) = e^x$. From Proposition 4.8, we have

$$g'(f(a)) = \frac{1}{f'(a)}, \quad \text{or} \quad g'(e^a)) = \frac{1}{e^a}.$$

Writing $b = e^a$ gives the familiar formula $g'(b) = (\ln b)' = \frac{1}{b}$.   ◊

Exercises

1. We said in this section that $f(x) = x^n$ for any positive integer n, then $f'(x) = nx^{n-1}$. For one inductive proof (using the product rule) see Problem 14, page 51.

   Another approach is to work directly with the definition:

   $$f'(a) = \lim_{x \to a} \frac{f(x) - f(a)}{x - a} = \lim_{x \to a} \frac{x^n - a^n}{x - a}.$$

   (a) Note that $x^2 - a^2 = (x-a)(x+a)$ and $x^3 - a^3 = (x-a)(x^2+ax+a^2)$. Find (proof optional) an analogous formula for $x^n - a^n$.

   (b) Use your formula from (a) to prove the derivative rule.

2. Let f, g, and h be everywhere differentiable functions.

   (a) Use the product rule to explain why $(fgh)' = f'gh + fg'h + fgh'$.

   (b) State a "three-deep chain rule" for the composition $f \circ g \circ h$.

   (c) State a formula for the *second* derivative $(f \circ g)''(a)$.

   (d) State a "product-chain hybrid" rule for $(f \circ (gh))'(a)$.

   (e) Consider the special case where $f(x) = Ax + B$ for some constants A and B. What is $(f \circ g)'(a)$? What does this have to do with the linear combination rule (Theorem 4.5, page 219)?

3. Suppose f and g are both differentiable at $x = a$, with derivatives $f'(a)$ and $g'(a)$. Use properties of limits to show that the linear combination $h(x) = 3f(x) - 42g(x)$ is also differentiable at a. What is $h'(a)$?

4. Prove the claim about linear combinations in Theorem 4.5, page 219.

5. This problem outlines a proof—assuming both the product rule and the chain rule, which we proved separately—of the quotient rule part of Theorem 4.5, page 219. Assume throughout that f and g are differentiable at $x = a$ and that $g(a) \neq 0$.

   (a) Assume (it is easy to show) that the function $k(x) = 1/x$ is differentiable for all $x \neq 0$, with $k'(x) = -1/x^2$. Use this and the chain rule to find $h'(a)$, where $h(x) = 1/g(x)$.

   (b) Apply the product rule to $f(x)/g(x) = f(x) \cdot 1/g(x)$ to deduce the quotient rule as stated in Theorem 4.5.

6. Use Theorem 4.5 and the fact that $(\sin x)' = \cos x$ to find derivatives of the other five standard trigonometric functions.

7. Find derivatives of the following functions (assuming that derivatives of $\sin x$, $\cos x$, and e^x are known).

   (a) $\sin(\cos(e^x))$

   (b) $\sin x \cdot \cos x \cdot e^x$

   (c) $\dfrac{\sin x}{\cos x \cdot e^x}$

   (d) $\dfrac{\sin x}{\cos(e^x)}$

8. Consider the function $q(x) = \dfrac{3x^7 - 5x + 2}{x^3 - x}$. The quotient rule guarantees that q is differentiable except where the denominator is zero.

   (a) Use technology to plot q; notice the two vertical asymptotes. How are they related to derivatives?

   (b) The number one is a root of the denominator, but the graph of q has no vertical asymptote there. Why not?

   (c) The value $q(1)$ is undefined in the formula above. What value of $q(1)$ makes q differentiable at $x = 1$? Why?

9. Let $f : \mathbb{R} \to \mathbb{R}$ be everywhere differentiable, with $f(a) = 0$, and let $g = f^2$. Show that $g(a) = g'(a) = 0$.

10. Let $g : \mathbb{R} \to \mathbb{R}$ be everywhere differentiable, with $g'(n) = 0$ for all integers n.

(a) Let $f : \mathbb{R} \to \mathbb{R}$ be another everywhere differentiable function. Show that $(f \circ g)'(n) = 0$ for all integers n.

(b) Let $h(x) = \cos(\pi x) \cdot g(x)$. Show that $h'(n) = 0$ for all integers n.

11. Suppose throughout this problem that f is differentiable at $x = 17$. Let $g(x) = |f(x)|$.

(a) Show that $g(x)$ is continuous at $x = 17$.

(b) Give an example to show that $g(x)$ *need not* be differentiable at $x = 17$.

(c) Show that $g(x)^2$ *is* differentiable at $x = 17$.

12. Let k be a positive integer, like 3 or 4. A function $f : \mathbb{R} \to \mathbb{R}$ is said to vanish to order k at $x = a$ if

$$f(a) = 0 = f'(a) = f''(a) = f'''(a) = \cdots = f^{(k-1)}(a), \text{ but } f^{(k)}(a) \neq 0.$$

(a) Show that x^3 vanishes to order 3 at $x = 0$.

(b) To what order does $x^2 \sin(x)$ vanish at $x = 0$? Why?

(c) Show carefully that if f vanishes to order 2 at $x = a$ and g vanishes to order 3 at $x = a$, then fg vanishes to order 5 at $x = a$.

4.3   The Mean Value Theorem

The mean value theorem (MVT) is a marquee result of real analysis. As the intermediate value theorem (IVT) and the extreme value theorem (EVT) do for *continuous* functions, the mean value theorem addresses a natural and fundamental property of *differentiable* functions.

Theorem 4.9 (The mean value theorem). *Suppose $f : [a, b] \to \mathbb{R}$ is continuous on $[a, b]$ and differentiable on (a, b). Then there exists $c \in (a, b)$ such that*

$$f'(c) = \frac{f(b) - f(a)}{b - a}.$$

Like the IVT and the EVT, the MVT says that a certain function—f', this time—assumes a certain "mean" (or "average") value. Unexciting as this might seem, it turns out to be a big deal, with surprisingly broad and deep implications. We explore several before proving the theorem.

The graphical view. The MVT equation says something about *slopes* on the graph of f: At some point c between a and b, the *tangent* line is parallel to the *secant* line joining $(a, f(a))$ and $(b, f(b))$. Figure 4.5 illustrates this. The (linear) secant line function, labeled $L(x)$ in the picture, will play a role in the proof.

The horizontal case: Rolle's theorem. If we add to the other hypotheses the requirement that $f(a) = f(b)$, then the conclusion takes a slightly simpler form:

Theorem 4.10 (Rolle's theorem). *Suppose $f : [a, b] \to \mathbb{R}$ is continuous on $[a, b]$ and differentiable on (a, b), and that $f(a) = f(b)$. Then $f'(c) = 0$ for some $c \in (a, b)$.*

In graphical terms, the new hypothesis says that the graph has the same height at $x = a$ and at $x = b$; the conclusion says that the graph has a horizontal tangent somewhere in between. Figure 4.6 shows *two* such horizontal tangents.

More than one possible c is fine with Rolle.

Car talk. If $f(t)$ is the *position* of a car at time t, then $(f(b) - f(a))/(b - a)$ is the car's *average* velocity over the time interval $[a, b]$, and $f'(t)$ is the car's *instantaneous* velocity at time t. At some intermediate time c, says the MVT (and common sense), these two velocities must be equal. Rolle's version sounds even more intuitive: if a car starts and ends at the same position, then it must be *stopped* at some instant in between.

The units might be miles and hours.

When is a function constant? It is obvious, and easy to prove, that a constant function has zero derivative. The converse—a function with constant zero derivative is constant—is less obvious than it might seem. For example, the function

$$s(x) = \begin{cases} 1 & \text{if } x > 0 \\ -1 & \text{if } x < 0 \end{cases}$$

is nonconstant, but $s'(x) = 0$ for all $x \neq 0$. If we rule out such "disconnected" domains, the converse *is* true—but the proof requires the MVT. Here are the details.

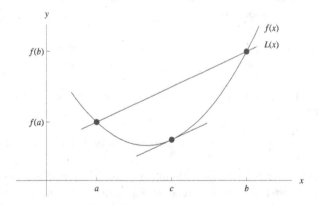

Figure 4.5. The mean value theorem illustrated.

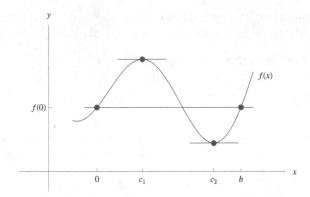

Figure 4.6. Rolle's theorem: two possible c-values.

Proposition 4.11. *Let I be an open interval and $f : I \to \mathbb{R}$ a function. If $f'(x) = 0$ for all $x \in I$, then f is constant on I.*

Proof: Let s and t be any two points in I. By the MVT, there is a point c between s and t for which

$$f'(c) = \frac{f(s) - f(t)}{s - t}.$$

Since the left side is zero, so is the right, which means $f(s) = f(t)$, as desired.□

Proving the Mean Value Theorem

Proving the MVT takes some effort—not unexpectedly, given the theorem's importance and usefulness. Fortunately, we've already done most of the hard work. We'll argue in three short steps:

- *Step 1:* Show that if $f : [a, b] \to \mathbb{R}$ is as hypothesized, and f has a local maximum or local minimum at $c \in (a, b)$, then $f'(c) = 0$.

- *Step 2:* Use Step 1—and another famous "value theorem"—to prove Rolle's theorem.

- *Step 3:* Apply Rolle's theorem in a clever way to deduce the MVT.

Find the largest rectangular pigpen . . . one side is a river . . . .

About Step 1. Our Step 1 claim should be familiar, for at least two reasons. First, it lurks behind most of those classic maximum–minimum problems of elementary calculus. Second, we've already proved it (and a bit more), as part (c) of Lemma 4.4, page 214. Let's move on.

About Step 2. We assume that $g : [a, b] \to \mathbb{R}$ is continuous on $[a, b]$ and differentiable on (a, b), and that $g(a) = g(b)$; we need to show that $g'(c) = 0$ for some $c \in (a, b)$.

We use g, not f, to avoid confusion in Step 3.

To apply Step 1, we invoke the EVT: Because g is *continuous* on $[a, b]$, it assumes both maximum and minimum values somewhere on that closed interval. If *both* the maximum and the minimum are assumed at endpoints, then the maximum and minimum values are equal, and so g is constant. In this case, $g'(c) = 0$ for *all* $c \in (a, b)$, and we're done already. The alternative is that at least one of the maximum and the minimum is attained somewhere in (a, b). In this case Step 1 applies, and again we're done.

The extreme value theorem; see page 175

About Step 3. Rolle's theorem is more than a junior-grade version of the MVT. With a little ingenuity, the latter can be deduced from the former. The trick is to apply Rolle's theorem not to the function f given in the MVT, but to another function g, cleverly contrived from f. We sketch the argument, leaving details to the exercises.

A common one in analysis.

1. Let $L(x)$ be the *linear* function whose graph joins the points $(a, f(a))$ and $(b, f(b))$; see Figure 4.5. Notice some properties of L:

   (i) $L(a) = f(a)$; (ii) $L(b) = f(b)$; (iii) $L'(x) = \dfrac{f(b) - f(a)}{b - a}$ for all x.

2. If we set $g(x) = f(x) - L(x)$, then $g(a) = 0 = g(b)$. By Rolle's theorem, there exists $c \in (a, b)$ with $g'(c) = 0$. This c satisfies the conclusion of the MVT, and the proof is complete.

Calculus Friends Revisited: Using the MVT

Standard ideas of elementary calculus link properties of a function f to those of its derivatives f' and f''. Using the MVT, we can prove many such facts—and some but not all of their converses. The following proposition offers some samples.

Proposition 4.12. *Suppose that $f : I \to \mathbb{R}$ is differentiable at all points of an open interval I.*

(i) *$f'(x) \geq 0$ for all $x \in I$ if and only if f is increasing on I.*

f need not be strictly increasing.

(ii) *If $f'(x) > 0$ for all $x \in I$, then f is* strictly *increasing on I.*

(iii) *If $f'(x) \neq 0$ for all $x \in I$, then f is monotone on I.*

Proof(s): For the "only if" part of (i), suppose that $f'(x) \geq 0$ for all $x \in I$. To show f is increasing, let s and t be in I, with $s < t$. By the MVT, we have, for some c between s and t,

$$\frac{f(t) - f(s)}{t - s} = f'(c) \quad \text{or, equivalently,} \quad f(t) - f(s) = f'(c)(t - s).$$

By hypothesis, the rightmost quantity is nonnegative. This implies that $f(t) \geq f(s)$, which means that f is increasing, as claimed.

For the "if" part of (i), suppose that f is increasing on I. For any $x \in I$, we have

$$f'(x) = \lim_{t \to x} \frac{f(t) - f(x)}{t - x} = \lim_{t \to x^+} \frac{f(t) - f(x)}{t - x},$$

See Proposition 3.6, page 157.

where the first equality holds by definition of $f'(x)$ and the second by a familiar property of limits. Because f is increasing, we have

$$\frac{f(t) - f(x)}{t - x} \geq 0$$

for all $t \in I$ with $t > x$. Thus $f'(x)$ is the limit of a *nonnegative* expression, and so must itself be nonnegative.

This completes the proof of (i); we leave remaining parts to the exercises. $\square$

Onward and Upward with the MVT

The mean value theorem has many and varied uses. Here we'll briefly glimpse two of them.

Darboux's theorem: derivatives and intermediate values. The intermediate value says that a continuous function $f : I \to \mathbb{R}$ defined on an interval I has the "intermediate value property":

> If s and t are in I and $f(s) < v < f(t)$, then $f(c) = v$ for some v between s and t.

It was widely believed in the nineteenth century that functions with the intermediate value property *must be* continuous.

This is false. In the late 1800s, Jean Gaston Darboux proved (essentially) the following theorem (a proof, based on the mean value theorem, is outlined in the exercises):

Theorem 4.13 (Darboux's theorem). *Suppose $f : I \to \mathbb{R}$ is continuous on $[a, b]$ and differentiable on (a, b). Then f' has the intermediate value property.*

Another well-known fact—also proved in exercises—is that a derivative function (f' in the notation above) need not be continuous.

EXAMPLE 1.   Consider the function $f(x) = x^2 \sin(1/x)$; set $f(0) = 0$. Discuss the derivative f'.

SOLUTION. For $x \neq 0$ the function is clearly differentiable; standard derivative rules apply. At $x = 0$, the function f is also differentiable, but the derivative f' is discontinuous. Details are explored in the exercises. $\Diamond$

Taylor's theorem: the MVT generalized. The mean value theorem equation can be rewritten to say that, under the appropriate hypotheses, we have

$$f(b) = f(a) + f'(c)(b - a)$$

for some number c between a and b. This equation can be thought of as saying that $f(b)$ approximates $f(a)$ with error no larger than the last term. Taylor's theorem, dating from around 1700 and named for the Scottish mathematician Brook Taylor, offers a more powerful version of such an approximation. Here is a special case:

Theorem 4.14 (Taylor's theorem, $n = 3$). *Suppose f, f', f'', and f''' all exist and are continuous on $[a, b]$. Then, for some input c in $[a, b]$, we have*

$$f(b) = f(a) + f'(a)(b - a) + \frac{f''(a)}{2}(b - a)^2 + \frac{f'''(c)}{6}(b - a)^3.$$

A proof is outlined in the exercises. Here is an application.

EXAMPLE 2. What does Taylor's theorem say about $f(x) = \cos x$ for $a = 0$ and $b = 1$?

SOLUTION. Applying Taylor's formula gives

$$\cos 1 = \cos(0) - \sin(0) \cdot 1 - \frac{\cos(0)}{2} \cdot 1^2 + \frac{\sin(c)}{6} \cdot 1^3 = 1 - \frac{1}{2} + \frac{\sin(c)}{6}.$$

Now, for $c \in [0, 1]$ we have $\sin c \leq \sin 1 \approx 0.84$, so the last summand above can't exceed 0.14. This means that $\cos 1$ differs from 0.5 by less than 0.14, which is easily checked to be true. (In fact, $\cos 1 \approx 0.54$.) So Taylor's theorem might help us calculate cosines in an emergency. $\Diamond$

Exercises

1. Let $f : R \to R$ be a differentiable function, and suppose that f has n real roots. Use Rolle's theorem to show that f' has *at least* $n - 1$ real roots. Give an example to show that f' may have more than $n - 1$ real roots.

2. (a) Use algebra to show that $p(x) = x^3 + x$ has exactly one (real) root.

   (b) Use Rolle's theorem to show (again) that $p(x) = x^3 + x$ has exactly one root. (Hint: If there are two roots, Rolle's theorem leads to a contradiction.)

(c) Show that $q(x) = x^{13} + x^{11} + x^9 + x^7 + x^5 + x^3 + x$ has exactly one root.

(d) Show that $r(x) = x^{13} + x^{11} + x^9 + x^7 + x^5 + x^3 + x + 987,654,321$ has exactly one root.

3. Let $f(x) = x^3 + ax + b$, where a and b are fixed real numbers.

   (a) Explain why $f(x)$ must have *at least* one root, regardless of the values of a and b.

   (b) Give examples (i.e., specific values of a and b) to show that $f(x)$ can have one, two, or three roots. No formal proofs needed, but say briefly why your examples work.

   (c) Prove that f can have no more than three roots. (Hints: (i) Take it as given that a *quadratic* polynomial can have at most two roots; (ii) use Rolle's theorem.)

   (d) Show that if $a > 0$, then f has *exactly* one root.

4. Use Rolle's theorem to show by induction that a nonconstant polynomial p of degree n has at most n roots. (Hint: If p has degree n, then p' has degree $n - 1$.)

5. Suppose $f : [a, c] \to \mathbb{R}$ has derivatives f' and f'', both continuous on $[a, b]$, and assume that there exists b in (a, c) with $f(a) = f(b) = f(c)$.

   (a) Show that there exists x_1 and x_2 with $a < x_1 < x_2 < c$ and $f'(x_1) = f'(x_2) = 0$.

   (b) Show that there exists x_0 with $a < x_0 < c$ and $f''(x_0) = 0$.

   (c) State a generalization of these hypotheses that guarantees there exists x_0 with $f^{(42)}(x_0) = 0$.

6. Let $f : \mathbb{R} \to \mathbb{R}$ have continuous derivatives f' and f'', and suppose that $f(0) = 0$, $f(1) = 1$, and $f(2) = 2$. Show that $f''(x_0) = 0$ for some x_0 in $(0, 2)$.

7. Suppose that $f : \mathbb{R} \to \mathbb{R}$ is periodic with positive period a (i.e., $f(x + a) = f(x)$ for all x), and that f has continuous derivatives of all orders. Show that $f', f'', f''', \ldots$ all have infinitely many roots.

8. Suppose $f : \mathbb{R} \to \mathbb{R}$ has first and second derivatives $f' : \mathbb{R} \to \mathbb{R}$ and $f'' : \mathbb{R} \to \mathbb{R}$.

   (a) Show that if $f'(x) > M$ for all x in an interval I, then $f(b) - f(a) > M(b - a)$ for all inputs a and b in I with $b > a$.

(b) Suppose $f(0) = 0$, $f'(0) = 0$ and $f''(x) > 2$ for all $x > 0$. Show that $f(x) > x^2$ for all $x > 0$.

9. In each part, find all possible values of c as described in the MVT.

   (a) $f(x) = 3x + 5$; $[a, b] = [1, 7]$.

   (b) $f(x) = 3x^2 + 5x + 7$; $[a, b] = [1, 7]$.

   (c) $f(x) = x^{100}$; $[a, b] = [0, 1]$.

   (d) $f(x) = 3x^2 + 5x + 7$; $[a, b] = [a, b]$ (the answer involves a and b).

   (e) If f is any quadratic function and $[a, b]$ any interval, then the MVT equation holds with c taken to be the *midpoint* of $[a, b]$. Make this precise and prove it.

10. In each part following, decide whether a value of c can be found as described in the MVT. If so, find one. If not, say which of the MVT hypotheses is not satisfied.

   (a) $f(x) = |x|$; $[a, b] = [-1, 1]$.

   (b) $f(x) = |x^2 - 10x + 21|$; $[a, b] = [4, 6]$.

   (c) $f(x) = |x^2 - 10x + 21|$; $[a, b] = [2, 4]$.

11. Suppose $f(t)$ is the *eastward velocity* of a car at time t, with all quantities measured in appropriate units. What does Rolle's theorem say in this setting? (Hint: $f'(t)$ measures *eastward acceleration*.)

12. One hypothesis of the MVT is that $f : [a, b] \to \mathbb{R}$ be continuous on all of $[a, b]$. Give an example (as simple as possible) to show that this hypothesis is necessary—i.e., find a function $f : [a, b] \to \mathbb{R}$, differentiable throughout (a, b), for which the MVT fails.

13. Suppose f and g are functions continuous on $[a, b]$ and differentiable on (a, b).

   (a) Use results in this section to show that if $f'(x) = g'(x)$ for all $x \in (a, b)$, then $f(x) = g(x) + C$ for some constant C. (This is sometimes called an *identity theorem* for differentiable functions.)

   (b) Suppose that $f'(x) = g'(x) + 5$ for all $x \in (a, b)$. What can be said about $f(x)$ and $g(x)$? Why?

14. Antiderivative tables in elementary calculus books are full of formulas like $\int \cos x \, dx = \sin x + C$.

   (a) Why exactly is the C there? What does this have to do with the MVT?

Forget the C on a test and lose a point.

(b) A very fussy professor might find fault with the formula $\int \frac{dx}{x} = \ln|x| + C$. Explain. (Hint: Consider domains.)

15. Give a detailed proof of Step 3 in the proof of the MVT. In particular, give an explicit formula for $L(x)$ and use it to verify the other claims.

16. This problem is about various parts of Proposition 4.12, page 233.

    (a) State (don't prove) versions of (i) and (ii) that involve the word "decreasing."

    (b) Prove (ii) carefully; mimic the proof of (i).

    (c) Prove (iii).

17. Let $f : \mathbb{R} \to \mathbb{R}$ be differentiable for all x, with $f'(x) < 1$ for all x and $f(0) = 0$.

    (a) Show that $f(x) < x$ for all $x > 0$.

    (b) Does a similar inequality hold for $x < 0$? Explain.

18. Let $f : \mathbb{R} \to \mathbb{R}$ be differentiable for all x. Suppose that $f'(x) > 1$ for all x. Show that the graph of $y = f(x)$ can intersect the graph of $y = x$ no more than once.

19. Let $f : \mathbb{R} \to \mathbb{R}$ be differentiable for all x. Suppose that $f'(x) < 1$ for all x and that $f(0) = 0$. Show that $f(x) < x$ for all $x > 0$. Must $f(x) < x$ hold also for $x < 0$? Why or why not?

20. A function $f : \mathbb{R} \to \mathbb{R}$ is called a *contraction* if the inequality

    $$|f(x) - f(y)| \leq |x - y|$$

    holds for all real numbers x and y. (As the name suggests, a contraction "shrinks distances.")

    (a) Show that if f is a contraction, then f is uniformly continuous on D.

    (b) Suppose that $f : \mathbb{R} \to \mathbb{R}$ is differentiable, and that $|f'(x)| \leq 1$ for all x. Show that f is a contraction. (Hint: MVT.)

    (c) Is $f(x) = \sin x$ a contraction? Is $g(x) = e^x$ a contraction? Why?

21. Prove Darboux's theorem (Theorem 4.13) as follows:

    (a) Suppose first that $a < s < t < b$, and $f'(s) < 0 < f'(t)$. Show that f must attain a *minimum* at some point c strictly between s and t. At this c, $f'(c) = 0$.

(b) Now suppose that $a < s < t < b$, and v is *any* number such that $f'(s) < v < f'(t)$. Consider the new function $g(x) = f(x) - vx$. Then g satisfies the conditions of the previous part. Apply the result of (a) to complete the proof.

22. In the situation of Example 1, find a (two-part) formula for f', and show that f' is not continuous at $x = 0$.

23. In the special case $a = 0$ and $b = 1$, Taylor's theorem (Theorem 4.14) says that $f(1) = f(0) + f'(0) + \dfrac{f''(0)}{2} + \dfrac{f'''(c)}{6}$ for some $c \in (0, 1)$. Prove this in the following steps.

   (a) Consider the function $p_2(t) = f(0) + f'(0)t + \dfrac{f''(0)}{2}t^2$. Show that $f(0) = p_2(0)$, $f'(0) = p_2'(0)$, and $f''(0) = p_2''(0)$.

   (b) Consider the new function $g(t) = f(t) - p_2(t) + (p_2(1) - f(1))t^3$. Show that $g(0) = 0 = g(1)$; conclude that there is some t_1 in $(0, 1)$ with $g'(t_1) = 0$.

   (c) Show that $g'(0) = 0 = g'(t_1)$, so there exists t_2 in $(0, t_1)$ with $g''(t_2) = 0$.

   (d) Show that $g''(0) = 0 = g''(t_2)$, so there exists c in $(0, t_2)$ with $g'''(c) = 0$. Conclude that Taylor's theorem holds in the case at hand.

24. Prove Taylor's theorem (Theorem 4.14) without assuming $a = 0$ and $b = 1$, as we did in the preceding problem. (Hint: Given a function f as in Theorem 4.14, define a new function $h(x) = f(a + x(b - a))$. Apply the result of the preceding problem to h; calculate carefully to derive the general case.)

4.4   Sequences and Series of Functions

In Chapter 2, we studied convergence for a sequence $\{a_n\}$ of *numbers*—under appropriate conditions, we say that the numbers a_n converge to another *number a*.

A sequence $\{f_n\}$ of *functions* may also converge or diverge to a limit *function*. We explore some basic examples before giving formal definitions.

EXAMPLE 1. Consider the functions $f_1(x) = x$, $f_2(x) = x^2$, $f_3(x) = x^3$, $\ldots$, $f_n(x) = x^n$. Figure 4.7(a) shows graphs of several f_n on the interval $[0, 1]$; Figure 4.7(b) shows one typical function, f_{20}. Does the sequence $\{f_n\}$ converge to some limit function?

SOLUTION.   As Figure 4.7(a) hints, the behavior of a sequence $\{f_n\}$ depends strongly on the domain we specify. Viewed on the interval $(0, 0.8)$, for instance, the graphs of f_n appear to become flatter and flatter, approaching the function $f(x) = 0$. At $x = 1$, on the other hand, we have $f_n(1) = 1$ for all n; graphs of all the f_n pass through the point $(1, 1)$. On the domain interval $[0, 1]$, therefore, the f_n appear to converge—in some sense yet to be defined—to the function

$$f(x) = \begin{cases} 0 & \text{if } 0 \leq x < 1, \\ 1 & \text{if } x = 1. \end{cases}$$

We also observe that, although each f_n is continuous on $[0, 1]$, the limit function f is discontinuous at $x = 1$. $\Diamond$

EXAMPLE 2.   Consider the functions

$$g_n(x) = x + \frac{\sin(nx)}{n}, \quad n = 1, 2, 3, \ldots.$$

Figure 4.8(a) shows several g_n on the domain $[-5, 5]$; Figure 4.8(b) shows only g_{20}. Does the sequence $\{g_n\}$ converge to a limit function?

SOLUTION.   The graphs suggest—correctly—that the g_n, although wavy for large n, approach the linear function $g(x) = x$. The formulas agree—for each n, the inequality

$$|g_n(x) - g(x)| = \left| \frac{\sin(nx)}{n} \right| \leq \frac{1}{n}$$

holds for all inputs x in $[-5, 5]$, the domain interval shown. By contrast with the situation in Example 1, moreover, the pictures suggest that the same sort of convergence occurs on *every* domain interval. $\Diamond$

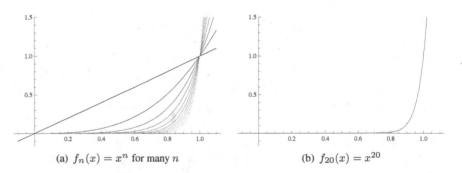

(a) $f_n(x) = x^n$ for many n (b) $f_{20}(x) = x^{20}$

Figure 4.7. Exploring the function sequence $\{f_n(x) = x^n\}$.

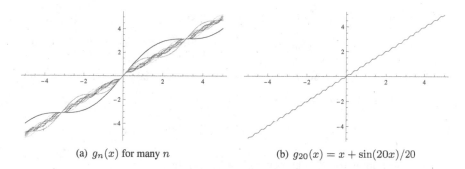

(a) $g_n(x)$ for many n (b) $g_{20}(x) = x + \sin(20x)/20$

Figure 4.8. Exploring the sequence $\{g_n(x) = x + \sin(nx)/n\}$.

What the examples show. The examples illustrate that a sequence of functions may behave quite differently on one domain than on another. On the domain $[0, 1/2]$, for example, the functions $\{f_n\}$ in Example 1 converge to the zero function. On the domain $[0, 1]$, too, these functions appear to converge, but to a limit function that is *discontinous* at $x = 1$. On the domain $[2, 3]$, by contrast, the same functions become larger and larger with n, and so have *no* sensible limit function. The functions $\{g_n\}$ in Example 1, on the other hand, appear to converge to $g(x) = x$ on *every* domain interval.

Not shown in Figure 4.7, but easily visualized.

Two Definitions

Let f_1, f_2, f_3, $\ldots$ be a sequence of real-valued functions, all defined on a fixed interval I. To sort out similarities and differences in the behaviors illustrated in Examples 1 and 2, we'll define *two* senses in which $\{f_n\}$ might converge to a limit function f, also defined on I.

Definition 4.15 (Pointwise convergence). The sequence $\{f_n\}$ converges *pointwise* to f on I if, for every fixed $x \in I$, we have $f_n(x) \to f(x)$ as $n \to \infty$.

Definition 4.16 (Uniform convergence). The sequence $\{f_n\}$ converges *uniformly* to f on I if for each $\epsilon > 0$ there exists N such that for every $n > N$ the inequality $|f_n(x) - f(x)| < \epsilon$ holds for all $x \in I$.

How do the sequences in Examples 1 and 2 stack up against these definitions?

- *Both sequences converge pointwise:* Both $\{f_n\}$ and $\{g_n\}$ converge *pointwise* to their limits, at least on the domain intervals shown. For $\{f_n\}$ we have

$$\lim_{n \to \infty} f_n(x) = \lim_{n \to \infty} x^n = 0 = f(x)$$

for all $x \in [0,1)$. If $x = 1$, by contrast, then

$$\lim_{n \to \infty} f_n(1) = \lim_{n \to \infty} 1^n = 1 = f(1).$$

Thus $\{f_n\}$ converges pointwise to f on $[0,1]$, as claimed.

For $\{g_n\}$, the squeezing inequality

$$x - \frac{1}{n} \le x + \frac{\sin(nx)}{n} \le x + \frac{1}{n},$$

which holds for all n and every fixed x, implies that

$$\lim_{n \to \infty} g_n(x) = \lim_{n \to \infty} \left(x + \frac{\sin(nx)}{n} \right) = x = g(x)$$

We could have used any
domain interval.

for all $x \in [-5,5]$. Thus $\{g_n\}$, too, converges pointwise to a limit function g, this time on the domain $[-5,5]$.

- *Only one sequence converges uniformly:* Both $\{f_n\}$ and $\{g_n\}$ converge *pointwise* to their limit functions; $\{g_n\}$ also converges *uniformly*. Showing this is straightforward. For given $\epsilon > 0$, the number $N = 1/\epsilon$ works: if $n > N$, then we have

$$|g_n(x) - g(x)| = \left| \frac{\sin(nx)}{n} \right| \le \frac{1}{n} < \frac{1}{N} = \epsilon,$$

and the inequality holds for *all* x in $[-5,5]$, as the definition requires.

The sequence $\{f_n\}$ does *not* converge uniformly to f on $[0,1]$. The problem, roughly speaking, is that the limit function f "jumps" abruptly at $x = 1$, while all the f_n increase smoothly. We explore this idea more carefully in the next example.

EXAMPLE 3.   The functions $f_n(x) = x^n$ converge *pointwise* on $[0,1]$ to the limit function f, with $f(x) = 0$ for $x \ne 1$ and $f(1) = 1$. Show that $\{f_n\}$ does *not* converge uniformly to f on $[0,1]$.

SOLUTION.   Suppose toward contradiction that $\{f_n\}$ *does* converge uniformly. Set $\epsilon = 0.1$, choose N as in Definition 4.16, and consider the function f_{N+1}. Then, by design,

$$|f_{N+1}(x) - f(x)| < 0.1$$

holds for all $x \in [0,1]$. Because $f(1) = 1$ and $f(x) = 0$ for $x \ne 1$, the preceding inequality boils down to two conditions:

(i) $|f_{N+1}(x)| < 0.1$ if $x \ne 1$;   (ii) $|f_{N+1}(1) - 1| < 0.1$.

In particular, we must have either $f_{N+1}(x) < 0.1$ or $f_{N+1}(x) > 0.9$ for *all* $x \in [0, 1]$. This is clearly absurd—the continuous function $f_{N+1}(x) = x^{N+1}$ must assume *every* value between zero and one. We've shown that uniform convergence fails.   ◊

On the other hand ...

EXAMPLE 4.   Show that the same sequence $\{f_n\}$ *does* converge uniformly on $[0, 0.99]$, with limit $f(x) = 0$.

This is the same limit function as before, but chopped off at $x = 0.99$.

SOLUTION.   Let $\epsilon > 0$ be given. Choose a number N such that $0.99^N < \epsilon$. Any such N works, because for all $n > N$ and all $x \in [0, 0.99]$, we have

Why can this be done? How big is N if $\epsilon = 0.001$?

$$|f_n(x) - f(x)| = |x^n - 0| = x^n \le 0.99^n < 0.99^N < \epsilon,$$

as desired.   ◊

Basic properties. Proposition 4.17 collects, as samples, some basic properties of pointwise and uniform convergence. Proofs and additional properties are left as exercises.

Proposition 4.17. *Let I be an interval, and let $\{f_n\}$ and $\{g_n\}$ be sequences of real-valued functions, all defined on I.*

- Uniform convergence is stronger: *If $f_n \to f$ uniformly, then $f_n \to f$ pointwise, too. The converse is false.*

- Algebraic combinations: *Suppose $f_n \to f$ and $g_n \to g$ pointwise on I, where f and g are functions. Then*

$$f_n \pm g_n \to f \perp g \quad and \quad f_n g_n \to fg,$$

pointwise on I in each case.

- Patching it together: *Let I_1 and I_2 be intervals with $I = I_1 \cup I_2$. If $f_n \to f$ uniformly (or pointwise) both on I_1 and on I_2, then $f_n \to f$ uniformly (or pointwise) on I.*

Properties of Limit Functions

Must the limit of a sequence "inherit" properties from members of the sequence? For a sequence $\{x_n\}$ of *numbers*, for instance, we know that if $x_n \ge 0$ for all n and $x_n \to L$, then $L \ge 0$, too. On the other hand, the limit of a strictly positive sequence need *not* be positive. Similar questions arise for sequences of functions.

Consider $\{1/n\}$, for instance.

EXAMPLE 5.   Let $\{f_n\}$ be a sequence of functions defined on an interval I, and suppose $f_n \to f$ pointwise on I. If $f_n(x) \geq 0$ for all $x \in I$ and for all $n > 0$, must $f(x) \geq 0$ for all $x \in I$? If all the f_n are continuous on I, must f be continuous on I, too?

SOLUTION.   The answers are respectively yes and no. If we fix any x in I, then $\{f_n(x)\}$ is a sequence of nonnegative *numbers*, and so its limit, $f(x)$, is also nonnegative, as claimed. As for continuity, we saw in Example 1 that the continuous functions $f_n(x) = x^n$ converge pointwise on $[0, 1]$ to a discontinuous limit—the function f defined $f(x) = 0$ if $x \neq 0$ and $f(1) = 1$.   ◊

And differentiable, for that matter.

So the *pointwise* limit of continuous functions need not be continuous. The situation is different for *uniform* limits:

Proposition 4.18. *Let $\{f_n\}$ be a sequence of functions, converging uniformly on an interval I to a function f. If all the f_n are continuous on I, then f is continuous on I, too.*

The proof idea. The formal proof is clever and slightly technical, but the basic idea is straightforward. Continuity of f at $x = a$ means, roughly, that if $x \approx a$, then $f(x) \approx f(a)$. This holds in the present case because we know (i) $f_n(x) \approx f(x)$ for large n and for *all* x in I; (ii) $f_n(x) \approx f_n(a)$ when $x \approx a$ because f_n is continuous at $x = a$. Putting these conditions together gives, for $x \approx a$,

$$f(x) \approx f_n(x) \approx f_n(a) \approx f(a).$$

Watch for three $\epsilon/3$'s.

We make this three-step approximation precise in the formal proof.

Proof: Let a be any point of I. To show that f is continuous at a, let $\epsilon > 0$ be given. To find a $\delta > 0$ that "works" for ϵ in the definition of continuity, we argue in three steps.

First, we choose any particular integer N such that

$$|f(x) - f_N(x)| < \frac{\epsilon}{3} \quad \text{for all } x \text{ in } I;$$

this is possible because $\{f_n\}$ converges uniformly to f. Second, because f_N is continuous at $x = a$, we can choose $\delta > 0$ such that

$$|f_N(x) - f_N(a)| < \frac{\epsilon}{3} \quad \text{whenever} \quad |x - a| < \delta.$$

Third, we note that, since $a \in I$, our choice of N guarantees

$$|f_N(a) - f(a)| < \frac{\epsilon}{3}.$$

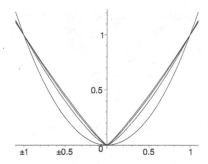

Figure 4.9. Smooth functions approaching the absolute value function.

Putting these inequalities together shows that our chosen δ does what we need it to do: if $|x - a| < \delta$, then

By a three-part triangle inequality.

$$|f(x) - f(a)| \le |f(x) - f_N(x)| + |f_N(x) - f_N(a)| + |f_N(a) - f(a)|$$
$$< \frac{\epsilon}{3} + \frac{\epsilon}{3} + \frac{\epsilon}{3} = \epsilon,$$

as desired. $\qquad\qquad\square$

Derivatives and limits. The uniform limit of a sequence of continuous functions is continuous, as just shown. Can anything be similar be said about *differentiable* functions?

As Example 1 illustrates, the *pointwise* limit of differentiable functions need not even be continuous, let alone differentiable. The following example shows that the *uniform* limit of differentiable functions need not be differentiable, either.

EXAMPLE 6. For each positive integer n, define $f_n(x) = |x|^{1+1/n}$. Figure 4.9 shows several of the f_n; the sequence appears to approach the absolute value function. What is happening with derivatives?

SOLUTION. The functions f_n behave as Figure 4.9 suggests. Writing $f(x) = |x|$, we can readily show:

- Each f_n is differentiable at all real x. This is clear for $x \ne 0$, and not hard to prove if $x = 0$.

See the exercises.

- $f_n \to f$ *uniformly* on $[-1, 1]$. This follows from the fact, readily shown, that
$$|f(x) - f_n(x)| < \frac{1}{n}$$
for all $x \in [-1, 1]$.

Again, see the exercises.

The point is that, although all the f_n are differentiable everywhere, their limit f is *not* differentiable at the origin. Thus, differentiability—unlike continuity—is not preserved by uniform limits.   ◊

Series of Functions

See Definition 2.21, page 120.

We saw in Section 2.5 the close relationship between series and sequences: Any given *series*

$$\sum_{k=1}^{\infty} a_k = a_1 + a_2 + a_3 + \ldots$$

determines the associated *sequence* $\{A_n\}$ of partial sums:

$$A_1 = a_1, \quad A_2 = a_1 + a_2, \quad \ldots, \quad A_n = a_1 + a_2 + \cdots + a_n, \quad \ldots$$

Convergence or divergence of the series $\sum a_k$ now reduces to the same questions for the sequence $\{A_n\}$. Per Definition 2.22, page 122, the series $\sum a_k$ converges to the sum A if and only if the sequence $\{A_n\}$ of partial sums converges to A.

Exactly the same approach works for series of functions.

Definition 4.19 (Convergence of function series). Let $g_1, g_2, g_3, \ldots$ be a sequence of functions, all defined on a fixed interval I. Define the partial sum functions $f_1, f_2, f_3, \ldots$ by

$$f_1(x) = g_1(x); \quad f_2(x) = g_1(x)+g_2(x); \quad f_3(x) = g_1(x)+g_2(x)+g_3(x); \quad \ldots$$

The function series $\sum_{n=1}^{\infty} g_n$ converges (pointwise or uniformly) to the sum function g on I if and only if the function sequence $f_1, f_2, f_3, \ldots$ converges (pointwise or uniformly) to g on I.

Power series. *Power series* are function series in which the summands are "monomials," like $3x^7$ and $x^5/120$, and the partial sums are polynomials—familiar, convenient, and well-behaved functions in their own right. Power series arise naturally and usefully in theory and applications. Taylor and Maclaurin series are crucial examples; we explore them briefly in the next section.

You've probably met these in calculus courses.

The topic of function series in general, and power series in particular, is vast. We take just a quick glance to suggest the topic's contours. The next example concerns the simplest and arguably most useful power series; note the close connection to geometric series already studied.

EXAMPLE 7.   Make sense of the function series

$$\sum_{k=0}^{\infty} x^k = 1 + x + x^2 + x^3 + \dots.$$

Does the power series converge to some function g on some interval I? Is the convergence pointwise? Uniform?

SOLUTION.   In the notation of Definition 4.19 we have $g_n(x) = x^n$ for $n \geq 0$, and the partial sum functions are

$$f_0(x) = 1; \quad f_1(x) = 1+x; \quad f_2(x) = 1+x+x^2; \, f_3(x) = 1+x+x^2+x^3; \quad \dots.$$

Here's the key point: For each x, $1 + x + x^2 + x^3 + \dots$ is a *geometric series* in powers of x, and and therefore converges if (but only if) $|x| < 1$. (Review geometric series around page 125, if necessary.) We even know the limit:

$$\sum_{k=0}^{\infty} x^k = 1 + x + x^2 + x^3 \dots = \frac{1}{1-x} \quad \text{if } |x| < 1;$$

the series diverges for $|x| \geq 1$. In the language of function series: the geometric power series *converges pointwise* for $x \in I = (-1, 1)$ to the function $g(x) = 1/(1-x)$.

Convergence in this case is not uniform on all of $I = (-1, 1)$, but it is uniform if we restrict x to smaller closed intervals, such as $[-0.8, 0.8]$. To see why, let $\epsilon > 0$ be given. Because $\sum 0.8^n$ converges, we can choose N such that $\sum_{k=N+1}^{\infty} 0.8^k < \epsilon$. This is another way of saying that $f_n(x) = \sum_{k=0}^{N} 0.8^k$ differs from the $\sum_{k=0}^{\infty} 0.8^k$ (which is $g(0.8)$) by less than ϵ. Similar reasoning applies to any $x \in [-0.8, 0.8]$: If $n > N$,

$$|g(x) - f_n(x)| = \left| \sum_{k=n+1}^{\infty} x^k \right| \leq \sum_{k=N+1}^{\infty} 0.8^k < \epsilon,$$

as uniform convergence requires.   ◊

More on power series. We saw above and will see again below that the limits of convergent function sequences and series may be functions that are less well-behaved (e.g., not differentiable) than the terms or summands that approach those limits.

Power series turn out to be especially well-behaved in this and other respects. Following, without proofs, are some special properties of power series not necessarily shared by arbitrary function series.

Proofs are within our reach, but complicated and off our main path.

Proposition 4.20. *Consider a power series $\sum a_n x^n$, and the function f defined by*

$$f(x) = a_0 + a_1 x + a_2 x^2 + \cdots = \sum_{n=0}^{\infty} a_n x^n$$

for all inputs x for which the series converges.

(a) *The series obviously converges (to a_0) if $x = 0$. If it converges for some $x_0 \neq 0$, then it also converges for all x with $|x| < R = |x_0|$. The supremum (possibly) infinite of all such r is the* radius of convergence *of the power series.*

(b) *If a power series function $f(x)$ has radius of convergence R, and $r < R$, then the series converges uniformly to $f(x)$ for $x \in [-r, r]$.*

(c) *A power series function with radius of convergence $R > 0$ is differentiable, and can be differentiated "term by term":*

If $f(x) = a_0 + a_1 x + a_2 x^2 + a_3 x^3 + \ldots$ then $f'(x) = a_1 + 2 a_2 x + 3 a_3 x^2 + \ldots$.

The series for f and f' have the same radius of convergence.

EXAMPLE 8.   The power series function

$$f(x) = 1 + x + \frac{x^2}{2!} + \frac{x^3}{3!} + \cdots = \sum_{n=0}^{\infty} \frac{x^n}{n!}$$

is justly famous. Why?

And see Example 2 for a similar calculation.

SOLUTION.   Apply the ratio test (Theorem 2.32, page 138): Since

$$\lim_{k \to \infty} \frac{|x|^{k+1}/(k+1)!}{|x|^k/k!} = \lim_{k \to \infty} \frac{|x|}{k+1} = 0$$

we see that the series $f(x)$ converges for *all* $x \in \mathbb{R}$.

Do this for yourself; it's easy but striking.

Differentiating $f(x)$ term by term, as Proposition 4.20 permits, shows something more remarkable: f is *its own derivative*. Since we have both $f' = f$ and $f(0) = 1$, we conclude from basic calculus theory that $f(x)$ has another famous formula: $f(x) = e^x$ for all x, and the series formula for $f(x)$ converges uniformly to e^x on all x-intervals $[-r, r]$. $\Diamond$

We touch on some further aspects of power series in exercises. We mention Maclaurin and Taylor series in the next section, designed for self-discovery.

A function of Weierstrass: everywhere continuous, nowhere differentiable.
One use of function limits is in exploring how "good" and "bad" behaviors can
coexist in the same function. In Example 6, for instance, we obtained the absolute
value function—continuous everywhere but not differentiable at one point—as the
uniform limit of a sequence of differentiable functions.

Using similar but more sophisticated methods, the mathematician Weierstrass
constructed around 1872 a much stranger limit function: one that is continuous
everywhere but differentiable *nowhere* on $\mathbb{R}$. Weierstrass's construction starts
with a function sequence of the following form:

$$f_0(x) = \cos(\pi x)$$
$$f_1(x) = \cos(\pi x) + a\,\cos\left(b\pi x\right)$$
$$f_2(x) = \cos(\pi x) + a\,\cos\left(b\pi x\right) + a^2\,\cos\left(b^2\pi x\right)$$

$$\cdots$$

$$f_n(x) = \sum_{k=0}^{n} a^k \cos\left(b^k \pi x\right).$$

Here a is a constant with $0 < a < 1$, small enough to assure that the f_n converge
uniformly; b is a positive integer, large enough to create rapid oscillation in the
cosines. We omit computational details, but here are the main points:

- The f_n converge uniformly on $\mathbb{R}$ to some function f.

- All the f_n are differentiable (and therefore continuous).

- The limit function f is continuous *everywhere*, as Proposition 4.18 guarantees.

- Thanks to rapid oscillation in the summands, the limit function is differentiable *nowhere*.

Figure 4.10 shows the first few f_n, with $a = 0.5$ and $b = 13$.

Exercises

1. In each part, show that the given sequence $\{f_n\}$ converges pointwise to the
   given limit f on the given interval I. Is the convergence also uniform? Why
   or why not?

   (a) $f_n(x) = \frac{1}{n}$ for all n; $f(x) = 0$; $I = \mathbb{R}$.

   (b) $f_n(x) = \frac{x}{n}$ for all n; $f(x) = 0$; $I = \mathbb{R}$.

   (c) $f_n(x) = \frac{\sin x}{n}$ for all n; $f(x) = 0$; $I = \mathbb{R}$.

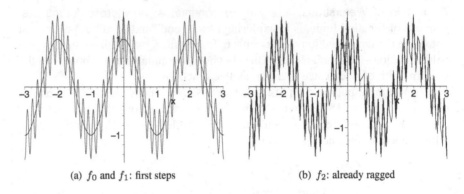

(a) f_0 and f_1: first steps   (b) f_2: already ragged

Figure 4.10. Steps toward Weierstrass's function.

2. Consider the sequence $\{f_n\}$ defined by $f_n(x) = \begin{cases} n & \text{if } x \geq n, \\ 0 & \text{if } x < n \end{cases}$.

   (a) Show that $\{f_n\}$ converges pointwise on $\mathbb{R}$ to the zero function $f(x) = 0$. Is the convergence also uniform?

   (b) Show that $\{f_n\}$ converges uniformly on $I = [-1000, 1000]$ to the zero function $f(x) = 0$.

3. Let $f : \mathbb{R} \to \mathbb{R}$ be a function. Define new functions $f_1, f_2, \ldots$ by $f_n(x) = f(x + 1/n)$.

   (a) Suppose f is continuous on $\mathbb{R}$. Show that $f_n \to f$ pointwise on $\mathbb{R}$.

   (b) Give an example to show that "continuous" is necessary in the preceding part.

   (c) Suppose f is uniformly continuous on $\mathbb{R}$. Show that $f_n \to f$ uniformly on $\mathbb{R}$.

   (d) Let f be defined by $f(x) = 1$ if $x \in \mathbb{Q}$ and $f(x) = 0$ otherwise. Does $f_n \to f$ pointwise? Uniformly?

   (e) Find a continuous function $f : \mathbb{R} \to \mathbb{R}$ such that $f_n \to f$ pointwise but not uniformly.

4. In the situation of Proposition 4.17, page 243, suppose that $f_n \to f$ and $g_n \to g$, both pointwise on I. Suppose that for all $x \in I$ we have $g(x) \neq 0$ and $g_n(x) \neq 0$ for all n. Show that $f_n/g_n \to f/g$ pointwise on I.

5. Let $\{f_n\}$ a sequence of real-valued functions defined on an interval I. Show that $\{f_n\}$ converges pointwise to a function f defined on I if and only if,

for each $x \in I$, the sequence $\{f_n(x)\}$ is Cauchy. (Such a sequence is said to satisfy a *Cauchy criterion*.)

6. In the situation and notation of Proposition 4.17, page 243, show that if $f_n \to f$ and $g_n \to g$, both pointwise on I, then $f_n + g_n \to f + g$ pointwise on I. Show also that the same result holds if "pointwise" is replaced with "uniformly."

7. For old times' sake, prove informally or disprove (with a counterexample) the following statements about limits of sequences of *numbers*.

   (a) If $x_n \in \mathbb{Z}$ for all n and $x_n \to L$, then $L \in \mathbb{Z}$, too.

   (b) If $x_n \notin \mathbb{Q}$ for all n and $x_n \to L$, then $L \notin \mathbb{Q}$, too.

   (c) If $x_n \in [-3, 42]$ for all n and $x_n \to L$, then $L \in [-3, 42]$, too.

8. This problem is about Example 6, page 245; consider the functions f_n defined there.

   (a) Let n be any positive integer. Show that $f_n(x) = |x|^{1+1/n}$ is differentiable at $x = 0$, with $f_n'(0) = 0$. Hint: Look at both left- and right-hand limits of the difference quotient for $f_n'(0)$.

   (b) Let n be any positive integer. It's a bit tricky, but we can show using elementary calculus that $|f(x) - f_n(x)| < \frac{1}{n}$ for all $x \in [-1, 1]$. Try it.

   (c) Use the inequality in the preceding part to show that $f_n \to f$ uniformly on $[-1, 1]$.

9. Consider the situation and notation of Proposition 4.17, page 243.

   (a) Show that if $f_n \to f$ and $g_n \to g$, both pointwise on I, then $f_n g_n \to fg$ pointwise on I.

   (b) Show that if $f_n \to f$ and $g_n \to g$, both uniformly on I, then $\{f_n g_n\}$ need not converge uniformly I to fg. Hint: One possibility is to consider $f_n(x) = g_n(x) = x + 1/n$ on $I = \mathbb{R}$.

10. Let $\{f_n\}$ be a sequence of functions with $f_n \to f$ pointwise on an interval I. Consider the new sequence $\{g_n\}$ defined by $g_n(x) = \sin(f_n(x))$ and the function $g(x) = \sin(f(x))$. Show that $g_n \to g$ pointwise on I.

11. Let $\{f_n\}$ be a sequence of functions with $f_n \to f$ uniformly on an interval I, and let $h : \mathbb{R} \to \mathbb{R}$ be uniformly continuous on $\mathbb{R}$. Consider the new sequence $\{g_n\}$ defined by $g_n(x) = h(f_n(x))$ and the function $g(x) = h(f(x))$. Show that $g_n \to g$ uniformly on I.

12. In Example 8, page 248, we observed that the power series $f(x)$ converges *uniformly* to e^x on bounded and closed intervals, like $I = [-1, 1]$. If $\epsilon = 1$, for instance, then $N = 1$ satisfies the conditions of Definition 4.16. (Use technology to plot e^x and $f_2(x) = 1 + x + x^2/2$ for $-1 \leq x \leq 1$ to see for yourself.)

   (a) Let $I = [-1, 1]$ and $\epsilon = 0.01$. Use plotting technology to find a value of N that works for this I and ϵ.

   (b) Let $I = [-2, 2]$ and $\epsilon = 1$. Use plotting technology to find a value of N that works for this I and ϵ.

13. In the spirit of Example 8, page 248, consider the power series

$$f(x) = x - \frac{x^3}{3!} + \frac{x^5}{5!} - \frac{x^7}{7!} + \cdots = \sum_{k=1}^{\infty} (-1)^{k+1} \frac{x^{2k-1}}{(2k-1)!}.$$

   (a) Use the ratio test to show that the series $f(x)$ converges (in absolute value) for all x.

   (b) Differentiate f term by term—twice—to see that $f'' = -f$; notice also $f(0) = 0$ and $f'(0) = 1$. What famous calculus function has these properties?

   (c) Differentiate f term by term to find a power series for f'. What famous function is this? How do you know?

   (d) Use technology to plot some partial sums of f and f' for $-\pi \leq x \leq \pi$. Do the results look familiar?

14. This problem is about the power series $g(x) = x + x^2/2 + x^3/3 + x^4/4 + \cdots$.

   (a) Use the ratio test to show that $g(x)$ converges for $-1 < x < 1$.

   (b) Explain why $g'(x) = 1/(1 - x)$ for $-1 < x < 1$. Note that $g(0) = 0$, too.

   (c) Explain why $g(x) = -\ln(1 - x)$ for $-1 < x < 1$.

   (d) The 5th partial sum of $g(x)$ is $f_5(x) = x + x^2/2 + x^3/3 + x^4/4 + x^5/5$, and $f_5(1/2) = 661/960$. Interpret this as an estimate to the natural logarithm of something.

4.5 Taylor Series and Taylor's Theorem: A Guided Discovery

This brief section, designed as an outline for self-study, explores an important application of Taylor's theorem, which we met earlier as a generalization of the

See Theorem 4.14, page 235.

mean value theorem. Here is a slightly modified form of the version from Section 4.3:

Theorem 4.21 (Taylor's theorem, $n = 3$). *Suppose f, f', f'', and f''' all exist and are continuous on $[a, b]$. Then, for some input c in $[a, b]$, we have*

$$f(b) - \left(f(a) + f'(a)(b - a) + \frac{f''(a)}{2}(b - a)^2 \right) = \frac{f'''(c)}{6}(b - a)^3.$$

As we observed earlier, Taylor's formula says that if the right-hand quantity is small, then the error committed in the approximation

$$f(b) \approx f(a) + f'(a)(b - a) + \frac{f''(a)}{2}(b - a)^2.$$

See Example 2, page 235, for more on this perspective.

Taylor polynomials: good approximations. The expression $f(a) + f'(a)(b - a) + \frac{f''(a)}{2}(b - a)^2$ in the preceding theorem points to a natural definition:

Definition 4.22 (Taylor polynomials). Let f be a function defined on an interval containing $x = a$, such that the values and derivatives $f(a)$, $f'(a)$, $f''(a)$, ..., $f^{(n)}(a)$ all exist. Then the polynomial function

$$P_n(x) = f(a) + f'(a)(x-a) + \frac{f''(a)}{2}(x-a)^2 + \frac{f'''(a)}{3!}(x-a)^3 + \cdots + \frac{f^{(n)}(a)}{n!}(x-a)^n$$

makes sense, and is called the *Taylor polynomial* of order n for f based at $x = a$. If $a = 0$ we refer to *Maclaurin polynomials*.

Let's work out some basic examples.

1. Let $f(x) = e^x$ and $a = 0$. Find the Maclaurin polynomials for f for orders up to 5.

2. Let $f(x) = e^x$ and $a = 1$. Find the Taylor polynomials through order 5 based at $a = 1$.

3. Let $f(x) = \dfrac{1}{1 - x}$ and $a = 0$. Find the Maclaurin polynomials for f for orders up to 5.

4. Let $f(x) = a + bx + cx^2 + dx^3$. Find the Maclaurin polynomials of f through order 3.

5. Find derivatives of $P_n(x)$ to show that

$$f(a) = P_n(a); \quad f'(a) = P_n'(a); \quad f''(a) = P_n''(a); \quad \cdots \quad f^{(n)}(a) = P_n^{(n)}(a).$$

As the examples show, the function f and the polynomial P_n have the same value and first n derivatives at $x = a$. In this sense P_n could reasonably be described as the "best polynomial approximation" to f at $x = a$. The version of Taylor's theorem above says something about how closely $P_2(b)$ approximates $f(b)$ when calculations are based at $x = a$.

An inequality form of Taylor's theorem. The version of Taylor's theorem above pertains specifically to the second-order Taylor polynomial P_2, and involves an *equality*, though with an unknown value c. The following more general version, which involves an *inequality*, works well for our purposes. We state it for $a = 0$ for simplicity:

Theorem 4.23 (Taylor's theorem, inequality version). *Suppose f, f', f'', ..., $f^{(n+1)}$ all exist for x in some interval $I = [-b, b]$, and that for some constant K, $\left| f^{(n+1)}(x) \right| < K$ for $x \in [-b, b]$. Then*

$$|f(x) - P_n(x)| \leq K \frac{|x|^{n+1}}{(n+1)!} \leq K \frac{|b|^{n+1}}{(n+1)!}$$

for $x \in [-b, b]$.

Again, some examples.

1. What does the theorem say for $f(x) = e^x$, $n = 4$, and $b = 2$? Use plotting technology to check that the claimed inequality does indeed hold.

2. What does the theorem say for $f(x) = x^4$, $n = 4$, and $I = [-2, 2]$? What if $f(x) = x^5$?

Toward a proof. Complete the following ingredients of a proof of Theorem 4.23 for the case $0 \leq x \leq b$. (The case $-b \leq x \leq 0$ is similar.)

1. Let $g(x) = f(x) - P_n(x)$ for $x \in [-b, b]$. Show that $0 = g(0) = g'(0) = g''(0) = \ldots g^{(n)}(0)$, and $g^{(n+1)}(x) = f^{(n+1)}(x)$.

2. Suppose h and k are functions on $[-b, b]$ with (i) $h(0) = k(0)$, and (ii) $h'(x) \leq k'(x)$ for $0 \leq x \leq b$. Then $h(x) \leq k(x)$ for $0 \leq x \leq b$. Hint: The mean value theorem or a corollary may be useful.

3. Our hypothesis is that

$$-K \leq g^{(n+1)}(x) \leq K \quad \text{for } x \in [0, b].$$

By the preceding fact, we can antidifferentiate:

$$-Kx \leq g^{(n)}(x) \leq Kx \quad \text{for } x \in [0, b].$$

For the same reason, antidifferentiating gives

$$-K\frac{x^2}{2} \le g^{(}(n-1))(x) \le K\frac{x^2}{2} \quad \text{for } x \in [0, b].$$

Repeating this process a total of $n + 1$ times gives

$$-K\frac{x^{n+1}}{(n+1)!} \le g(x) \le K\frac{x^{n+1}}{(n+1)!}$$

which is what we wanted to show.

Taylor series. If f has derivatives of *all* orders at $x = a$, then we can write its Taylor *series*:

$$S(x) = f(a) + f'(a)(x-a) + \frac{f''(a)}{2}(x-a)^2 + \cdots = \sum_{k=0}^{\infty} \frac{f^{(k)}(a)}{k!}(x-a)^k.$$

Note that the partial sums of S are the P_n.

Like any power series, $S(x)$ may or may not converge for particular x. Even if $S(x)$ does converge, there's no automatic guarantee that it converges to $f(x)$, as we'd hope. But under good conditions, Taylor's theorem can assure us that all is well.

Let's explore one poster-child example.

1. Write the Taylor (i.e., Maclaurin) series for $f(x) = e^x$ centered at $a = 0$.

2. Use the ratio test to show that the Maclaurin series for $f(x) = e^x$ converges for all x.

3. Consider the interval $[-b, b] = [-10, 10]$. What does Theorem 4.23 say about P_n and f in this situation? Hint: All derivatives of f are the same, and are bounded by e^{10} on $[-10, 10]$.

4. Conclude from the preceding part that $S(x)$ converges uniformly to $f(x) = e^x$ on $[-10, 10]$.

5. Use technology to plot f and some of the P_n for $-10 < x < 10$.

CHAPTER 5

Integrals

5.1 The Riemann Integral: Definition and Examples

Familiar Ideas Revisited

Expressions and calculations like

$$\int_0^1 f(x)\,dx, \quad \int_0^\pi \sin x\,dx, \quad \text{and} \quad \int_1^5 x^2\,dx = \left.\frac{x^3}{3}\right]_1^5 = \frac{124}{3}$$

are familiar from elementary calculus. In this chapter we interpret the integral, define it rigorously, and explore some of its properties—including the fundamental theorem of calculus, which justifies calculations like the one above.

Notes on notation. If $f(x) = x^2$, then the expressions

$$\int_0^1 f(x)\,dx, \quad \int_0^1 f(p)\,dp, \quad \int_0^1 x^2\,dx, \quad \text{and} \quad \int_0^1 f$$

look a little different, but all four mean the same thing. The first two forms use different variable *names*, but these choices are arbitrary—all four expressions have the same numerical value. For simplicity and economy, we'll often drop the variable name entirely, and use the last form. We'll use the other forms when we want to emphasize a variable name or, as in the third form, when no specific function name is given.

Riemann's and other integrals. In this book (and in elementary calculus) "integral" means "Riemann integral," after the German mathematician G. F. B. Riemann (1826–1866), who first defined integrability rigorously. Riemann's mathematical accomplishments, despite his short life, ranged across the discipline. One of his conjectures in number theory, now known as the *Riemann hypothesis*, remains after 150 years among mathematics' most important unsolved problems. And not for lack of trying: A $1 million prize for its solution, offered by the Clay Mathematics Institute, lies unclaimed.

The phrase "Riemann integral" honors a person, but it also distinguishes one particular approach to integration from several others, each with its own features and (depending on one's viewpoint) bugs. Among important alternatives to Riemann's integral are the *Lebesgue* and the *Henstock* integrals, developed around 1900 and 1950, respectively. These integrals "agree with" the Riemann integral for all the standard functions of calculus, but they are more general in the sense that they handle larger classes of functions. More advanced courses in analysis treat such integrals carefully, but we will stick with Riemann's version.

Areas and integrals. The most familiar view of integrals from elementary calculus involves area: If $f(x) \geq 0$ for $x \in [a, b]$, then $\int_a^b f(x)\, dx$ measures the area above the x-axis, below the curve $y = f(x)$, and between the vertical lines $x = a$ and $x = b$. If $f(x) < 0$ for some inputs, then area below the x-axis is involved, and counts as negative.

> Draw your own pictures to illustrate the possibilities.

Thinking of integrals as areas will often be helpful for us, too. This view suggests—correctly—that

$$\int_1^7 3\, dx = 18, \qquad \int_0^3 (1 + x)\, dx = \frac{15}{2}, \quad \text{and} \quad \int_0^{3.14} \cos(x)\, dx \approx 0,$$

> Is the last integral slightly positive or slightly negative? Why?

assuming, as we do for now, that all three integrals exist. But extra care will be needed for other functions, whose graphs may be ragged or broken, not smooth.

Defining the Integral

What do statements like

$$\int_0^3 2x\, dx = 9, \qquad \int_0^\pi \sin x\, dx = 2, \quad \text{and} \quad \int_a^b f = -7$$

> Draw your own pictures for the first two integrals.

really mean? For the first integral, the integral-as-area view is enough—9 is the only reasonable answer. For the second integral, the area view suggests 2 is possible, but is it exact? Why not 2.034? Or 1.957? For the third integral, the negative result and elementary calculus intuition suggest that $f(x)$ is in some sense more negative than positive for x in the interval $[a, b]$. But in what sense is the answer exactly -7?

Resolving all these questions asks a lot of the integral, and so it is no surprise that the recipe is complicated. Here, first, we assemble the ingredients:

- *Integrand and interval:* The definition requires an *integrand*—a real-valued function $f : [a, b] \to \mathbb{R}$ defined for *all* inputs x in a closed and bounded interval $[a, b]$. Note, in particular, that although integral equations like

$$\int_1^\infty \frac{1}{x^2}\, dx = 1 \quad \text{and} \quad \int_0^1 \frac{1}{\sqrt{x}}\, dx = 2$$

are sometimes seen in elementary calculus, the integrals in question are *not* of the type considered here.

That's why we call them "improper."

- *Partitions, subintervals, and norms:* A *partition* $\mathcal{P}$ of an interval $[a, b]$ is any choice of $n + 1$ points $\{x_0, x_1, \ldots, x_n\}$, arranged in increasing order from a to b:

$$a = x_0 < x_1 < x_2 < \cdots < x_{n-1} < x_n = b.$$

We'll often just write $\mathcal{P} = \{x_0, x_1, x_2, \ldots, x_n\}$ to emphasize that $\mathcal{P}$ is a *set*; the smallest-to-biggest ordering is understood.

The object of choosing $\mathcal{P}$ is to chop $[a, b]$ into n smaller closed intervals, meeting only at their endpoints:

"Partition" is the polite term.

$$[a, b] = [x_0, x_1] \cup [x_1, x_2] \cup \cdots \cup [x_{i-1}, x_i] \cup \cdots \cup [x_{n-1}, x_n].$$

We'll call $[x_{i-1}, x_i]$ the *i-th subinterval*; its *length* is the positive number $\Delta x_i = x_i - x_{i-1}$. If the points of $\mathcal{P}$ happen to be regularly spaced, then the partition is called *regular*, and $\Delta x_i = (b - a)/n$ for all i. For any partition, regular or not, we have

$$\Delta x_1 + \Delta x_2 + \cdots + \Delta x_n = x_n - x_0 = b - a.$$

The *norm* of a partition $\mathcal{P}$, denoted $\|\mathcal{P}\|$, is the *largest* of the Δx_i. Requiring that $\|\mathcal{P}\| < \delta$, as we will in the definition, ensures that *every* subinterval is narrow.

Also known as the mesh.

- *Sampling points:* Given any partition $\mathcal{P} = \{x_0, x_1, x_2, \ldots, x_n\}$ of an interval $[a, b]$, we can choose *sampling points*

$$s_1 \in [x_0, x_1], \quad s_2 \in [x_1, x_2], \quad \ldots, \quad s_n \in [x_{n-1}, x_n],$$

one from each subinterval of $\mathcal{P}$, to form a full set $\mathcal{S} = \{s_1, s_2, \ldots, s_n\}$ of sampling points from $\mathcal{P}$.

There are many, many ways to choose a sample $\mathcal{S}$ from any given partition $\mathcal{P}$. We could choose each subinterval's left endpoint, its right endpoint, or its midpoint, or any combination. Or we could throw an imaginary dart into each subinterval to choose a sampling point at random. A partition endpoint, say x_{42}, might even serve *twice* as a sample point, representing both $[x_{41}, x_{42}]$ and $[x_{42}, x_{43}]$.

- *Riemann sums:* Given all the data (a function f, an interval $[a, b]$, a partition $\mathcal{P}$, and samples $\mathcal{S}$) just described, a *Riemann sum* is an expression of the form

$$\sum_{i=1}^{n} f(s_i)\Delta x_i = f(s_1)\Delta x_1 + f(s_2)\Delta x_2 + \cdots + f(s_n)\Delta x_n.$$

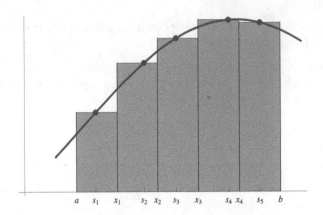

Figure 5.1. A generic Riemann sum.

Even the shorthand is slightly cumbersome.

As a shorthand we'll sometimes use the notation $\text{RS}(f, \mathcal{P}, \mathcal{S})$ to denote a Riemann sum with all these data. When all the data are given explicitly, $\text{RS}(f, \mathcal{P}, \mathcal{S})$ reduces to a *number*. Under good conditions—as specified in the coming definition—we expect this number to approximate an integral.

A Riemann sum is a sort of "representative sample" of the function f on $[a, b]$: We chop $[a, b]$ into smaller pieces, evaluate f at one point of each piece, and add up the results, "weighting" each value $f(s_i)$ with the "size" of the subinterval it represents. Figure 5.1 illustrates the familiar idea.

We don't know "officially" yet that the integral exists. But it does.

EXAMPLE 1.   Explore some Riemann sums for $\int_0^2 \sin(x)\, dx$.

SOLUTION.   The integral involves the integrand $f(x) = \sin x$ and the interval $[0, 2]$. The simplest and least interesting partition of $[0, 2]$ involves *no* chopping: $\mathcal{P} = \{x_0, x_1\} = \{0, 2\}$ has just one subinterval, and $\|\mathcal{P}\| = 2$. Here we need

Chosen at random by Mathematica.

just one sampling point, say, $s_1 = 1.3657$, and the corresponding Riemann sum is simply

$$f(s_1)\Delta x_1 = \sin(1.3657) \cdot 2 \approx 1.9581.$$

With so little work or thought invested, we can't expect much return from the answer.

Let's work (just) a little harder. With

$$\mathcal{P} = \{0, 0.8, 1.0, 1.7, 2\} \quad \text{and} \quad \mathcal{S} = \{0, 0.81, 1.69, 1.73\},$$

we have $\|\mathcal{P}\| = 0.8$ and the Riemann sum $\mathrm{RS}(f, \mathcal{P}, \mathcal{S})$ becomes

$$\sum_{i=1}^{4} f(s_i)\Delta x_i = \sin(0) \cdot 0.8 + \sin(0.81) \cdot 0.2 + \sin(1.69) \cdot 0.7 + \sin(1.73) \cdot 0.3$$

$$\approx 1.57.$$

With the same partition, but new samples $\mathcal{S} = \{0.8, 0.8, 1.7, 1.7\}$, the Riemann sum becomes

$$\sin(0.8) \cdot 0.8 + \sin(0.8) \cdot 0.2 + \sin(1.7) \cdot 0.7 + \sin(1.7) \cdot 0.3 \approx 2.14,$$

which isn't all that close to earlier estimates.

Now let's get serious. With a regular, 20-piece partition (so $\|\mathcal{P}\| = 0.1$) and *midpoint* samples, we have

$$\mathcal{P} = \{0, 0.1, 0.2, \ldots, 1.9, 2\} \quad \text{and} \quad \mathcal{S} = \{0.05, 0.15, \ldots, 1.95\},$$

and the Riemann sum works out to

Thanks, Mathematica.

$$\sin(0.05) \cdot 0.1 + \sin(0.15) \cdot 0.1 + \cdots + \sin(1.95) \cdot 0.1 \approx 1.42.$$

With *right endpoint* samples $\mathcal{S} = \{0.01, 0.02, \ldots, 2.0\}$ from the same partition the result is not much different; Mathematica gives about 1.421. These numbers, and all the extra work, deserve more credibility.   ◇

Rightly so. The exact value, as we will be able to prove soon, is $1 - \cos 2 \approx 1.41615$.

We are ready at last for the formal definition.

Definition 5.1. Let $[a, b]$ be a closed interval, $f : [a, b] \to \mathbb{R}$ a function, and I a real number. Then f is *integrable* on $[a, b]$, and we write $\int_a^b f = I$, if for every given $\epsilon > 0$ there exists $\delta > 0$ such that

$$|\mathrm{RS}(f, \mathcal{P}, \mathcal{S}) - I| < \epsilon$$

whenever $\mathcal{P} = \{x_0, x_1, \ldots, x_n\}$ is a partition of $[a, b]$ with $\|\mathcal{P}\| < \delta$ and $\mathcal{S} = \{s_1, \ldots, s_n\}$ is any set of samples from $\mathcal{P}$.

Illustrating the Definition

The definition is undeniably complicated, but simple examples will help us unpack it, and reveal some pleasant surprises.

EXAMPLE 2.   Let $f(x) = 3$ for all x and $[a, b] = [1, 7]$. Thinking of area suggests that $\int_1^7 f = 18$. Does the definition agree? Can the idea be generalized?

SOLUTION.   Yes, and yes.

If $\mathcal{P}$ is *any* partition of $[1, 7]$ and $\mathcal{S}$ is *any* set of sampling points, then

$$\sum_{i=1}^{n} f(s_i)\Delta x_i = \sum_{i=1}^{n} 3\Delta x_i = 3 \sum_{i=1}^{n} \Delta x_i = 3 \cdot (7 - 1) = 18,$$

which is *exactly* the desired answer. Thus, for a given $\epsilon > 0$ we can choose *any* positive δ, say $\delta = 42$, and the definition is satisfied. $\Diamond$

There's nothing special, of course, about the data in the preceding example.

Proposition 5.2. *If $f(x) = k$ is any constant function and $[a, b]$ any interval, then f is integrable on $[a, b]$, and*

$$\int_a^b k\, dx = k \cdot (b - a).$$

Nonnegative integrands, nonnegative integrals. Following is another basic property of integrals that follows directly from the definition.

Proposition 5.3. *Suppose $f(x) \geq 0$ for all $x \in [a, b]$. If $\int_a^b f$ exists, $\int_a^b f \geq 0$.*

Proof: The idea is simple: since all Riemann sums for $I = \int_a^b f$ are nonnegative, I itself must be nonnegative, too. The formal proof takes some care.

Let $\epsilon > 0$ be given. Choose $\delta > 0$ that works for this ϵ in the sense of Definition 5.1. If we choose *any* Riemann sum $\mathrm{RS}\,(f, \mathcal{P}, \mathcal{S})$ with $\|\mathcal{P}\| < \delta$, then we must have $|\mathrm{RS}\,(f, \mathcal{P}, \mathcal{S}) - I| < \epsilon$. Because $f(x) \geq 0$ for all $x \in [a, b]$, it is clear that $\mathrm{RS}\,(f, \mathcal{P}, \mathcal{S}) \geq 0$, too, and so we must have $I \geq -\epsilon$. Since this inequality holds for *all* positive ϵ, we must have $I \geq 0$. $\square$

Stranger integrands. In Proposition 5.3 we *assumed* that the integral exists. Deciding which integrals exist takes some work, and we start slowly. The following integrand is discontinuous, but only at one point. Does it matter?

EXAMPLE 3.   Let

$$f(x) = \begin{cases} 1 & \text{if } x = 1, \\ 0 & \text{if } x \neq 1. \end{cases}$$

Does $\int_0^2 f$ exist?

SOLUTION.   Thinking about area suggests $\int_0^2 f = 0$; the graph of f seems too skinny to bound appreciable area above the x-axis.

Draw your own.

Proving this is not difficult. Let $\epsilon > 0$ be given, set $\delta = \epsilon/2$, and let $\mathcal{P} = \{x_0, x_1, \ldots, x_n\}$ be any partition with $\|\mathcal{P}\| < \delta$. We need to show that if $\mathcal{S} = \{s_1, s_2, \ldots, s_n\}$ is *any* set of sampling points for $\mathcal{P}$, then the associated Riemann sum satisfies

$$-\epsilon < f(s_1)\Delta x_1 + \cdots + f(s_n)\Delta x_n < \epsilon.$$

Since all of the $f(s_i)\Delta x_i$ are nonnegative, it is enough to prove the right-hand inequality.

The left-hand inequality is obvious.

Now $f(s_i) = 0$ unless $s_i = 1$, in which case $f(s_i) = 1$. Thus, each summand $f(s_i)\Delta x_i$ is either 0 or Δx_i. Because the "offender point" $x = 1$ can lie in *at most two* subintervals, say $[x_{i-1}, x_i]$ and $[x_i, x_{i+1}]$, our Riemann sum can have at most two nonzero summands. Adding everything up gives

Very likely, all summands are zero.

$$\sum_{i=1}^n f(s_i)\Delta x_i = f(s_i)\Delta x_i + f(s_{i+1})\Delta x_{i+1}$$

$$\leq \Delta x_i + \Delta x_{i+1} < \frac{\epsilon}{2} + \frac{\epsilon}{2} = \epsilon,$$

as desired.   ◇

Example 3 shows that *one* discontinuity is forgivable in an integral. What about many?

EXAMPLE 4.   Recall our old friend

$$g(x) = \begin{cases} 1 & \text{if } x \in \mathbb{Q}, \\ 0 & \text{if } x \notin \mathbb{Q}. \end{cases}$$

Does $\int_0^1 g$ exist?

Can you guess anything from the graph of g?

SOLUTION.   No. The problem, roughly speaking, is that for *every* partition $\mathcal{P} = \{x_0, x_1, x_2, \ldots, x_n\}$ of $[0, 1]$, no matter how "fine," there are both rational and irrational numbers inside every subinterval $[x_{i-1}, x_i]$. If all of the sample points s_i are chosen to be rational, then $g(s_i) = 1$ for all i, and the corresponding Riemann sum works out to

$$\text{RS}(f, \mathcal{P}, \mathcal{S}) = \sum_{i=1}^n g(s_i)\Delta x_i = \sum_{i=1}^n 1 \cdot \Delta x_i = 1.$$

If, instead, all the s_i are irrational, then we get

$$\text{RS}(f, \mathcal{P}, \mathcal{S}) = \sum_{i=1}^n g(s_i)\Delta x_i = \sum_{i=1}^n 0 \cdot \Delta x_i = 0.$$

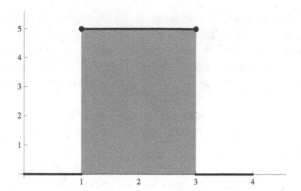

Figure 5.2. A discontinuous integrand.

This situation—widely varying Riemann sums even for "fine" partitions—is in-compatible with Definition 5.1.   ◊

The last example involves another type of discontinuity.

EXAMPLE 5.   Figure 5.2 shows the function

$$f(x) = \begin{cases} 5 & \text{if } x \in [1,3], \\ 0 & \text{if } x \notin [1,3]; \end{cases}$$

a region of interest is shaded. Show that f is integrable on $[0,4]$, with $\int_0^4 f = 10$.

SOLUTION.   We know already from the discussion in Example 2 that f is inte-grable on $[1,3]$, and that $\int_1^3 f = 10$. We need now to show that f has the same integral over a larger interval. This should seem reasonable if we think in terms of area; the (slightly laborious) proof makes this intuition precise.

Suppose, then, that $\epsilon > 0$ is given. We'll show that $\delta = \epsilon/10$ works in the sense of Definition 5.1. To see this, let $\mathcal{P} = \{x_0, \ldots, x_n\}$ be any partition of $[0,4]$ with $\|\mathcal{P}\| < \delta$, let $\mathcal{S} = \{s_1, \ldots, s_n\}$ be any choice of sampling points within $\mathcal{P}$, and let $\text{RS}(f, \mathcal{P}, \mathcal{S})$ be the corresponding Riemann sum. To prove our claim, we need to show that

$$10 - \epsilon < \text{RS}(f, \mathcal{P}, \mathcal{S}) < 10 + \epsilon.$$

We'll tackle the left-hand inequality and leave the right-hand one as an exercise.

Because $\mathcal{P}$ is a partition of $[0,4]$, we can choose among the partition points the *first* x_i with $x_i \geq 1$ and the *last* x_i with $x_i \leq 3$. If, say, these points are x_{42}

Why the denominator of ten? Read on.

and x_{237}, then we have

$$\cdots < x_{41} < 1 \leq x_{42} < x_{43} < \cdots < x_{236} < x_{237} \leq 3 < x_{238} < \cdots.$$

This implies that $1 \leq s_i \leq 3$ for $i = 43, 44, \ldots, 237$, and so $f(s_i) = 5$ for all these i. Moreover, because $\|\mathcal{P}\| < \delta$, we know that

$$x_{42} < 1 + \delta \quad \text{and} \quad 3 - \delta < x_{237},$$

and so $x_{237} - x_{42} > 2 - 2\delta$. Now we can calculate:

$$
\begin{aligned}
\mathrm{RS}\,(f, \mathcal{P}, \mathcal{S}) &= f(s_1)\Delta x_1 + \cdots + f(s_n)\Delta x_n \\
&\geq f(s_{43})\Delta x_{43} + \cdots + f(s_{237})\Delta x_{237} \qquad \text{all terms are nonnegative} \\
&= 5\,\Delta x_{43} + \cdots + 5\,\Delta x_{237} \\
&= 5\,(x_{237} - x_{42}) \qquad \text{the } \Delta\,x_i \text{ "collapse"} \\
&> 5\,(2 - 2\delta) = 10 - 10\delta = 10 - \epsilon,
\end{aligned}
$$

as we aimed to show. $\diamond$

Lessons from the examples. The examples illustrate one pleasant and one less pleasant property of integrals. The good news is that the Riemann integral is relatively forgiving of minor misbehavior, such as occasional discontinuities, in an integrand. Less convenient is the fact that proofs and calculations using the definition, even for quite simple integrands, can be messy and technical. The moral is that we'd like to have both simpler tests for integrability and more efficient methods of calculating integrals. We'll find some in the following sections.

As we just saw in Example 5.

Which Functions are Riemann Integrable?

Deciding precisely which functions have integrals and which don't turns out to be a very hard problem; we won't tackle it in full generality. Partial answers, however, are interesting and accessible—and sometimes surprising. We will show, for instance, that every *continuous* function (and hence every *differentiable* function) on an interval $[a, b]$ is indeed integrable. We'll see, too, that every *monotone* function $f : [a, b] \to \mathbb{R}$ is also integrable—even though such an f might have infinitely many discontinuities on $[a, b]$.

Try drawing a monotone-but-nasty function.

We will study the question of integrability more carefully in Section 5.3. Meanwhile, here's one useful criterion.

Proposition 5.4. *If f is integrable on $[a, b]$, then f is bounded on $[a, b]$.*

About the proof. We sketch the proof of a special case, leaving generalities to the exercises. Suppose, say, that $\int_0^3 f = 42$. Now let $\epsilon = 1$. Choose $\delta > 0$ that works for this ϵ in the sense described in Definition 5.1, and let $\mathcal{P} = \{x_0, x_1, x_2, \ldots, x_n\}$ be any partition of $[1, 3]$ with $\|\mathcal{P}\| < \delta$. If we can show that f is bounded on each of the n subintervals $[x_{i-1}, x_i]$, we can conclude that f is bounded on *all* of $[1, 3]$.

t is the only variable; all the x_i are constants.

To see why f is bounded on, say, $[x_6, x_7]$, consider the expression

$$S(t) = f(x_1)\Delta x_1 + \cdots + f(x_6)\Delta x_6 + f(t)\Delta x_7 + f(x_8)\Delta x_8 + \cdots + f(x_n)\Delta x_n.$$

Here is the key idea: For each $t \in [x_6, x_7]$, $S(t)$ is a Riemann sum, based on the partition $\mathcal{P}$, for the integral $\int_0^3 f$, and so Definition 5.1 guarantees that

$$41 < S(t) < 43,$$

which means that $S(t)$ is bounded for $t \in [x_6, x_7]$. This implies, as desired, that $f(t)$ is bounded for $t \in [x_6, x_7]$.

Exercises

1. Let $f(x) = x^2$ and consider the integral $I = \int_0^3 f$, (We know from elementary calculus—and we'll prove soon—that $I = 9$.)

   (a) Use the partition $\mathcal{P} = \{0, 0.8, 1.0, 1.7, 2, 2.9, 3\}$ and the sample set $\mathcal{S} = \{0, 1, \sqrt{2}, \sqrt{3}, 2, 3\}$ to evaluate the Riemann sum $\mathrm{RS}(f, \mathcal{P}, \mathcal{S})$. What is $\|\mathcal{P}\|$?

   (b) Using the same partition $\mathcal{P} = \{0, 0.8, 1.0, 1.7, 2, 2.9, 3\}$, choose another sample set $\mathcal{S}$ for which the $\mathrm{RS}(f, \mathcal{P}, \mathcal{S})$ is as *small* as possible. How small can $\mathrm{RS}(f, \mathcal{P}, \mathcal{S})$ be?

   (c) Using the same partition $\mathcal{P} = \{0, 0.8, 1.0, 1.7, 2, 2.9, 3\}$, choose a sample set $\mathcal{S}$ for which $\mathrm{RS}(f, \mathcal{P}, \mathcal{S})$ approximates I (the exact value is nine) as *badly* as possible. How much error can $\mathrm{RS}(f, \mathcal{P}, \mathcal{S})$ commit?

   (d) Using the (one-piece) partition $\mathcal{P} = \{0, 3\}$, choose a (one-member) sample set $\mathcal{S}$ for which $\mathrm{RS}(f, \mathcal{P}, \mathcal{S}) = 9$.

2. Let $f(x) = x$ and consider the integral $I = \int_0^4 f$, (We know from elementary calculus—and we'll prove soon—that $I = 8$. See also Problem 12 below.)

   (a) Using the partition $\mathcal{P} = \{0, 1, 2, 3, 4\}$, find a sample set $\mathcal{S}$ for which $\mathrm{RS}(f, \mathcal{P}, \mathcal{S}) = 8$.

   (b) Using the partition $\mathcal{P} = \{0, 1, 2, 3, 4\}$, find a sample set $\mathcal{S}$ for which $\mathrm{RS}(f, \mathcal{P}, \mathcal{S}) = 9$.

(c) For $\mathcal{P} = \{0, 4\}$ find a sample set $\mathcal{S}$ so that $\text{RS}(f, \mathcal{P}, \mathcal{S}) = 0$.

(d) For $\mathcal{P} = \{0, 4\}$ find a sample set $\mathcal{S}$ so that $\text{RS}(f, \mathcal{P}, \mathcal{S}) = 8$.

(e) Is there a two-piece partition $\mathcal{P} = \{0, x_1, 4\}$ and a sample set $\mathcal{S}$ for which $\text{RS}(f, \mathcal{P}, \mathcal{S}) = 0$? Why or why not?

(f) Describe a 100-piece partition $\mathcal{P} = \{0, x_1, \ldots, x_{99}, 4\}$ and a sample set $\mathcal{S}$ for which $\text{RS}(f, \mathcal{P}, \mathcal{S}) < 0.01$.

3. Suppose that f is integrable on $[0, 2]$ and that $f(x) \geq 3$ for all $x \in [0, 2]$. Show that $\int_0^2 f \geq 6$. (Mimic the proof of Proposition 5.3, page 262.)

4. Use Definition 5.1 to prove Proposition 5.2: If $f(x) = k$ is a constant function, then f is integrable and $\int_a^b k\, dx = k \cdot (b - a)$.

5. Let $f(x) = 0$ for $x \neq 0$ and $f(0) = 42$. We explore the proof that f is integrable on $[0, 1]$ and $\int_0^1 f = 0$.

(a) Let $\mathcal{P} = \{0, 0.001, 0.002, \ldots, 1\}$ be a 1000-piece partition of $[0, 1]$, and let $\mathcal{S} = \{s_1, s_2, \ldots, s_{1000}\}$ be any corresponding set of samples. What are the possible values of $\text{RS}(f, \mathcal{P}, \mathcal{S})$?

(b) Let $\mathcal{P} = \{0, x_1, \ldots, x_{n-1}, 1\}$ be any n-piece partition of $[0, 1]$, and $\mathcal{P} = \{s_1, \ldots, s_n\}$ any set of samples. Explain why we must have either $\text{RS}(f, \mathcal{P}, \mathcal{S}) = 0$ or $\text{RS}(f, \mathcal{P}, \mathcal{S}) = 42x_1$.

(c) Let $\epsilon > 0$ be given and set $\delta = \epsilon/42$. Show that if $\|\mathcal{P}\| < \delta$ and $\mathcal{S} = \{s_1, \ldots, s_n\}$ is any set of samples from $\mathcal{P}$, then $0 \leq \text{RS}(f, \mathcal{P}, \mathcal{S}) < \epsilon$. Conclude that $\int_0^1 f = 0$.

6. Let $f(x) = 0$ for $x \neq 0$ and $f(1) = 42$. Prove as follows that f is integrable on $[0, 2]$ and $\int_0^2 f = 0$.

(a) Let $\mathcal{P} = \{0, 0.001, 0.002, \ldots, 1.999, 2\}$ be a 2000-piece partition of $[0, 1]$, and $\mathcal{S} = \{s_1, s_2, \ldots, s_{2000}\}$ any corresponding set of samples. What are the three possible values of $\text{RS}(f, \mathcal{P}, \mathcal{S})$?

(b) Let $\epsilon > 0$ be given and set $\delta = \epsilon/84$. Show that if $\|\mathcal{P}\| < \delta$ and $\mathcal{S} = \{s_1, \ldots, s_n\}$ is any set of samples from $\mathcal{P}$, then $0 \leq \text{RS}(f, \mathcal{P}, \mathcal{S}) < \epsilon$. Conclude that $\int_0^2 f = 0$.

7. Suppose g is integrable on $[a, b]$, with $\int_a^b g = 0$. Suppose $0 \leq f(x) \leq g(x)$ for all x. Show that f is also integrable, with $\int_a^b f = 0$.

8. What is the converse of Proposition 5.3? Is it true? If so, why? If not, give a counterexample.

9. Prove that the value of I in Definition 5.1 is unique. (In other words, if both I_1 and I_2 satisfy the definition, then $I_1 = I_2$.)

10. Suppose that $I = \int_a^b f$ exists. Show that the following "Cauchy condition" holds: for all $\epsilon > 0$ there exists $\delta > 0$ such that if $\mathcal{P}_1$ and $\mathcal{P}_2$ are two partitions of $[a, b]$ with $\|\mathcal{P}_1\| < \delta$ and $\|\mathcal{P}_2\| < \delta$, and $\mathcal{S}_1$ and $\mathcal{S}_2$ are corresponding sample sets, then $|\mathrm{RS}(f, \mathcal{P}_1, \mathcal{S}_1) - \mathrm{RS}(f, \mathcal{P}_2, \mathcal{S}_2)| < \epsilon$.

11. Suppose that f is continuous on $[a, b]$. We will show later that f is also integrable on $[a, b]$; here are some steps in that direction.

    (a) Explain why the following condition holds: for any given $\epsilon > 0$, there exists some $\delta > 0$ such that if s and t are in $[0, 1]$ and $|s - t| < \delta$, then $|f(s) - f(t)| < \epsilon$.

    (b) Let ϵ and δ be as in the preceding part, let $\mathcal{P} = \{x_0, x_1, x_2, \ldots, x_n\}$ be any partition with $\|\mathcal{P}\| < \delta$, and let $\mathcal{S} = \{s_1, s_2, \ldots, s_n\}$ and $\mathcal{T} = \{t_1, t_2, \ldots, t_n\}$ be two different sample sets for $\mathcal{P}$. Show that

    $$|\mathrm{RS}(f, \mathcal{P}, \mathcal{S}) - \mathrm{RS}(f, \mathcal{P}, \mathcal{T})| < \epsilon(b - a).$$

12. Show as follows that $f(x) = x$ is integrable on $[0, 1]$, and that $\int_0^1 f = 1/2$.

    (a) Let $\mathcal{P} = \{x_0, x_1, x_2, \ldots, x_n\}$ be *any* partition of $[0, 1]$ (so $x_0 = 0$ and $x_n = 1$). Let $\mathcal{M} = \{m_1, m_2, \ldots, m_n\}$ be the sample set whose elements are the *midpoints* of their respective subdivisions. (For example, $m_7 = (x_7 + x_6)/2$.) Show that $\mathrm{RS}(f, \mathcal{P}, \mathcal{M}) = 1/2$.

    (b) Let $\epsilon > 0$ be given. Show that $\delta = \epsilon$ works in Definition 5.1 to show that $I = \int_0^1 f = 1/2$. (Use the preceding fact and part (b) of Problem 11.)

13. Generalize ideas in Problem 12 to show that $f(x) = x$ is integrable on every interval $[a, b]$, and that $\int_a^b x\, dx = (b^2 - a^2)/2$.

14. Let $f(x) = 1/x$ if $x \neq 0$ and $f(0) = 0$. Proposition 5.4 says that f is not integrable on $[0, 1]$. To illustrate this, show that, for *any* partition $\mathcal{P} = \{0, x_1, \ldots, x_{n-1}, 1\}$, there is some sample set $\mathcal{P} = \{s_1, s_2, \ldots, s_n\}$ such that $\mathrm{RS}(f, \mathcal{P}_2, \mathcal{S}_2) > 1000$.

5.2   Properties of the Integral

Basic properties of the definite integral are familiar from elementary calculus—where they may be taken for granted. Using the formal definition, we can readily give rigorous proofs.

Algebra with Integrals

It is standard operating procedure in calculus to break complicated integrands apart, integrate the pieces separately, and then reassemble the results. For example, we can write

Use basic calculus to check the last identity, if you like.

$$\int_0^1 \left(\frac{7}{x+1} - \frac{3}{x^2+1} \right) dx = 7 \int_0^1 \frac{1}{x+1} dx - 3 \int_0^1 \frac{1}{x^2+1} dx$$
$$= 7 \ln 2 - 3 \frac{\pi}{4},$$

assuming (for now) that all three integrals in question exist. In elementary calculus we may skip over some theoretical niceties, but now we can do things right. The next theorem guarantees that integrable functions (like differentiable functions, continuous functions, and other objects we have studied) "behave well" with respect to addition and scalar multiplication.

Theorem 5.5. *Suppose that functions f and g are integrable on $[a, b]$, and let k be any number. Then the new functions kf and $f + g$ are also integrable on $[a, b]$, with*

$$\int_a^b kf = k \int_a^b f \quad and \quad \int_a^b (f+g) = \int_a^b f + \int_a^b g.$$

Integration is linear. Before proving the theorem we observe that we can apply it repeatedly to handle more complicated linear combinations. If, say, f, g, and h are all integrable on $[2, 5]$, then the following equation makes good sense—and it's true:

$$\int_2^5 \left(3f - 7g + \frac{\pi}{4}h \right) = 3 \int_2^5 f - 7 \int_2^5 g + \frac{\pi}{4} \int_2^5 h.$$

A fancier way to state these ideas uses the language of linear algebra. The set V of integrable functions on $[a, b]$ is a *vector space*, and integration on $[a, b]$ is a *linear transformation* from V to $\mathbb{R}$ (another vector space!).

Proving Theorem 5.5. We'll sketch the proof for sums and leave the rest to the exercises. For brevity we write I_f for $\int_a^b f$ and I_g for $\int_a^b g$.

Let $\epsilon > 0$ be given. Since I_f and I_g exist, there are positive numbers δ_f and δ_g such that, for any sampling points $s_1, \ldots, s_n$,

$$\|\mathcal{P}\| < \delta_f \implies \left| \sum_{i=1}^n f(s_i)\Delta x_i - I_f \right| < \frac{\epsilon}{2};$$

$$\|\mathcal{P}\| < \delta_g \implies \left| \sum_{i=1}^n g(s_i)\Delta x_i - I_g \right| < \frac{\epsilon}{2}.$$

Now let $\delta = \min\{\delta_f, \delta_g\}$; we'll show that this δ "works" in the sense of Definition 5.1 for the sum function $f + g$. To do so, consider any partition $\mathcal{P} = \{x_0, x_1, x_2, \ldots x_n\}$ with $\|\mathcal{P}\| < \delta$, and let $\mathcal{S} = \{s_1, s_2, \ldots s_n\}$ be any corresponding choice of sampling points. Now we calculate:

Watch for the triangle inequality.

$$\left| \sum_{i=1}^{n} (f(s_i) + g(s_i)) \, \Delta x_i - (I_f + I_g) \right|$$

$$= \left| \sum_{i=1}^{n} f(s_i)\Delta x_i - I_f + \sum_{i=1}^{n} g(s_i)\Delta x_i - I_g \right|$$

$$\leq \left| \sum_{i=1}^{n} f(s_i)\Delta x_i - I_f \right| + \left| \sum_{i=1}^{n} g(s_i)\Delta x_i - I_g \right|$$

$$< \frac{\epsilon}{2} + \frac{\epsilon}{2} = \epsilon.$$

This shows what we claimed: $f + g$ is integrable on $[a, b]$, with integral $I_f + I_g$.

Bigger integrands, bigger integrals. Theorem 5.5 has a simple and natural corollary; we leave the proof to the exercises.

Corollary 5.6. *Let f and g be integrable functions on $[a, b]$.*

$$\text{If } f(x) \leq g(x) \text{ for all } x \in [a, b], \text{ then } \int_a^b f \leq \int_a^b g.$$

Combining ideas from Example 3, page 262, and Theorem 5.5 produces some possible surprises.

EXAMPLE 1. Suppose

$$f(1) = 1, \quad f(2) = 2, \quad f(3) = 3, \quad \ldots, \quad f(1000) = 1000$$

and $f(x) = 0$ for other x. Does $\int_0^{1000} f$ exist?

SOLUTION. Yes, and $\int_0^{1000} f = 0$. Example 3, page 262, slightly modified, shows that functions like $f_{17} : [0, 1000] \to \mathbb{R}$, given by

$$f_{17}(x) = \begin{cases} 17 & \text{if } x = 17, \\ 0 & \text{otherwise} \end{cases}$$

are all integrable on $[0, 1000]$, and that $\int_0^{1000} f_{17} = 0$. Our given f satisfies

$$f = f_1 + f_2 + \cdots + f_{1000},$$

and so Theorem 5.5 gives

$$\int_0^{1000} f = \int_0^{1000} f_1 + \int_0^{1000} f_2 + \cdots + \int_0^{1000} f_{1000}$$
$$= 0 + 0 + \cdots + 0 = 0,$$

as desired. $\Diamond$

EXAMPLE 2.   Suppose f is integrable on $[a, b]$, with $\int_a^b f = 42$, and suppose $g(x) = f(x)$ for all but finitely many x in $[a, b]$. What can be said about $\int_a^b g$?

SOLUTION.   In this case, $\int_a^b g = 42$, too. To see why, consider the difference function $g - f$. Since $g(x) - f(x) = 0$ for all but finitely many x, the method of Example 1 shows that $\int_a^b (g - f) = 0$. By Theorem 5.5,

$$\int_a^b g = \int_a^b f + (g - f) = \int_a^b f + \int_a^b (g - f) = 42 + 0 = 42.$$

Here is another way to express the result: If $\int_a^b f = I$, then altering $f(x)$ at finitely many points in $[a, b]$ leaves the integral unchanged. $\Diamond$

Joinery

Another way to build new integrals from old is to stick "pieces" of various functions together. Interestingly, the pieces need not fit together continuously. In this sense, the integral is more forgiving than the derivative, which requires smoother joinery.

EXAMPLE 3.   Figure 5.3 shows a function f formed—clumsily—from three separate pieces. What can be said about $\int_0^4 f$?

SOLUTION.   Despite the breaks at $x = 1$ and $x = 3$, the resulting function *is* integrable on $[0, 4]$, and can be integrated piece by piece:

$$\int_0^4 f = \int_0^1 f + \int_1^3 f + \int_3^4 f = 2 + 2 + 3 = 7.$$

Justifying such calculations rigorously takes a little effort. The main idea is in the next lemma. $\Diamond$

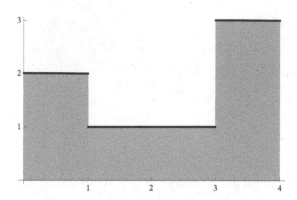

Figure 5.3. Clumsy joinery.

Lemma 5.7. *Suppose that $f : [a, b] \to \mathbb{R}$ is integrable, with $\int_a^b f = I$. Assume $c > b$, so $[a, b] \subset [a, c]$, and consider the "extended" function $\bar{f} : [a, c] \to \mathbb{R}$ defined by*

$$\bar{f}(x) = \begin{cases} f(x) & \text{if } x \in [a, b], \\ 0 & \text{if } x > b. \end{cases}$$

Then $\bar{f}$ is integrable on $[a, c]$, and

$$\int_a^c \bar{f} = \int_a^b f = I.$$

The live issue, by the way, is *integrability* of $\bar{f}$. Once we know that, it is no surprise that extending f in the way described doesn't change the *value* of the integral I. The proof—like others that involve the definition of the integral—is a bit messy, but it is worth sticking with as an example of its type. The main idea is that Riemann sums for $\int_a^c \bar{f}$ can't differ by much from Riemann sums for $\int_a^b f$.

Proof: Let $\epsilon > 0$ be given. We need to find $\delta > 0$ so that if $\mathcal{P}$ is any partition of $[a, c]$ with $\|\mathcal{P}\| < \delta$, and $\mathcal{S}$ is any choice of sampling points for $\mathcal{P}$, then

$$\left| \sum_{i=1}^n \bar{f}(s_i)\Delta x_i - I \right| < \epsilon.$$

Let's use what we know about the original integral $\int_a^b f$ to produce such a δ.

First, since $\int_a^b f$ exists, we can choose $\delta_f > 0$ so that

$$\left| \sum_{i=1}^m f(s_i)\Delta x_i - I \right| < \frac{\epsilon}{2}$$

Draw your own picture of a typical f and $\bar{f}$.

The $\epsilon/2$ will come in handy later.

holds whenever $\mathcal{P}' = \{x_0, x_1, \dots x_m\}$ is a partition of $[a, b]$ with $\|\mathcal{P}'\| < \delta_n$, and the sampling set $\mathcal{S}' = \{s_1, s_2, \dots s_m\}$ is drawn from $\mathcal{P}'$. Second, since $\int_a^b f$ exists, we know f must be bounded on $[a, b]$, so we have $|f(x)| \le M$ for some positive number M and all $x \in [a, b]$.

Here $a = x_0$ and $b = x_m$.

So says Proposition 5.4, page 5.4.

Our goal is now visible in the distance. We'll show that the positive number

Strange as it may look at first glance.

$$\delta = \min\left\{\delta_f, \frac{\epsilon}{4M}\right\}$$

has the desired properties. To this end, let $\mathcal{P} = \{x_1, x_2, \dots x_n\}$ be any partition of $[a, c]$ with $\|\mathcal{P}\| < \delta$, let $\mathcal{S}$ be any selection of sampling points from $\mathcal{P}$, and consider the corresponding Riemann sum

$$S = \sum_{i=1}^{n} \bar{f}(s_i)\Delta x_i = \bar{f}(s_1)\Delta x_1 + \bar{f}(s_2)\Delta x_2 + \cdots + \bar{f}(s_n)\Delta x_n.$$

We need to show that S lies within ϵ of I. To do so, let x_{i_0} be the *largest* member of $\mathcal{P}$ that does not exceed b; that is, $x_{i_0} \le b < x_{i_0+1}$. Now for $i > i_0 + 1$ we know $\bar{f}(s_i) = 0$, while for $i \le i_0$ we have $\bar{f}(s_i) = f(s_i)$. With these facts in mind, we can rewrite S:

$$S = f(s_1)\Delta x_1 + f(s_2)\Delta x_2 + \cdots + f(s_{i_0})\Delta x_{i_0} + \bar{f}(s_{i_0+1})\Delta x_{i_0+1}.$$

This form reveals that S is *almost* a Riemann sum for $\int_a^b f$. Indeed, the similar sum

$$S' = f(s_1)\Delta x_1 + f(s_2)\Delta x_2 + \cdots + f(s_{i_0})\Delta x_{i_0} + f(b)\,(b - x_{i_0})$$

is a Riemann sum for $\int_a^b f$, corresponding to the new partition $\mathcal{P}' = \{a, x_1, x_2, \dots, x_{i_0}, b\}$; note that $\|\mathcal{P}'\| < \delta$. (If $x_{i_0} = b$ then the last summand in S' vanishes.)

To finish the proof, we observe that, by our hypothesis,

$$|S' - I| < \frac{\epsilon}{2}. \qquad (**)$$

Also, S and S' differ by (at most) two terms, each of which is small:

$$\begin{aligned}|S - S'| &= \left|\bar{f}(s_{i_0+1})\Delta x_{i_0+1} - f(b)\,(b - x_{i_0})\right| \\ &\le \left|\bar{f}(s_{i_0+1})\Delta x_{i_0+1}\right| + |f(b)\,(b - x_{i_0})| \\ &< M\delta + M\delta \le 2M\frac{\epsilon}{4M} = \frac{\epsilon}{2}.\end{aligned}$$

Combining this with inequality $(**)$ gives

$$|S - I| \le |S - S'| + |S' - I| < \frac{\epsilon}{2} + \frac{\epsilon}{2} = \epsilon,$$

as desired. $\qquad\qquad\square$

Lemma 5.7 lets us prove the theorem we really want.

Theorem 5.8. *Let $f : [a, b] \to \mathbb{R}$ be a function, and let c be a number between a and b. If both $\int_a^c f$ and $\int_c^b f$ exist, then $\int_a^b f$ exists, too, and*

$$\int_a^b f = \int_a^c f + \int_c^b f.$$

Proof: Define f_1 and f_2 on $[a, b]$ by

$$f_1(x) = \begin{cases} f(x) & \text{if } a \le x \le c, \\ 0 & \text{if } c < x \le b; \end{cases} \qquad f_2(x) = \begin{cases} 0 & \text{if } a \le x < c, \\ f(x) & \text{if } c \le x \le b. \end{cases}$$

Note that $f(x) = f_1(x) + f_2(x)$ for all $x \in [a, b]$. Now Lemma 5.7 says that both f_1 and f_2 are integrable on $[a, b]$, and that

$$\int_a^b f_1 = \int_a^c f \quad \text{and} \quad \int_a^b f_2 = \int_c^b f.$$

To finish up, we add up integrals, as Theorem 5.5 permits:

$$\int_a^b f = \int_a^b f_1 + \int_a^b f_2 = \int_a^c f + \int_c^b f,$$

as claimed. $\square$

Theorem 5.8 needs an important caveat: The add-up-the-pieces strategy is guaranteed to work only if we *know in advance* that the "smaller" integrals $\int_a^c f$ and $\int_c^b f$ exist. (We *do* know this in Example 3, in which f is constant on each smaller subinterval.) In particular, we have *not* proved the tempting (and true) fact that if f is integrable on $[a, b]$, then it is also integrable on any smaller interval $[a, c]$. We will prove this in the next section, where we consider more carefully the question of which functions are integrable.

Antiderivatives and Integrals: Toward a Fundamental Theorem

Elementary calculus experience suggests, almost irresistibly, that integrals and antiderivatives are closely connected. So they are, and the following result, a junior-grade version of the familiar fundamental theorem of calculus, begins the connection.

But there's more to say: a stronger version (Theorem 5.19, page 293) appears in the next section.

Theorem 5.9 (Fundamental theorem of calculus, version 0). *Let $f : [a, b] \to \mathbb{R}$ be a function, and suppose*

(i) *f is integrable on $[a, b]$;*

(ii) *there is a function $F : [a, b] \to \mathbb{R}$, continuous on $[a, b]$ and differentiable on (a, b), such that $F'(x) = f(x)$ for all $x \in (a, b)$.*

Then $\int_a^b f = F(b) - F(a)$.

Proof: By our hypotheses, both of the numbers $I = \int_a^b f$ and $F(b) - F(a)$ exist; we need to show they're equal. We'll do so by proving that

$$|F(b) - F(a) - I| < \epsilon$$

for every positive ϵ, no matter how small.

Let $\epsilon > 0$ be given. Because I satisfies Definition 5.1, we can choose $\delta > 0$ as described there. Now let

$$\mathcal{P} = \{x_0 = a, x_1, x_2, \dots x_n = b\}$$

be *any* partition with $\|\mathcal{P}\| < \delta$.

Finding such a $\mathcal{P}$ is easy. One is to chop $[a,b]$ into n equal pieces, where $[a,b]/n < \delta$.

The proof uses a clever idea: If we choose the sampling points $s_1, s_2, \dots, s_n$ just right, the associated Riemann sum adds up, after some fancy telescoping, to none other than $F(b) - F(a)$. Once this is done, Definition 5.1 guarantees that $F(b) - F(a)$, like *every* Riemann sum associated with $\mathcal{P}$, lies within ϵ of I. Since ϵ is arbitrary, we must have $F(b) - F(a) = I$, as claimed.

To choose these special sampling points, we apply the mean value theorem (MVT) to our antiderivative function F, once on each of the n partition subintervals. Hypothesis (ii) assures that the MVT applies. For each subinterval $[x_{i-1}, x_i]$, the MVT says that for some s_i in (x_{i-1}, x_i),

$$F(x_i) - F(x_{i-1}) = F'(s_i)(x_i - x_{i-1}) = f(s_i)\Delta x_i.$$

Adding all n of these results gives the desired Riemann sum:

Watch the summands collapse.

$$\sum_{i=1}^n f(s_i)\Delta x_i = \sum_{i=1}^n F(x_i) - F(x_{i-1})$$
$$= F(x_1) - F(x_0) + F(x_2) - F(x_1) + \cdots + F(x_n) - F(x_{n-1})$$
$$= F(x_n) - F(x_0) = F(b) - F(a).$$

This completes the proof. $\square$

Good news—and some cautions. Theorem 5.9 should look familiar. It allows many standard integral calculations of elementary calculus, like this one:

$$\int_1^5 x^2 \, dx = \frac{x^3}{3}\bigg]_1^5 = \frac{5^3}{3} - \frac{1^3}{3} = \frac{124}{3}.$$

Avoiding all that fuss over partitions, norms, and Riemann sums seems—and is— a big advantage in calculating a lot of integrals. But some sticky questions remain:

- *Can we find an antiderivative?* In the preceding calculation, with $f(x) = x^2$, it was easy to find (or just to know) that $F(x) = x^3/3$ is a suitable antiderivative. For other functions it can be much harder, or even impossible, to find antiderivative formulas. For instance, neither of the harmless-looking functions

$$f(x) = \cos(x^2) \quad \text{and} \quad g(x) = \cos(x)\ln(x)$$

has an "elementary antiderivative"—a function built from standard function "elements": polynomials, trigonometric functions, logarithms, etc. Without a suitable antiderivative F, Theorem 5.9 is useless for calculation.

- *Is f integrable?* We assumed in the theorem—and used crucially in the proof—that $\int_a^b f$ exists. Elementary calculus courses often skirt the question of integrability, perhaps forgivably both because the matter is subtle and because the basic functions of elementary calculus turn to *be* integrable on intervals within their domains.

Ask Mathematica or Maple to antidifferentiate these functions; notice the strange ingredients in the answer.

Theorem 5.9 dodges the tough questions above by simply *assuming*, as hypotheses, that all is well. Doing so simplifies the proof, but it weakens the theorem. In the next section, we'll grapple more seriously with the question of which functions are integrable. A key result (which applies to all basic functions of elementary calculus) is that every function f *continuous* on a closed interval $[a, b]$ is also *integrable* there.

Using the Fundamental Theorem

Definition 5.1, page 261

Among the drawbacks to using the (complicated!) basic definition to prove that an integral I exists is the need to know (or guess) a specific numerical value for I. This number may not be obvious, but Theorem 5.9 sometimes helps.

EXAMPLE 4. Does $\int_0^\pi (2x + \sin(x)) \, dx$ exist? If so, how big is it?

We will see later that f, being continuous, is integrable.

SOLUTION. The function $f(x) = 2x + \sin(x)$ has antiderivative $F(x) = x^2 - \cos(x)$, and so $F(\pi) - F(0) = \pi^2 + 2 \approx 11.87$. Theorem 5.9 does *not* guarantee that the integral exists, but it does say that $\pi^2 + 2$ is the only possible value. ◇

EXAMPLE 5.   Show that $\int_0^1 2x\, dx = 1$; use Theorem 5.9 and Definition 5.1.

SOLUTION.   The easy part is finding a value for the integral. Since the integrand $f(x) = 2x$ has antiderivative $F(x) = x^2$, and $F(1) - F(0) = 1$, Theorem 5.9 says that one is the only possible value.

We could have thought about areas, too.

The tricky bit is showing that the integral exists. We have, for now, only Definition 5.1 to work with, so we start as usual with a given positive ϵ. Let $\delta = \epsilon$; we'll use some clever algebra and a nice collapsing sum to show that this δ "works."

Let $\mathcal{P} = \{x_0, x_1, x_2, \ldots x_n\}$ be any partition of $[0,1]$ with $\|\mathcal{P}\| < \delta$. We claim that, for any samples $\mathcal{S} = \{s_1, s_2, \ldots, s_n\}$ drawn from $\mathcal{P}$,

$$1 - \epsilon < \sum_{i=1}^n f(s_i)\Delta x_i = \sum_{i=1}^n 2s_i \Delta x_i < 1 + \epsilon.$$

We will prove just the second inequality, leaving the first as an (easy) exercise.

The key observation is that, since $s_i \le x_i$ for all i,

$$\sum_{i=1}^n 2s_i \Delta x_i \le \sum_{i=1}^n 2x_i \Delta x_i.$$

It suffices, therefore, to show that the right-hand sum above can't exceed $1 + \epsilon$. For this we use an algebraic trick. If we write

$$2x_i = (x_i + x_{i-1}) + (x_i - x_{i-1}) = x_i + x_{i-1} + \Delta x_i$$

for each i, then substitution and a little algebra give

$$\sum_{i=1}^n 2x_i \Delta x_i = \sum_{i=1}^n \left(x_i + x_{i-1} + \Delta x_i \right) \Delta x_i$$

$$= \sum_{i=1}^n (x_i + x_{i-1}) \Delta x_i + \Delta x_i{}^2$$

$$= \sum_{i=1}^n \left(x_i^2 - x_{i-1}^2 \right) + \sum_{i=1}^n \Delta x_i{}^2 = S_1 + S_2.$$

We will handle S_1 and S_2 separately. A telescoping argument helps with S_1:

$$S_1 = \left(x_1^2 - x_0^2 \right) + \left(x_2^2 - x_1^2 \right) + \cdots + \left(x_n^2 - x_{n-1}^2 \right)$$
$$= x_n^2 - x_0^2 = 1 - 0.$$

The sum S_2 is small. Since $\Delta x_i < \delta = \epsilon$ for each i, we get

$$S_2 = \sum_{i=1}^n \Delta x_i{}^2 < \sum_{i=1}^n \epsilon\, \Delta x_i = \epsilon \sum_{i=1}^n \Delta x_i = \epsilon \cdot 1 = \epsilon.$$

Putting our results together gives

$$\sum_{i=1}^{n} 2s_i \, \Delta x_i \leq \sum_{i=1}^{n} 2x_i \, \Delta x_i = S_1 + S_2 < 1 + \epsilon,$$

which completes the proof. $\Diamond$

Exercises

1. This problem is about Corollary 5.6, page 270.

   (a) Consider the function $h(x) = g(x) - f(x)$ on $[a, b]$. Why must h be integrable? Why is $\int_a^b h \geq 0$?

   (b) Prove Corollary 5.6.

2. Suppose that f is integrable on $[a, b]$ and that $m \leq f(x) \leq M$ for all $x \in [a, b]$. Explain carefully why

$$m \, (b - a) \leq \int_a^b f \leq M \, (b - a).$$

3. Suppose f is integrable on $[1, 7]$ and $-1 \leq f(x) \leq 2$ for all $x \in [1, 7]$. Find upper and lower bounds for $\int_1^7 f$.

4. Use the result of Problem 2 to find upper and lower bounds on each of the following integrals; assume (it's true!) that they exist.

   (a) $\displaystyle\int_{100}^{200} \left(5 + \frac{\sin(x^2)}{100} \right) \, dx$

   (b) $\displaystyle\int_{100}^{200} \left(5 + \frac{\sin^2 x}{100} \right) \, dx$

   (c) $\displaystyle\int_{100}^{200} \frac{1}{100 + \sin^2 x} \, dx$

5. It is a fact (we will prove it later, but just assume it here) that if f is *continuous* on $[a, b]$ then f is also *integrable* on $[a, b]$.

   (a) Suppose that f is continuous on $[a, b]$ and that $f(x) > 0$ for all $x \in [a, b]$. Show that $\int_a^b f > 0$, too. (Hint: Use the extreme value theorem and Corollary 5.6, page 270.)

(b) Suppose f is continuous on $[a, b]$ and that $\int_a^b f = 0$. Show that $f(c) = 0$ for some $c \in [a, b]$. (This fact is sometimes known as the *mean value theorem for integrals*.)

(c) Give an example to show that the result in (b) need not hold if f is discontinuous on $[a, b]$.

6. Suppose that f is integrable on $[a, b]$. Show (in the spirit of the proof of Theorem 5.5, page 269) that $3f$ is also integrable and that $\int_a^b 3f = 3 \int_a^b f$.

7. Suppose that both $\int_a^b f$ and $\int_a^b |f|$ exist. Prove the "triangle-like" inequality

$$\left| \int_a^b f \right| \le \int_a^b |f|.$$

8. Consider the function f defined by $f(n) = \sin n$ if $n \in \mathbb{Z}$ and $f(x) = 0$ otherwise. Show that f is integrable on every interval $[-a, a]$, where $a > 0$. What is $\int_{-a}^a f$?

9. Consider the "floor function" $f(x)$ defined by

$$f(x) = \sup\{n \in \mathbb{N} \mid n \le x\}.$$

(a) Explain (cite an appropriate theorem) why f is integrable on every interval $[a, b]$.

(b) What is $\int_{-1}^1 f$? Why?

(c) What is $\int_{-n}^n f$? Why?

10. We showed in and near Problem 12, page 268, that $f(x) = x$ is integrable on $[a, b]$, and that $\int_a^b x \, dx = (b^2 - a^2)/2$.

(a) Use the formula above and Theorem 5.5 to find $\int_1^5 (3x + 7) \, dx$.

(b) A function g has the W-shaped graph formed by connecting the dots at $(0, 2)$, $(1, 0)$, $(2, 1)$, $(3, 0)$, and $(4, 2)$ in the xy-plane. Explain why the integral $\int_0^4 g$ exists and find its value.

11. Assume (it's true) that the integral in each part following exists. Use Theorem 5.9 to find the value.

(a) $\int_0^1 x^{42} \, dx$

(b) $\int_0^\pi \sin x \, dx$

(c) $\int_a^b \left(C + Dx + Ex^2 + F(x^3) \right)\, dx$

(d) $\int_0^1 xe^{x^2}\, dx$

12. Consider the function f defined by $f(1/n) = 1$ for $n \in \mathbb{N}$ and $f(x) = 0$ otherwise. Assuming that f is integrable (as we'll see in the next section), complete the following proof sketch to show that $I = \int_0^1 f = 0$.

    (i) Clearly $I \geq 0$, so it is enough to show $I \leq \epsilon$ for all $\epsilon > 0$. (ii) For given ϵ, define $f_\epsilon(x) = 1$ if $0 \leq x \leq \epsilon$, and $f_\epsilon(x) = 0$ otherwise. Then f_ϵ is integrable, and $\int_0^1 f_\epsilon(x) = \epsilon$. (iii) We have $f(x) \leq f_\epsilon(x)$ for all but finitely many x, so $\int_0^1 f \leq \int_0^1 f_\epsilon(x) = \epsilon$.

13. Suppose that both $\int_a^b f$ and $\int_a^b g$ exist, and that $f(x) \leq g(x)$ for all but finitely many x in $[a, b]$. Show that $\int_a^b f \leq \int_a^b g$.

5.3   Integrability

How can we decide whether a given function $f : [a, b] \to \mathbb{R}$ is integrable on $[a, b]$? The question is obviously important: a useful integral should apply to many functions. The question is also difficult: deciding which functions satisfy a complicated definition naturally takes some work.

So far we've seen only piecemeal results:

See Proposition 5.4, page 265.

- If f is integrable, then f is bounded.

See Example 3, page 262, and Example 4, page 263.

- A discontinuous function f may or may not be integrable.

See Example 4, page 276.

- We've also said—but not proved—that every continuous function is integrable.

In this section we approach these matters rigorously, and identify some important classes of integrable functions. Our main tool, detailed in Theorem 5.14, is the *box sum*, a useful and practically usable criterion for integrability.

Criteria for Integrability

Our definition of the Riemann integral is relatively technical, so we expect that detecting integrability (or its absence) will be challenging, too. An immediate hurdle is the fact the definition requires a "candidate" I for the value of a given integral. It turns out that we can skirt this problem with a "Cauchy criterion" for Riemann sums.

We did something similar—for similar reasons—for sequences.

Lemma 5.10. *Let $f[a, b] \to \mathbb{R}$ be a bounded function. Suppose that for every $\epsilon > 0$ there exists $\delta > 0$ such that*

$$|\mathrm{RS}(f, \mathcal{P}_1, \mathcal{S}_1) - \mathrm{RS}(f, \mathcal{P}_2, \mathcal{S}_2)| < \epsilon$$

whenever $\mathcal{P}_1$ and $\mathcal{P}_2$ are partitions of $[a, b]$ such that both $\|\mathcal{P}_1\| < \delta$ and $\|\mathcal{P}_2\| < \delta$, and samples $\mathcal{S}_1$ and $\mathcal{S}_2$ come from $\mathcal{P}_1$ and $\mathcal{P}_2$. Then the integral $\int_a^b f$ exists.

Proof: The idea is to concoct a certain Cauchy sequence $\{I_n\}$ of *numbers*, whose limit I will turn out to be the desired integral.

Every Cauchy sequence has one.

To get started, for each $n \in \mathbb{N}$ we use the hypothesis to choose a positive number δ_n such that

$$|\mathrm{RS}(f, \mathcal{P}_1, \mathcal{S}_1) - \mathrm{RS}(f, \mathcal{P}_2, \mathcal{S}_2)| < \frac{1}{n}$$

for all Riemann sums based on partitions $\mathcal{P}_1$ and $\mathcal{P}_2$ with both $\|\mathcal{P}_1\| < \delta_n$ and $\|\mathcal{P}_2\| < \delta_n$. For technical reasons we choose the δ_n to be decreasing: $\delta_1 \geq \delta_2 \geq \delta_3 \geq \dots$.

Convince yourself this is possible.

Next, for each n we choose any particular partition $\mathcal{P}_n$ with $\|\mathcal{P}_n\| < \delta_n$ and any particular set $\mathcal{S}_n$ of samples from $\mathcal{P}_n$. The associated Riemann sum $\mathrm{RS}(f, \mathcal{P}_n, \mathcal{S}_n)$ is then a number; let's call it I_n for short. This process produces a numerical sequence $\{I_n\}$, which is readily shown to be Cauchy—and hence converges to some limit I. (The proof that $\{I_n\}$ is Cauchy uses the fact that the δ_n decrease; see the exercises.)

A regular partition works fine.

Last, we use Definition 5.1, page 261, to show that I is the sought-after integral. For given $\epsilon > 0$, we first choose any positive integer N for which both

$$|I_N - I| < \frac{\epsilon}{2} \quad \text{and} \quad \frac{1}{N} < \frac{\epsilon}{2},$$

and we set $\delta = \delta_N$ as chosen above. This δ does what Definition 5.1 asks. If $\mathcal{P}$ is any partition with $\|\mathcal{P}\| < \delta_N$, $\mathcal{S}$ is any set of samples, and $\mathrm{RS}(f, \mathcal{P}, \mathcal{S})$ is the corresponding Riemann sum, then the triangle inequality and our choices give

$$|\mathrm{RS}(f, \mathcal{P}, \mathcal{S}) - I| \leq |\mathrm{RS}(f, \mathcal{P}, \mathcal{S}) - I_N| + |I_N - I|$$
$$\leq \frac{\epsilon}{2} + \frac{\epsilon}{2} = \epsilon,$$

as Definition 5.1 requires. $\qquad\qquad\square$

Not quite there. Lemma 5.10 will prove useful, but it has a serious practical drawback: showing that the hypothesized inequality holds for *all* suitable partitions $\mathcal{P}$ and *all* corresponding sample sets $\mathcal{S}$ seems difficult. The *box-sum* criterion, which turns out to be equivalent to the hypothesis of Lemma 5.10, will prove much easier to work with.

There are infinitely many $\mathcal{P}$ for each δ, and infinitely many $\mathcal{S}$ for each $\mathcal{P}$.

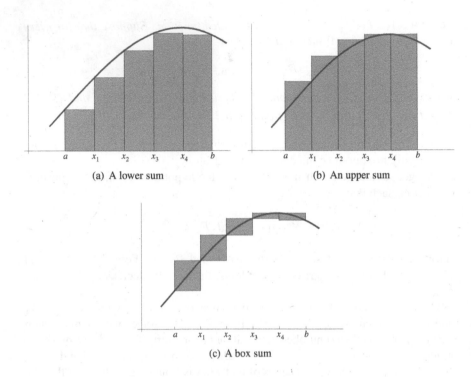

(a) A lower sum   (b) An upper sum

(c) A box sum

Figure 5.4. Lower, upper, and box sums for an integral.

Integrability and Box Sums

Figure 5.4 shows (shaded) the *lower sum*, *the upper sum*, and the corresponding *box sum*—the difference between upper and lower sums—for a well-behaved integrand f and a five-piece partition of $[a, b]$.

The picture suggests the appropriate definitions, but some extra care is needed for discontinuous integrands. The fine print follows.

Upper and lower sums. For any given partition $\mathcal{P}$ of an interval $[a, b]$, there are infinitely many ways of choosing samples $\mathcal{S}$ compatible with $\mathcal{P}$, and hence infinitely many possible Riemann sums associated to $\mathcal{P}$. To get upper and lower bounds on the values of all these Riemann sums, it is helpful to consider *upper sums* and *lower sums*.

If the integrand f is continuous on $[a, b]$, as in Figure 5.4, then upper and lower sums are ordinary Riemann sums, with sampling points chosen to *maximize* or *minimize* f over each separate subinterval. (A continuous function *has* maximum and minimum values on each subinterval, by the extreme value theorem.) In this case, therefore, upper and lower sums are simply the *largest and smallest possible*

Riemann sums for a given partition.

A *discontinuous* integrand f, on the other hand, need not attain maximum and minimum values on a given subinterval. If f is *bounded* on $[a, b]$, however, then both

$$m_i = \inf\{\, f(x) \mid x \in [x_{i-1}, x_i]\,\} \quad \text{and} \quad M_i = \sup\{\, f(x) \mid x \in [x_{i-1}, x_i]\,\}$$

exist for each i, and so the following definitions make sense.

The completeness axiom guarantees this.

Definition 5.11. Let $f : [a, b] \to \mathbb{R}$ be a bounded function, and $\mathcal{P} = \{x_0, x_1, \ldots, x_n\}$ a partition of $[a, b]$. With m_i and M_i as above, we define *upper* and *lower sums* as follows:

$$\mathrm{US}\,(f, \mathcal{P}) = \sum_{i=1}^{n} M_i \Delta x_i = M_1 \Delta x_1 + M_2 \Delta x_2 + \cdots + M_n \Delta x_n;$$

$$\mathrm{LS}\,(f, \mathcal{P}) = \sum_{i=1}^{n} m_i \Delta x_i = m_1 \Delta x_1 + m_2 \Delta x_2 + \cdots + m_n \Delta x_n.$$

Upper and lower sums may not *be* Riemann sums, but they have useful connections to Riemann sums and to the integrals they approximate. We collect two such facts in the following proposition, using RS, US, and LS to denote Riemann, upper, and lower sums.

They are Riemann sums if f happens to be continuous.

Proposition 5.12. *Let $f : [a, b] \to \mathbb{R}$ be a bounded function, and $\mathcal{P}$ any partition of $[a, b]$.*

Upper and lower sums don't depend on samples, and so their notations don't involve an $\mathcal{S}$.

(i) *If $\mathcal{S}$ is set of samples from $\mathcal{P}$, then*

$$\mathrm{LS}\,(f, \mathcal{P}) \leq \mathrm{RS}\,(f, \mathcal{P}, \mathcal{S}) \leq \mathrm{US}\,(f, \mathcal{P}).$$

(ii) *If f is integrable, with $\int_a^b f = I$, then*

$$\mathrm{LS}\,(f, \mathcal{P}) \leq I \leq \mathrm{US}\,(f, \mathcal{P}).$$

About proofs. The inequality in (i) follows immediately from the definitions of upper and lower sums. Part (ii) is slightly subtler; see the exercises for both parts.

Clamping down: box sums. Proposition 5.12 suggests that it is a good thing for integrability when upper and lower sums are close together. Box sums measure this closeness.

Definition 5.13 (The box sum). Let $f : [a, b] \to \mathbb{R}$ be a bounded function and $\mathcal{P} = \{x_0, x_1, \ldots x_n\}$ a partition of $[a, b]$. The associated *box sum* is the difference

$$\mathrm{US}\,(f, \mathcal{P}) - \mathrm{RS}\,(f, \mathcal{P}) = (M_1 - m_1)\Delta x_1 + \cdots + (M_n - m_n)\Delta x_n.$$

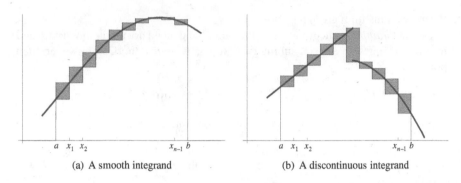

(a) A smooth integrand   (b) A discontinuous integrand

Figure 5.5. Two more box sums.

Box sums have a nice geometric interpretation, as seen already in Figure 5.4(c). Figure 5.5(a) shows another box sum for the same integrand as before, but here with a finer partition. Figure 5.5(b) shows another ten-element box sum, this time for a *discontinuous* integrand.

Here is the key point: For *both* integrands in Figure 5.5, the total shaded area—the box sum—can be made *as small as we wish* by requiring that the partition have small norm. This seems clear enough for the smooth integrand in Figure 5.5(a): For a partition with tiny norm, the shaded area becomes smaller and smaller, approaching a skinny "tube" around the graph of f.

Less obvious, but equally important, is the fact that the total shaded area can also be made small for the *discontinuous* integrand in Figure 5.5(b). Figure 5.6 suggests why. Any partition $\mathcal{P}$ may generate one or two *tall* box sum elements, but if $\mathcal{P}$ has small norm, then all boxes are narrow and their *areas* don't amount to much. We formalize these ideas in a theorem.

Theorem 5.14 (The box-sum criterion). *A function* $f : [a, b] \to \mathbb{R}$ *is integrable on* $[a, b]$ *if and only if for each* $\epsilon > 0$ *there exists some partition* $\mathcal{P}$ *of* $[a, b]$ *with box sum less than* ϵ.

Many uses. We outline the rather technical proof at the end of this section; it involves some meticulous bookkeeping. First we illustrate the theorem's uses and advantages—including the fact that for a given ϵ we need only *one* suitable partition to prove integrability. Observe also that the box-sum criterion works both ways: it detects both integrability and non-integrability. Example 1 illustrates both of these uses.

EXAMPLE 1.   Consider the functions

$$f(x) = \begin{cases} 0 & \text{if } x < 0.5, \\ 1 & \text{if } x \geq 0.5; \end{cases} \qquad g(x) = \begin{cases} 0 & \text{if } x \notin \mathbb{Q}, \\ 1 & \text{if } x \in \mathbb{Q}. \end{cases}$$

We have shown by other methods that f is integrable on $[0, 1]$ but g is not. What do box sums say?

SOLUTION.   For f, consider the three-piece partition

$$\mathcal{P} = \{0, 0.499, 0.501, 1\},$$

which "isolates" the jump discontinuity at $x = 0.5$ in the skinny interval $[0.499, 0.501]$. The box sum—only 0.002 in this case—can be made even smaller narrowing the middle interval even further. Thus f is indeed integrable.

With integral 0.5.

For g the conclusion is negative: for *any* partition $\mathcal{P}$ of $[0, 1]$, *every* box element has height one, and so the box sum is one. $\Diamond$

Draw your own picture.

The box-sum criterion is exactly what we need to prove some familiar and important—but otherwise elusive—results.

Theorem 5.15. *If f is continuous on $[a, b]$, then f is integrable, too.*

The proof idea. Continuity means that f cannot rise or fall very much over a small domain interval. Thus, for a partition with small norm, all box sum elements must be *short*. Since the total *width* of boxes is only $(b - a)$, their total *area* must also be small.

Proof: Let $\epsilon > 0$ be given. Because f is continuous on $[a, b]$, it is also *uniformly* continuous there, according to Theorem 3.22, page 184. This means we can choose $\delta > 0$ such that

$$|f(s) - f(t)| < \frac{\epsilon}{b - a} \quad \text{whenever} \quad |s - t| < \delta \text{ and } s, t \in [a, b].$$

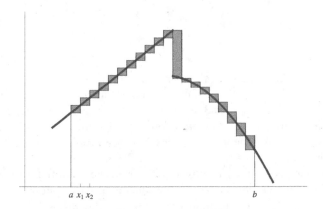

Figure 5.6. Fine partitions, small box sums.

A regular partition will do.

Now let $\mathcal{P}$ be *any* partition with $\|\mathcal{P}\| < \delta$. Since f is continuous on each subinterval $[x_{i-1}, x_i]$, it attains maximum and minimum values there, so there exist s_i and t_i in $[x_{i-1}, x_i]$ with

$$f(s_i) = M_i = \sup\{\, f(x) \mid x \in [x_{i-1}, x_i]\,\};$$
$$f(t_i) = m_i = \inf\{\, f(x) \mid x \in [x_{i-1}, x_i]\,\}.$$

Our uniform continuity condition gives $M_i - m_i < \epsilon/(b-a)$ for all i, and so

$$\text{box sum} = \sum_{i=1}^{n} (M_i - m_i)\, \Delta x_i$$
$$< \sum_{i=1}^{n} \frac{\epsilon}{b-a} \Delta x_i = \frac{\epsilon}{b-a} \sum_{i=1}^{n} \Delta x_i = \epsilon,$$

as desired. $\square$

More on integrability. We can use the box-sum criterion to prove other familiar, reasonable-seeming properties of the integral. Proposition 5.16 gives two samples.

Proposition 5.16. *Suppose that $f : [a, b] \to \mathbb{R}$ is integrable.*

(i) *If $a < c < b$, then $\int_a^c f$ and $\int_c^b f$ exist, and*

$$\int_a^b f = \int_a^c f + \int_c^b f.$$

(ii) *The function $|f|$ is integrable, and*

$$\left| \int_a^b f \right| \leq \int_a^b |f|.$$

Proof (sketch): The main proof challenge for both (i) and (ii) turns out to be integrability: We need to show that if f is integrable on $[a, b]$, then it is also integrable on the smaller intervals $[a, c]$ and on $[c, b]$, and that $|f|$ is also integrable on $[a, b]$. Once all the integrals in question are known to exist, the rest is easy. The equation in (i) is essentially Theorem 5.8, page 274. See Problem 7, page 279, for the inequality in (ii).

We don't even have candidates for the values of the integrals.

Proving our integrability claims directly from the definition of integrability would be difficult. With the box-sum criterion, the proof is routine.

Let $\epsilon > 0$ be given. By Theorem 5.14, applied to the integral $\int_a^b f$, there exists a partition $\mathcal{P}$ of $[a, b]$ with box sum less than ϵ.

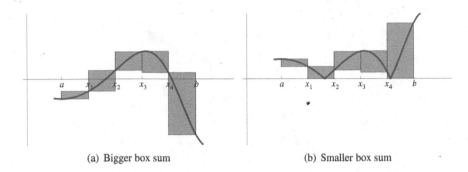

(a) Bigger box sum   (b) Smaller box sum

Figure 5.7. Comparing box sums for f and for $|f|$.

Let's show first that $\int_a^c f$ and $\int_c^b f$ exist. We may as well assume that $c \in \mathcal{P}$; if not, we can add c to $\mathcal{P}$ without increasing the box sum. Thus, $\mathcal{P}$ has the form

Sketch the situation to convince yourself.

$$\mathcal{P} = \{a = x_0, x_1, \ldots, x_m = c, x_{m+1}, \ldots, x_n = b\},$$

and therefore

$$\mathcal{P}_1 = \{a = x_0, x_1, \ldots, x_m = c\} \quad \text{and} \quad \mathcal{P}_2 = \{c = x_m, x_{m+1}, \ldots, x_n = b\}$$

are, respectively, partitions of $[a, c]$ and $[b, c]$. Because $\mathcal{P}_1$ and $\mathcal{P}_2$ are subsets of $\mathcal{P}$, and all box summands are nonnegative, the box sums for $\mathcal{P}_1$ and $\mathcal{P}_2$ are clearly *smaller* than that for $\mathcal{P}$. By Theorem 5.14, f is indeed integrable on each smaller interval.

To show that $|f|$ is also integrable on $[a, b]$, we compare box sums for f and $|f|$. Figure 5.7 illustrates the nice answer: For *any* partition $\mathcal{P}$ of $[a, b]$,

$$\text{box sum for } |f| \leq \text{box sum for } f.$$

In Figure 5.7(a), box-sum elements that "straddle" the x-axis become smaller for $|f|$, as shown in Figure 5.7(b). Thus any partition for f with box sum less than ϵ works for $|f|$, too. $\square$

Proposition 5.17. *If $f : [a, b] \to \mathbb{R}$ is monotone, then f is integrable.*

The proof idea. Figure 5.8 illustrates the proof idea for an increasing integrand, with m and M as lower and upper bounds.

The pictured integrand has one discontinuity, but a monotone function can have many.

If $\mathcal{P}$ is a regular partition with norm δ, then the box sum elements can be "stacked" vertically as shown, to give total area less than $(M - m)\delta$. By choosing δ small, we can keep the box sum as small as we like.

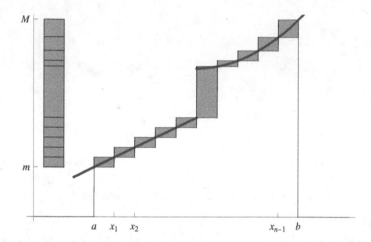

Figure 5.8. Stack the boxes: why every monotone function is integrable.

EXAMPLE 2. What does Proposition 5.17 say about integrability of some favorite functions?

SOLUTION. Proposition 5.17 guarantees that integrals like

$$\int_0^{100} x^{10}\, dx, \quad \int_0^{100} \sqrt{x}\, dx, \quad \int_0^{100} \ln(x+1)\, dx, \quad \int_0^{100} \frac{1}{\ln(x+1)}\, dx$$

all exist, because all integrands are monotone on $[0, 100]$. With help from Theorem 5.8, page 274, we can conclude that all of

$$\int_0^{100} \cos x\, dx, \quad \int_0^{100} \ln(\cos x + 2)\, dx, \quad \int_0^{100} e^{\cos x}\, dx$$

also exist. Even though the integrands are not monotone on $[0, 100]$, in each case we can break $[0, 100]$ up into finitely many smaller subintervals, on each of which the given integrand *is* monotone.   ◊

Proving the Box-Sum Criterion

Theorem 5.14 has both an "if" and an "only if" part. We start with the latter—easier—assertion.

Why the box-sum criterion is necessary. Suppose f is integrable on $[a, b]$; let $I = \int_a^b$. For any given $\epsilon > 0$, we'll find a partition $\mathcal{P}$ of $[a, b]$ with box sum less than ϵ. Since f is integrable, we can choose $\delta > 0$ such that

$$I - \frac{\epsilon}{4} < \mathrm{RS}\,(f, \mathcal{P}, \mathcal{S}) < I + \frac{\epsilon}{4}$$

for every Riemann sum $\mathrm{RS}\,(f, \mathcal{P}, \mathcal{S})$ based on a partition $\mathcal{P}$ with $\|\mathcal{P}\| < \delta$ and samples $\mathcal{S}$ drawn from $\mathcal{P}$. Now let $\mathcal{P}$ be *any* such partition; this $\mathcal{P}$ will do what we want.

A regular partition will do.

To see why, note first that the upper sum $\mathrm{US}\,(f, \mathcal{P})$, although perhaps not a Riemann sum itself, is the *supremum* of all possible Riemann sums based on $\mathcal{P}$. In particular, we can choose a set of samples $\mathcal{S}$ from $\mathcal{P}$ so that

$$\mathrm{RS}\,(f, \mathcal{P}, \mathcal{S}) > \mathrm{US}\,(f, \mathcal{P}) - \frac{\epsilon}{4},$$

and so we have

$$\mathrm{US}\,(f, \mathcal{P}) < \mathrm{RS}\,(f, \mathcal{P}, \mathcal{S}) + \frac{\epsilon}{4} < I + \frac{\epsilon}{4} + \frac{\epsilon}{4} = I + \frac{\epsilon}{2}.$$

A similar argument shows that

$$\mathrm{LS}\,(f, \mathcal{P}) > I - \frac{\epsilon}{2},$$

and so

$$\text{box sum} = \mathrm{US}\,(f, \mathcal{P}) - \mathrm{LS}\,(f, \mathcal{P}) < \epsilon,$$

as we wanted to show.

Why the box-sum criterion is sufficient. The plan of the proof is to use the box-sum condition to obtain the "Cauchy criterion" of Lemma 5.10, page 281, which in turn implies integrability. We show, in fact, that if for given $\epsilon > 0$ we can find *some* partition $\mathcal{P}_0$ with small box sum, then there exists some (very small!) $\delta > 0$ such that *every* partition $\mathcal{P}$ with $\|\mathcal{P}\| < \delta$ must also have small box sum, and this does the trick.

One picky technical lemma will prove useful. We leave its straightforward proof as an exercise.

Draw a simple picture to get started.

Lemma 5.18. *Let $f : [a, b] \to \mathbb{R}$ be a bounded function, with $m \le f(x) \le M$ for $x \in [a, b]$. Let $\mathcal{P}$ be a partition of $[a, b]$ with $\|\mathcal{P}\| < \delta$, and let $\mathrm{US} = \mathrm{US}\,(f, \mathcal{P})$ and $\mathrm{LS} = \mathrm{LS}\,(f, \mathcal{P})$ be the corresponding upper and lower sums.*

Adding one *more point to $\mathcal{P}$ produces new upper and lower sums, $\mathrm{US}_{\mathrm{new}}$ and $\mathrm{LS}_{\mathrm{new}}$, such that*

(i) $\mathrm{US}_{\mathrm{new}} \le \mathrm{US}$ *and* $\mathrm{LS}_{\mathrm{new}} \ge \mathrm{LS}$;

(ii) $\mathrm{US} - \mathrm{US}_{\mathrm{new}} < (M - m)\delta$ *and* $\mathrm{LS}_{\mathrm{new}} - \mathrm{LS} < (M - m)\delta$.

In words: Adding one point to $\mathcal{P}$ *decreases* upper sums *and increases* lower *sums—but by no more than* $(M - m)\delta$.

Now we can prove that the box-sum condition implies integrability. For given $\epsilon > 0$, we first choose any particular partition

$$\mathcal{P}_0 = \{w_0, w_1, \ldots, w_N\},$$

of $[a, b]$ with N partition points and box sum less than $\epsilon/2$; that is,

$$\mathrm{US}\,(f, \mathcal{P}_0) - \mathrm{LS}\,(f, \mathcal{P}_0) < \frac{\epsilon}{2}.$$

Next we set

$$\delta = \frac{\epsilon}{4(N - 1)(M - m)},$$

where M and m are, respectively, upper and lower bounds for f on $[a, b]$. This δ

Unlikely as that may seem in advance.

will turn out to work in Lemma 5.10.

Now let $\mathcal{P}$ be any partition of $[a, b]$ with $\|\mathcal{P}\| < \delta$. Consider the new partition $\mathcal{P}'$ formed from $\mathcal{P}$ by adding in the $N - 1$ partition points $w_1, w_2, \ldots, w_{N-1}$. (Since $w_0 = a$ and $w_N = b$, adding them to $\mathcal{P}$ has no effect.)

Now Lemma 5.18(i) implies that

$$\mathrm{US}\,(f, \mathcal{P}') \leq \mathrm{US}\,(f, \mathcal{P}_0),$$

because $\mathcal{P}'$ "refines" $\mathcal{P}_0$. Because $\mathcal{P}'$ has at most $N - 1$ more points than $\mathcal{P}$, Lemma 5.18(ii) says that

$$\mathrm{US}\,(f, \mathcal{P}') > \mathrm{US}\,(f, \mathcal{P}) - (N - 1) \cdot (M - m) \cdot \delta$$
$$= \mathrm{US}\,(f, \mathcal{P}) - \frac{\epsilon}{4}.$$

Putting these inequalities together gives

$$\mathrm{US}\,(f, \mathcal{P}) < \mathrm{US}\,(f, \mathcal{P}') + \frac{\epsilon}{4} \leq \mathrm{US}\,(f, \mathcal{P}_0) + \frac{\epsilon}{4}.$$

A similar argument shows that

$$\mathrm{LS}\,(f, \mathcal{P}) < \mathrm{LS}\,(f, \mathcal{P}_0) + \frac{\epsilon}{4}.$$

All these inequalities boil down saying that, for any partition $\mathcal{P}$ with $\|\mathcal{P}\| < \delta$, both the upper and lower sums $\mathrm{LS}\,(f, \mathcal{P})$ and $\mathrm{US}\,(f, \mathcal{P})$ lie in the interval

$$\left(\mathrm{LS}\,(f, \mathcal{P}_0) - \frac{\epsilon}{4}, \ \mathrm{US}\,(f, \mathcal{P}_0) + \frac{\epsilon}{4} \right),$$

which has length less than ϵ. In particular, every *Riemann* sum $\text{RS}(f, \mathcal{P}, \mathcal{S})$ is between $\text{LS}\,(f, \mathcal{P})$ and $\text{US}\,(f, \mathcal{P})$, and therefore lies in the same interval.

Finally, we observe that if both $\mathcal{P}_1$ and $\mathcal{P}_2$ are partitions of $[a, b]$ with both $\|\mathcal{P}_1\| < \delta$ and $\|\mathcal{P}_2\| < \delta$, and if samples $\mathcal{S}_1$ and $\mathcal{S}_2$ come from $\mathcal{P}_1$ and $\mathcal{P}_2$, respectively, then both $\text{RS}(f, \mathcal{P}_1, \mathcal{S}_1)$ and $\text{RS}(f, \mathcal{P}_2, \mathcal{S}_2)$ lie in the same interval of length less than ϵ, and are thus within ϵ of each other. Thus the hypothesis of Lemma 5.10 is satisfied, and the integral $\int_a^b f$ exists.

Exercises

1. Suppose f is continuous on $\mathbb{R}$. Explain why the functions defined by $f^2(x)$, $\sin(f(x))$, $f(\sin x)$, and $\ln(\sin f(x) + 2)$ are all integrable on every interval $[a, b]$.

2. Consider the integral $\int_0^{10} \sin x\, dx$. Find a partition $\mathcal{P}$ of $[0, 10]$ that gives box sum less than 0.01. (Hint: Use the fact that inequality $|\sin x - \sin y| \leq |x - y|$ holds for all $x, y \in \mathbb{R}$.)

3. Consider the function f defined by $f(n) = 0$ if $n \in \mathbb{Z}$ and $f(x) = \pi$ otherwise, and let $\epsilon > 0$. Find a partition $\mathcal{P}$ of $[0, 10]$ that gives box sum less than ϵ. (Hint: Isolate the "offender points" in small subintervals.) What is $\int_0^{10} f$?

4. Use box sums to show carefully that any function of your choice is *not* integrable on $[0, 1]$.

5. Show using box sums and Theorem 5.14 that the function $f(x) = x^2$ is integrable on $[0, 1]$.

6. Show using box sums and Theorem 5.14 that the function $f(x) = x^3$ is integrable on $[0, 1]$.

7. Suppose that f is integrable on $[0, 2]$. Show using box sums and Theorem 5.14 that f is integrable on $[0, 1]$, too.

8. We said in Example 2, page 288, that $\int_0^{100} \cos x\, dx$ can be shown to exist using Proposition 5.17, page 287 and Theorem 5.8, page 274. Give the details.

9. State and prove the converse of Lemma 5.10.

10. Show that the sequence $\{I_n\}$ defined in the proof of Lemma 5.10 is Cauchy.

11. Show that if $f : [a, b] \to \mathbb{R}$ is a bounded function, $\mathcal{P}$ is a partition of $[a, b]$, and $\mathcal{S}$ is any sampling set from $\mathcal{P}$, then

$$\text{LS} (f, \mathcal{P}) \leq \text{RS} (f, \mathcal{P}, \mathcal{S}) \leq \text{US} (f, \mathcal{P}) .$$

Note: This is (i) of Proposition 5.12; it says that for any partition $\mathcal{P}$, the corresponding upper and lower sums are upper and lower bounds, respectively, for the set of all Riemann sums associated with $\mathcal{P}$.

12. Give a "picture proof" of Lemma 5.18, page 289. (Hint: Adding one more point to a partition changes only one box sum element. Convince yourself that this change is no more than the claimed amount.)

13. Show that if $\int_a^b f = I$ exists, and $\mathcal{P}$ is any partition of $[a, b]$, then

$$\text{LS} (f, \mathcal{P}) \leq I \leq \text{US} (f, \mathcal{P}) .$$

Note: This is (ii) of Proposition 5.12; it says that, for any partition $\mathcal{P}$, the corresponding upper and lower sums "trap" the exact value of I from above and below.

(Hint: To show that $\text{US} (f, \mathcal{P})$ overestimates I, consider the "stair-step" function f_{big} defined by

$$f_{\text{big}}(x) = \begin{cases} M_1 & \text{if } a \leq x < x_1 \\ M_2 & \text{if } x_1 \leq x < x_2 \\ \cdots & \cdots \\ M_n & \text{if } x_{x_{n-1}} \leq x \leq b \end{cases}$$

Why is f_{big} integrable? What is its integral? How does $\int_a^b f_{\text{big}}$ compare to $\int_a^b f$?)

14. Consider the function f given by $f(1/n) = 1$ if $n \in \mathbb{N}$ and $f(x) = 0$ otherwise. Show, using box sums and Theorem 5.14, that $I = \int_0^1 f$ exists. (We know already—see Problem 12, page 280—that if the integral exists, then $I = 0$.)

5.4   Some Fundamental Theorems

We met a useful but preliminary version of the fundamental theorem of calculus (Theorem 5.9, page 275) in Section 5.2. Using Theorem 5.15, page 285, we can now tie up a serious loose end—existence of the integral—in connecting integrals and derivatives. Here is a new, improved version:

Theorem 5.19 (Fundamental theorem of calculus, version 1). *Let $f : [a, b] \to$ $\mathbb{R}$ be continuous, and suppose $F'(x) = f(x)$ for all $x \in [a, b]$. Then f is integrable on $[a, b]$, and*

$$\int_a^b f = F(b) - F(a).$$

Proof: We've already done all the hard work. By Theorem 5.15, the integral $\int_a^b f$ exists, and Theorem 5.9 guarantees that the integral's value is indeed $F(b) - F(a)$. $\square$

Elementary calculus revisited. Theorem 5.19 makes precise a sense in which integration and differentiation are inverse processes. Now that we know it, we can assert standard calculations of elementary calculus, like this one—

$$\int_0^1 x^2 \, dx = \left. \frac{x^3}{3} \right]_0^1 = \frac{1}{3} - 0,$$

with perfect confidence.

EXAMPLE 1.   Find $\int_0^\pi \sin x \, dx$ exactly.

SOLUTION.   Nothing could be easier. If $f(x) = \sin x$, then $F(x) = -\cos(x)$ is an antiderivative, and so

$$\int_0^\pi \sin x \, dx = -\cos(\pi) + \cos(0) = 2.$$

Having worked very hard to build the machine, we now only need to turn the crank. $\Diamond$

EXAMPLE 2.   Find $\int_0^\pi \sin(x^2) \, dx$.

SOLUTION.   The problem superficially resembles the one in Example 1, and it would be nice to proceed in the same way. But there is a problem: $f(x) = \sin(x^2)$ has no nice antiderivative formula $F(x)$ into which we can plug endpoints. We can always *estimate* the answer numerically, but that is less satisfying. $\Diamond$ It's about 0.773.

Antiderivatives *do* exist. The difficulty in Example 2 notwithstanding, the integrand $f(x) = \sin(x^2)$ there *does* have an antiderivative—it just doesn't happen to be expressible in simple symbolic form. Indeed, *every* continuous function has antiderivatives, and another version of the fundamental theorem of calculus says why.

Theorem 5.20 (Fundamental theorem of calculus, version 2). *Let* $f : [a, b] \to \mathbb{R}$ *be continuous. Define a new function* F *by*

$$F(x) = \int_a^x f \quad \text{for all } x \in [a, b].$$

Then $F'(c) = f(c)$ *for all* $c \in (a, b)$.

Proof: Note first that $F(x)$ is defined for all $x \in [a, b]$. Part (i) of Proposition 5.16, page 286. says so, as does the fact that f is continuous on all subintervals $[a, x]$.

Fix $c \in (a, b)$; we'll show $F'(c) = f(c)$. By definition,

$$F'(c) = \lim_{h \to 0} \frac{F(c+h) - F(c)}{h}.$$

if the limit exists. We'll show that

$$\lim_{h \to 0^+} \frac{F(c+h) - F(c)}{h} = f(c);$$

the proof for the left-hand limit (where $h \to 0^-$) is similar.

For any $h > 0$, we have, from basic properties of the integral,

$$\frac{F(c+h) - F(c)}{h} = \frac{\int_a^{c+h} f - \int_a^c f}{h} = \frac{\int_c^{c+h} f}{h}.$$

Since f is continuous on $[c, c + h]$, it assumes maximum and minimum values, say M_h and m_h, on this interval. Because $m_h \leq f(x) \leq M_h$ for $x \in [c, c + h]$, it follows that

$$m_h \cdot h = \int_c^{c+h} m_h \, dx \leq \int_c^{c+h} f(x) \, dx \leq \int_c^{c+h} M_h \, dx = M_h \cdot h,$$

and therefore

$$m_h \leq \frac{\int_c^{c+h} f}{h} \leq M_h.$$

This means, in turn, that the middle quantity is "intermediate" between m_h and M_h for the continuous function f. Thus, by the intermediate value theorem, there exists some input x_h with $c < x_h < c + h$ and

$$\frac{\int_c^{c+h} f}{h} = f(x_h).$$

As $h \to 0^+$, we have $x_h \to c$, and since f is continuous at c, we have

$$\lim_{h \to 0^+} \frac{\int_c^{c+h} f}{h} = \lim_{h \to 0^+} f(x_h) = f(c),$$

as desired. $\square$

Nice to know, but .... Theorem 5.20 assures us that every continuous function f, no matter how ill-behaved, *has* an antiderivative F. (Being an antiderivative, F is automatically differentiable and hence also continuous.) This is nice to know in the abstract—but not helpful for *calculating* integrals like the one in Example 2. Indeed, many useful and harmless-looking calculus functions, including

Some are very ill-behaved.

$$\cos(x^2), \quad \exp(x^2), \quad \text{and} \quad \frac{\sin x}{x},$$

turn out not to have *elementary* antiderivatives, and are said not to be "integrable in closed form."

Elementary functions are nice combinations of the familiar calculus-style functions.

Average values, and another mean value theorem. Integration has a natural connection to averaging.

Definition 5.21. If f is integrable on $[a, b]$, then the average value of f on $[a, b]$ is given by

$$\text{average value} = \frac{\int_a^b f}{b - a}.$$

Embedded in the proof of Theorem 5.20 is another theorem (and its proof) of independent interest.

Theorem 5.22 (Mean value theorem for integrals). *If f is continuous on $[a, b]$, then there exists c in (a, b) for which*

$$\int_a^b f = f(c) \cdot (b - a).$$

In words: a continuous function on $[a, b]$ assumes its average value.

We explore some implications in the exercises.

Exercises

1. Show that if $\int_a^b f = 0$ and f is continuous on $[a, b]$, then $f(c) = 0$ for some $c \in (a, b)$. Give an example to show that the conclusion need not hold if f is not continuous.

2. Let f and g be continuous on $[a, b]$ and suppose $\int_a^b f = \int_a^b g$. Show that $f(c) = g(c)$ for some c in $[a, b]$.

3. Suppose f has average value 3 on $[a, b]$. What is the average value of $5f + 7$ on $[a, b]$? Why?

4. In each part, find $I = \int_0^2 f$, the average value of f on $[0, 2]$, and a value of c at which the average value is achieved (in the sense of Theorem 5.22).

(a) $f(x) = \sqrt{x}$.

(b) $f(x) = x$.

(c) $f(x) = x^2$.

(d) $f(x) = x^{42}$.

5. In each part, find an interval $[0, b]$ on which f has average value 1.

(a) $f(x) = \sqrt{x}$.

(b) $f(x) = x$.

(c) $f(x) = x^2$.

(d) $f(x) = x^{42}$.

6. Let $h : [a, b] \to \mathbb{R}$ be continuous on $[a, b]$. Suppose that $h(x) \geq 0$ for all $x \in [a, b]$ and that $\int_a^b h = 0$. Use Theorem 5.20 to show that $h(x) = 0$ for all $x \in [a, b]$. (Hint: Use properties of the function $H(x) = \int_a^x h$ to derive the result.)

7. Let f be continuous on $[a, b]$. Suppose that the average value and the maximum value of f on $[a, b]$ are equal. Show that f is constant. Must the same result hold if f is not continuous? Why?

8. Let $f : [0, \infty) \to \mathbb{R}$ be continuous everywhere, and suppose that f has average value 3 on every interval $[0, b]$. Find a formula for f.

9. Let $f : [0, \infty) \to \mathbb{R}$ be continuous everywhere, and suppose that f has average value $b/2$ on every interval $[0, b]$. Find a formula for f.

10. Let $f : \mathbb{R} \to \mathbb{R}$ be a *linear* function. Show that the average value of f on any interval $[a, b]$ is $f(m)$, where m is the midpoint of $[a, b]$.

11. Find an interval $[a, b]$ such that the average value of $f(x) = x^2$ occurs at the midpoint.

12. We used the mean value theorem for integrals to prove Theorem 5.20. Use Theorem 5.20 and the mean value theorem for derivatives to prove the mean value theorem for integrals. (Assume that f is continuous, of course.)

13. (This problem refers to material in Section 4.4.) Consider the functions h_n defined by $h_n(x) = n$ if $x \in (0, 1/n)$ and $h_n(x) = 0$ otherwise. Let h be the constant function $h(x) = 0$.

(a) Show that the sequence $\{h_n\}$ converges pointwise to h on $[0, 1]$. Does $\int_0^1 h_n$ converge to $\int_0^1 h$? Explain.

(b) Show that the sequence $\{h_n\}$ converges *uniformly* to h on $[0.1, 1]$. Does $\int_{0.1}^{1} h_n$ converge to $\int_{0.1}^{1} h$? Explain.

14. (This problem refers to material in Section 4.4.) Consider the functions $f_n(x) = x^n$ for $n = 1, 2, 3, \ldots$ and the limit function f given by $f(x) = 0$ if $x \in [0, 1)$ and $f(1) = 1$.

   (a) We showed in Section 4.4 that the sequence $\{f_n\}$ converges *pointwise* to f on $[0, 1]$. Does $\int_0^1 f_n$ converge to $\int_0^1 f$? Explain.

   (b) The sequence $\{f_n\}$ converges *uniformly* to f on $[0, 0.9]$. Does $\int_0^{0.9} f_n$ converge to $\int_0^{0.9} f$? Explain.

15. (This problem refers to material in Section 4.4.) Let $\{f_n\}$ be a sequence of continuous functions on $[0, 1]$, and suppose $\{f_n\}$ converges uniformly on $[0, 1]$ to a function f.

   (a) Explain why f must be integrable on $[0, 1]$.

   (b) Prove that $\int_0^1 f_n$ converges to $\int_0^1 f$.

Selected Solutions

1.1 Numbers 101: The Very Basics

1. (a) The claim makes sense and is true.

    (b) The claim makes no sense; $\sqrt{8}$ isn't a subset.

    (c) The claim makes sense and is true.

    (d) The claim makes sense but is false; consider $a = 0$ and $b = \sqrt{2}$.

    (e) The claim makes sense and is true.

    (f) The claim makes sense but is false: consider $a = 0$.

3. The number $1/a$ is an integer only if $a = \pm 1$. The number $1/a$ is rational for all nonzero integers a. The equation $1/a = a$ holds only if $a = \pm 1$.

5. (a) $1 \in S_1$ but $-1 \notin S_1$

    (b) $2 \in S_2$ but $1/2 \notin S_2$

    (c) $\sqrt{2} \in S_3$ but $1/\sqrt{2} = \sqrt{2}/2 \notin S_3$

    (d) $\pi \in S_4$ but $\pi^2 \notin S_4$

7. All of xy, $x + y$, $x - y$ and x/y can be either rational or irrational. Examples are easy to find.

9. Assume toward contradiction that $\sqrt{3} = a/b$ for integers a and b, where a/b is in reduced form. Then squaring both sides gives $3b^2 = a^2$. This implies (essentially as in the proof of Theorem 1.2) that 3 divides both a and b, which contradicts the assumption that a/b is in reduced form.

11. Parts (i) and (ii) follow from the fact that $1 < a/b < 2$. For part (iii), note that

$$\frac{a'^2}{b'^2} = \frac{(2b - a)^2}{(a - b)^2} = \frac{4b^2 - 4ab + a^2}{a^2 - 2ab + b^2},$$

    and substituting $a^2 = 2b^2$ shows that the last fraction is 2.

    This all shows that if $\sqrt{2} = a/b$ holds for *any* positive integers a and b, then we can find a new fraction a'/b' with $\sqrt{2} = a'/b'$ and $b' < b$, which is absurd.

13. Matrix addition in $M_{2 \times 2}$ is commutative, but multiplication is not; examples are easy to find. Every matrix A in $M_{2 \times 2}$ has an additive inverse $-A$, but multiplicative inverses exist only for some nonzero matrices (those with nonzero determinant); again, examples are easy to find. Distributivity does indeed hold in $M_{2 \times 2}$.

15. Since $\ln \ln \ln n$ tends to infinity, it must exceed two for large n. The well-ordering property guarantees that a smallest such n_0 exists. (Using a calculator we can see that n_0 has about 703 decimal digits.)

1.2   Sets 101: Getting Started

1. (a) $D \subset I; D \in C$.

    (b) $B = \{m \in A \mid m \text{ has 31 days}\}$.

    (c) $A \times D$ is the set of ordered pairs (January, 2), (February, 2), ..., (December, 2), (January, 3), (February, 3), ..., (December, 3). There are 24 such pairs.

    (d) $A\backslash B = \{\text{February, April, June, September, November}\}$; $B\backslash A = \emptyset$; $A\cap C = \{\text{November}\}$; $B \cap A = B$; $D \cap I = D$; $D \cup I = I$.

3. (a) $\mathbb{R} \setminus A = (-\infty, 1) \cup (3, \infty)$

    (c) $\mathbb{R} \setminus A = (-\infty, 1]) \cup [2, 3] \cup [4, \infty)$

    (e) $\mathbb{R} \setminus A = \{0\}$

5. To say that a is in $\mathbb{R} \setminus (\mathbb{R} \setminus A)$ means that a is *not* in $\mathbb{R} \setminus A$; this means, in turn that $a \in A$.

7. We know $\mathbb{R} \setminus A_1 = (-\infty, 1] \cup [3, \infty)$ and $\mathbb{R} \setminus A_2 = (-\infty, 2] \cup [5, \infty)$. Also, $\mathbb{R} \setminus (A_1 \cap A_2) = (-\infty, 2] \cup [3, \infty)$ and $\mathbb{R} \setminus (A_1 \cup A_2) = (-\infty, 1] \cup [5, \infty)$.

    It's easy to see that, as claimed, $\mathbb{R} \setminus (A_1 \cap A_2) = (-\infty, 2] \cup [3, \infty) = (-\infty, 1] \cup [3, \infty) \cup (-\infty, 2] \cup [5, \infty) = (\mathbb{R} \setminus A_1) \cup (\mathbb{R} \setminus A_2)$. Similarly, $\mathbb{R} \setminus (A_1 \cup A_2) = (-\infty, 1] \cup [5, \infty) = ((-\infty, 1] \cup [3, \infty)) \cap ((-\infty, 2] \cup [5, \infty)) = (\mathbb{R} \setminus A_1) \cup (\mathbb{R} \setminus A_2))$.

9. If $x \in T'$ then $x \notin T$. Since $T \supset S$, we have $x \notin S$, which means $x \in S'$, as desired.

11. (a) $I = (-42, 0)$ and $J = (0, \infty)$ work.

    (b) $I = (-42, 0)$ and $J = [0, \infty)$ work.

    (c) The given conditions (draw a picture) mean that $a < c < 0 < b < d$, so $I \cup J = (a, d)$ which is indeed an open interval.

13. It's easy for I and $\mathbb{R} \setminus I$ to be intervals. For instance, if $I = (-\infty, 0)$, then $\mathbb{R} \setminus I = [0, \infty)$ is another interval. I and $\mathbb{R} \setminus I$ cannot both be bounded intervals; two bounded intervals can't "add up" to the unbounded set $(-\infty, \infty)$.

15. No. Suppose a and b are rational numbers in I, with $a < b$. Consider $c = a + (b - a)/\sqrt{2}$. Note that $c \in \mathbb{R} \setminus \mathbb{Q}$ and that $a < c < b$. If I were an interval, we'd have $c \in I$, which is impossible.

17. (a) The complement of $\{1, 2, 3\}$ consists of four open intervals.

    (b) $\mathbb{R} \setminus \mathbb{Z}$ is the union of all open intervals of the form $(n, n + 1)$, where $n \in \mathbb{Z}$.

    (c) If $\mathbb{Q}$ were open, we could find for each rational q an open interval I with $q \in I \subseteq$. But $I \subseteq \mathbb{Q}$ is impossible.

    (d) If $(0, 1]$ were open there would be an open interval I with $1 \in I \subseteq (0, 1]$. That's impossible—every such I includes points "to the right" of $x = 1$. A similar argument with $0 \in \mathbb{R} \setminus (0, 1]$ shows that $\mathbb{R} \setminus (0, 1]$ isn't open, so $(0, 1]$ isn't closed.

19. (a) There are $10 \times 9 \times 8 = 720$ ways to choose 3 different elements *in order*. There are six ways to reorder each such choice, so the answer is that S had $720/6 = 120$ elements.

   (b) T has $10 \times 10 \times 10 = 1000$ elements.

   (c) S_{10} has $10! = 10 \times 9 \times 8 \times \cdots \times 1 = 3628800$ elements.

   (d) The sets N_{10}, S, T, S_{10} have no elements in common.

21. We have $S \in A_{42}$ if and only if $N_{100} \setminus S \in A_{58}$, so there is a one-to-one correspondence between A_{42} and A_{58}, which therefore have the same number of elements.

23. (a) The picture is a diagonal stripe from upper left to lower right.

   (b) The black squares can be described by the set $\{(x,y) \mid x + y \text{ is even}\}$.

   (c) The element $(2,3,\text{black})$ corresponds to a black square at position $(2,3)$. The set $G \times \{\text{black}, \text{white}\}$ represents all possible ways to choose a square and color it.

   (d) A picture is, in effect, a *subset* of squares to be colored black. Thus $P(G)$ corresponds to the full set of possible pictures.

1.3   Sets 102: The Idea of a Function

1. There are many possibilities; following are some.

   (a) Let A be the set of all humans who have ever lived, and B the set of all women who have ever lived. The function is not injective, because siblings have the same mother. The function is not surjective, either, since some women are not mothers.

   (b) Let A be the set of all mothers of sons, and B the set of all male humans. Then FIRSTBORNSON : $A \to B$ is one-to-one but not onto.

   (c) Let A be the set of all humans and B the set of all colors. Then EYECOLOR : $A \to B$ is neither one-to-one nor onto, since several people have blue eyes, and nobody has silver eyes.

   (d) Let A be the set of all US citizens and $B = \{\text{January } 1, \text{January } 2, \dots, \text{December } 31\}$. Then BIRTHDAY : $A \to B$ is onto (every day is someone's birthday) but not one-to-one (several people have the same birthday).

3. (a) The natural domain is all of $\mathbb{R}$.

   (b) The natural domain is all of $\mathbb{R}$ except for points x at which $\cos x = 0$.

   (c) The natural domain is $(-\infty, -1] \cup [1, \infty)$.

5. (a) $f(x) = x^2 + 1$ works

   (b) $f(x) = (x-2)^2 = x^2 - 4x + 4$ works

   (c) $f(x) = (x+5)^2 = x^2 + 10x + 25$ works

7. (a) Horizontal lines intersects the graph at most once.

(b) Every horizontal line $y = a$, where $0 \le a \le 1$, intersects the graph.

(c) $f(x) = 1 - x$, $f(x) = 1 - x^2$, $f(x) = 1 - x^{42}$ all work. So do many others.

(d) Here $f(1/\sqrt{2}) = f(1/2) = 1/2$, so f is not one-to-one. But f is onto, since every irrational number in $[0, 1]$ has an irrational square root, also in $[0, 1]$.

9. It's reasonable to use $f : (0, \infty) \to (-\infty, \infty)$ and $g : (-\infty, \infty) \to (0, \infty)$. Graphs of f and g are mirror images across the line $y = x$. We have $f(g(x)) = x$ and $g(f(x)) = x$ for x in the appropriate domains.

11. (a) The domain is the set $A = \{a, e, i, o, u\}$; a good codomain is $B = \{1, 2, 3, 4, 5\}$.

(b) f is one-to-one because no two elements of G have the same second coordinate. f is onto because every element of B is the second coordinate of an element of G.

(c) The graph of f^{-1} is the set $\{(1, a), (2, e), (3, i), (4, o), (5, u)\}$.

(d) We might call these functions VowelNumber and NumberOfVowl, or something similar.

13. (a) One approach is to look at the graph. Or note that $L(x) = L(y)$ means that $3x + 2 = 3y + 2$, and this can happen only if $x = y$. Thus L is one-to-one. Also, for any $b \in \mathbb{R}$ we can set $a = (b - 2)/3$, and see that $f(a) = b$. Thus L is onto.

15. (a) f is one-to-one and onto for all *odd* positive integers, and not for even positive integers.

(b) Here f is one-to-one and onto for *all* integers, positive or negative.

17. ℓ is both one-to-one and onto; make a table of some early values to see why. This might seem weird since $\mathbb{N}$ might seem "smaller" than $\mathbb{Z}$. In fact (as we'll see) they're the same "size" in the sense at hand.

19. Use domain $[-\pi, \pi]$ and codomain $[-1, 1]$ for the sine function; and codomain $[-\pi, \pi]$ and domain $[-1, 1]$ for the arcsine function. Then the arcsine becomes an honest inverse for the sine.

21. (a) That Mod5 is reflexive and symmetric is obvious. To see it's transitive, suppose xMod5y and yMod5z. This means that 5 divides both $x - y$ and $y - z$. But then 5 also divides the sum $(x - y) + (y - z) = x - z$, which means xMod5z.

(b) The equivalence class $[0]$ is the set $\{\ldots, -10, -5, 0, 5, 10, \ldots\}$ of multiples of 5. The equivalence class $[1]$ is the set $\{\ldots, -9, -4, 1, 6, 11, \ldots\}$ of numbers one more than a multiple of 5. Similarly, $[2] = \{\ldots, -8, -3, 2, 7, 12, \ldots\}$ and so on.

1.4   Proofs 101: Proofs and Proof-Writing

1. Statement P is false, R is true, and Q is nonsense. R is the negation of P.

3. (a) If a function f is increasing for all x, then $f'(x) \ge 0$ for all x.

(b) If a function f has a maximum at $x = a$, then either $f'(a) = 0$ or $f'(a)$ does not exist.

(c) If $f''(x) > 0$ for all x, then the graph of f is concave up.

(d) If a series $\sum a_n$ converges, then $\lim a_n = 0$. (This is known as the nth term test.)

5. (a) The statement is false; no integer is greater than *all* real numbers.

(b) The statement is true; $x = 0.1$ works, for instance.

7. (a) Given: Raining implies not sunny (true). Converse: Not sunny implies raining (false). Contrapositive: Sunny implies not raining (true).

(b) Given: Raining implies clouds (true). Converse: Cloudy implies raining (false). Contrapositive: Not cloudy implies not raining (true).

(c) Given: Raining implies not sunny and cloudy (true). Converse: Not sunny and cloudy implies raining (false). Contrapositive: Sunny or not cloudy implies not raining (true).

(d) Given: Cloudy implies not sunny (true). Converse: Not sunny implies cloudy (false ... it could be night). Contrapositive: Sunny implies not cloudy (true).

9. R: **Negation:** $\exists n \in \mathbb{N}$ such that $\sqrt{n^2 + 6} \notin \mathbb{N}$. This is true.

S: **Negation:** $\forall \in \dot{\mathbb{N}}$, $\sqrt{n^2 + 6} \notin \mathbb{N}$. This is true.

T: **Negation:** $\exists x \in [0, 1]$ such that $\cos(x) < 0.6$. This is true; look at $x = 1$, for instance.

U: **Negation:** Some positive even integer n with $4 < n < 24$ is not the sum of two odd primes. This is false; one just checks.

11. (a) Negation: All of a, b, and c are negative.

(b) Negation: $f(x) > 3$ for some $x \in [2, 7]$.

(c) Negation: $\sin n$ is rational for some positive integer n.

(d) Negation: $\forall x \in \mathbb{R}$ we have $x^2 \neq -1$.

(e) Negation: $\exists x \in \mathbb{R}$ such that $\forall y \in \mathbb{R}$ we have $xy \neq 1$.

13. (a) Statement: If $x > 3$ then $x^2 > 9$. Converse: If $x^2 > 9$ then $x > 3$. Contrapositive: If $x^2 \leq 9$ then $x \leq 3$. These are true, false, and true, respectively.

(b) Statement: If $x > 3$ then $x^3 - 4x^2 + 3x > 0$. Converse: If $x^3 - 4x^2 + 3x > 0$ then $x > 3$. Contrapositive: If $x^3 - 4x^2 + 3x \leq 0$ then $x \leq 3$. The statements are true, false, and true, respectively. (It may be useful to plot the function.)

(c) Statement: If $a > 0$ and $b > 0$ then $|a + b| = |a| + |b|$. Converse: If $|a+b| = |a|+|b|$, then $a > 0$ and $b > 0$. Contrapositive: If $|a+b| \neq |a|+|b|$, then $a \leq 0$ or $b \leq 0$. The statements are true, false, and true, respectively.

1.5   Types of Proof

1. (a) True; the simplest proof is indirect.

    (b) False; $x = -\sqrt{2}$ is a counterexample.

    (c) True; the simplest proof is indirect.

    (d) False; $p = 11$ is a counterexample.

    (e) False; $x = 0$ and $y = 1/2$ is a counterexample.

3. The claim is obvious for $n = 1$. Given $F = \{x_1, x_2, \ldots, x_n\}$, consider $F' = \{x_1, x_2, \ldots, x_{n-1}\}$. By the inductive hypothesis F' has a largest member, say x_{n-1}. The largest member of F is then x_n if $x_n > x_{n-1}$, and x_{n-1} otherwise.

5. (a) If $n = 1$, then both sides have value 1; if $n = 10$, then both sides are 100.

    (b) The *base case*, $P(1)$, holds as indicated in (a).

    For the inductive step we assume $P(n)$ and try to prove $P(n + 1)$, which is the claim that $1 + 2 + \cdots + (n+1) = (n+1)(n+2)/2$. We start by adding $(n + 1)$ to both sides of $P(n)$, and then do algebra:

    $$1 + 2 + 3 + \cdots + n + (n+1) = \frac{n(n+1)}{2} + (n+1)$$

    $$= (n+1)\left(\frac{n}{2} + 1\right) = \frac{(n+1)(n+2)}{2}.$$

    This shows $P(n + 1)$, and completes the proof.

    (c) Calculate with sums: $\displaystyle\sum_{k=1}^{n}(2n-1) = 2\sum_{k=1}^{n}n - \sum_{k=1}^{n}1 = 2\frac{n(n+1)}{2} - n = n^2.$

7. Prove this by induction. The base case $n = 1$ is clear. For the inductive step, suppose that $(1 + x)^k \geq 1 + kx$ for a particular k. Multiplying both sides by $(1 + x)$ and doing a little algebra shows that $(1 + x)^{k+1} \geq 1 + (k + 1)x$, as desired.

9. (a) Rewrite $f_1 f_2 f_3$ as $(f_1 f_2) f_3$ and use the ordinary product rule twice.

    (b) The formula is $(f_1 f_2 f_3 \cdots f_n)' = f_1' f_2 f_3 \cdots f_n + f_1 f_2' f_3 \cdots f_n + f_1 f_2 f_3' \cdots f_n + \cdots + f_1 f_2 f_3 \cdots f_n'$. We've proved the first few cases. For the inductive step, we write $f_1 f_2 f_3 \cdots f_{n+1} = (f_1 f_2 f_3 \cdots f_n) f_{n+1}$ and apply the ordinary product rule.

11. (a) Checking early cases suggests the answer $n(n + 1)(n + 2)/3$; this is readily proved by induction.

    (b) Notice that

    $$\sum_{j=1}^{n} j(j + 1) = \sum_{j=1}^{n} j^2 + \sum_{j=1}^{n} j = \frac{n(n+1)(2n+1)}{6} + \frac{n(n+1)}{2}.$$

    The result now follows from basic algebra.

13. Note first that $P(n)$ is obviously true for the base case $n = 4$; just check both sides. To complete the inductive proof, we assume $P(n)$ (for $n \geq 4$) and show $P(n+1)$. By $P(n)$, we have $2^n < n!$. We know also that $2 < (n+1)$. Multiplying these inequalities gives

$$2^n < n! \implies 2^{n+1} < n!(n+1) = (n+1)!.$$

This shows that $P(n+1)$ holds, and we're done.

15. The base case is an assumption. (We'll prove it later in the course.) The inductive step follows readily from the product rule.

17. Among small amounts, it's clear we can pay out $3, $5, $6, $8, $9. In fact, we can pay *all* numbers greater than $9, too. Here's an informal inductive proof. Note first that the base case $10 is clearly possible. Suppose we can pay out all amounts from $10 to $k. To show $k + 1$ is also possible, note that $k + 1 - 3 = k - 2 \geq 8$, so $k - 2$ is payable. Adding one more $3 bill brings the total to $k.

1.6   Sets 103: Finite and Infinite Sets; Cardinality

1. Here's a sketch: The base case is given. For the inductive step, write $F_1 \cup F_2 \cup \cdots \cup F_n = (F_1 \cup F_2 \cup \cdots \cup F_{n-1}) \cup F_n$ and use the inductive hypothesis.

3. (a) $f(x) = 7x - 2$ is one-to-one: If $f(a) = f(b)$, then $7a - 2 = 7b - 2$, which implies $a = b$. $g(x) = x/(1+x)$ is also one-to-one: if $g(a) = g(b)$, then $a/(1+a) = b/(1+b)$. A little algebra (do it—keep in mind that a and b are positive) shows that $a = b$.

    (b) For any $b \in \mathbb{R}$ we can set $a = (b+2)/7$ and see that $f(a) = b$. If $b \in (-2, 5)$ then $b + 2 \in (0, 7)$ and $(b + 2)/7 \in (0, 1)$ as desired.

    (c) We know that both f and g are one-to-one and onto; it follows (from one of our theorems) that $f \circ g$ is one-to-one and onto. This means (among other things) that all of the sets under consideration here have the same cardinality.

5. f can be one-to-one but g cannot. The latter holds by the pigeonhole principle, since g tries, in effect, to put 43 pigeons into 42 holes.

7. (a) The function $f : \mathbb{N} \to \mathbb{N}$ given by $f(n) = n + 1$ works.

    (b) The function $g : \mathbb{R} \to \mathbb{R}$ given by $g(x) = \arctan x$ works.

    (c) Define $h : \mathbb{R} \to \mathbb{R}$ by $h(x) = x$ if $x \notin \mathbb{N}$ and $h(x) = x + 1$ if $x \in \mathbb{N}$.

9. Proof by contradiction. If the hypotheses hold and $A \setminus B$ is finite, then $A = (A \setminus B) \cup B$ is the union of two finite sets, and hence finite itself. This is absurd.

11. (a) Members of Q_3 correspond to (some of the) points (p, q) with integer coordinates inside the circle $p^2 + q^2 = 3$—there are nine such points (p, q). Since we're interested in fractions p/q, we may as well require $q > 0$. This reduces the eligible points to just three: $(-1, 1)$, $(0, 1)$, and $(1, 1)$, all of which give different reduced fractions p/q.

Note that Q_4 is the same set as Q_3.

For Q_{10}, similar reasoning finds 11 points (p, q) with $q > 0$ and $p^2 + q^2 < 10$. These 11 points correspond to 7 different rational numbers: $\pm 2, \pm 1, \pm 0.5$, and 0.

(b) If $p/q \in Q_i$, then clearly both $|p| < i$ and $|q| < i$. There are at most $2i - 1$ choices for each of p and q, and so no more that $(2i - 1)^2$ choices for both p and q. (In fact, the number of choices is less, but that's OK for present purposes.) An even rougher (but simpler-looking) upper bound is that Q_i has fewer than $4i^2$ elements.

(c) All parts are straightforward implications of the definitions.

13. As one possible bijection $f : [0, \infty) \to (0, \infty)$ we can define $f(x) = x + 1$ if $x \in \mathbb{Z}$ and $f(x) = x$ otherwise.

15. The set B is countably infinite. The set B_n of n-character books is clearly finite for each n, and B is the union of the (countably many) B_n.

17. Different approaches are possible. In the case discussed, it's fun to see that, in binary digit notation, $h(0.1\text{stuff}) = 0.01\text{stuff}$, $h(0.01\text{stuff}) = 0.1\text{stuff}$, $h(0.001\text{stuff}) = 0.0001\text{stuff}$, $h(0.0001\text{stuff}) = 0.001\text{stuff}$, etc.

1.7   Numbers 102: Absolute Values

1. (a) From elementary calculus we can see that $|\sin(x) + \cos(x)| \le \sqrt{2}$.

   (b) The RTI says that $|\sin(x) - \cos(x)| \ge ||\sin(x)| - |\cos(x)|| \ge 0$. We can't improve on this, since $|\sin(x) - \cos(x)| = 0$ when $x = \pi/4$.

   (c) Yes. A famous trigonometric identity says that $\sin(x)^2 + \cos(x)^2 = 1$ for all x.

3. For the first inequality, use the ordinary triangle inequality and the fact that $|x - y| = |x + (-y)|$. The second inequality is half of the RTI.

5. (a) $x \in (1, \infty)$

   (b) $x \in \left(-\sqrt{4.07}, -\sqrt{3.93}\right) \cup \left(\sqrt{3.93}, \sqrt{4.07}\right)$

   (c) $x \in (-3, 1)$

7. (a) Subtracting 4 from all parts of $4.96 < y < 5.04$ gives $0.96 < y - 4 < 1.04$; note also that $y - 4 = |y - 4|$.

   (b) Note that $2.93 < x < 3.07$ and $4.96 < y < 5.04$. Thus $|y - x| < 5.04 - 2.93 = 2.11 = K$.

   (c) Since $2.93 < x < 3.07$ and $4.96 < y < 5.04$, we have $|y - x| > 4.96 - 3.07 = 1.89 = L$.

9. The base case ($n = 2$) is the ordinary TI. For the inductive step, let's assume that the TI holds for n summands. To show it holds for $n + 1$ summands we calculate

just as in the problem, except that we use $n + 1$ summands instead of three.

$$|x_1 + \cdots + x_n + x_{n+1}| = |(x_1 + \cdots + x_n) + x_{n+1}| \leq |x_1 + \cdots + x_n| + |x_{n+1}|$$
$$\leq |x_1| + \cdots + |x_n| + |x_{n+1}|.$$

11. (a) Look separately at the cases $x \geq y$ (so $|x - y| = x - y$) and $x < y$ (so $|x - y| = y - x$).

(b) $\min\{x, y\} = \dfrac{x + y - |x - y|}{2}$.

(c) We can use $h(x) = \dfrac{f(x) + g(x) - |f(x) - g(x)|}{2}$.

(d) A good interval is something like $[-20, 2]$.

13. (a) $|x - 1| < 0.5$ implies $1 - x < 0.5$, or $x > 0.5$.

(b) Use the triangle inequality: $|c| = |c - x + x| \leq |c - x| + |x| < \frac{|c|}{2} + |x|$. This implies $|c| < \frac{|c|}{2} + |x|$, or $|x| > \frac{|c|}{2}$.

1.8 Bounds

1. Following are sketches.

(a) Say $m \leq t \leq M$ for all $t \in T$. If $s \in S$, then $s \in T$, too, so $m \leq s \leq M$. Thus S is bounded.

(c) If $|S|$ is bounded, then $|s| < K$ for all $s \in S$ and some $K > 0$. But then $-K < s < K$ for all $s \in S$, so S is bounded. The converse is similar.

3. (a) $I = [1, 3)$ works

(b) $I = (-\infty, 3)$ works

(c) $I = (1, 3)$ works

(d) No; if $\max(S)$ exists then $\max(S) = \sup(S)$.

5. (a) We have $1 = e^0 \leq f(x) = e^x \leq e^{10}$ and $-1 \leq g(x) = \sin(x) \leq 1$ for $x \in A$. We know from calculus that (i) $f(x) = e^x$ is bounded below by 0 but is unbounded above on $\mathbb{R}$, and (ii) $g(x) = \sin(x)$ is bounded below by -1 and above by 1 on all of $\mathbb{R}$.

(b) For $f \circ g$: When $x \in [0, 10]$ we have $0.37 \approx e^{-1} \leq e^{\sin(x)} \leq e^1 \approx 2.72$. For $g \circ f$: When $x \in [0, 10]$ we have $-1 \leq \sin(e^x) \leq 1$.

7. Since A is bounded we know that $-M \leq x \leq M$ for some $M > 0$ and all $x \in A$. But then also $-3M + 5 \leq 3x + 5 = f(x) \leq 3M + 5$ for all $x \in A$, as desired.

9. (a) $\inf(S) = \min(S) = 2$; $\sup(S) = \max(S) = 3$.

(b) $\inf(S) = \min(S) = 2$; $\sup(S)$ does not exist, as there are infinitely many primes.

    (c) Here $S = \{2\}$, so $\inf(S) = \min(S) = 2 = \sup(S) = \max(S)$

    (d) $\inf(S) = \min(S) = 41$; $\sup(S) = \max(S) = 43$.

    (e) $\inf(S) = -\sqrt{2}$; $\sup(S) = \sqrt{2}$. Maximum and minimum values don't exist, as $\sqrt{2} \notin S$.

    (f) $\inf(S) = \min(S) = -\sqrt{2}$; $\sup(S) = \max(S) = \sqrt{2}$.

11.   (a) The shortest English words have one letter, so 1 is a lower bound for f on EW. It's hard to identify the longest English word, but 1000 letters is probably a safe upper bound.

    (b) Words in this problem vary in length from 1 to 10 letters.

13.   (a) All subsets $S \subseteq \mathbb{R}$ have this property

    (b) The given property means that S is bounded.

    (c) For all $s \in S, |s| \geq 1$.

15.   (a) The maximum value is $f(0) = 1$; the minimum is $f(\pm K) = 1/(K+1)$.

    (b) The graph is a parabola with vertex at $x = -1/3$, so the maximum occurs at the right endpoint; $g(K) = 3K^2 + 2K - 7$. If $K \geq 1/3$, then $-2/3 \in [-K, K]$, and so the minimum value is $g(-1/3) = -22/3$. If $K < 1/3$ then the minimum occurs at $x = -K$, and the value is $g(-K) = 3K^2 - 2K - 7$.

    (c) Note that $h(x) = 1 + x$, so the function is largest at the right endpoint ($h(K) = 1 + K$) and smallest at the left ($h(-K) = 1 - K$).

17.   (a) $S = (1, \infty)$ is one possibility; there are many.

    (b) Let S be bounded away from zero. Then there exists $\delta > 0$ so $|s| > \delta$ for all $s \in S$. Thus, if $t \in T$ we have $|t| = \left| \dfrac{1}{s} \right| = \dfrac{1}{|s|} < \dfrac{1}{\delta}$, so T is bounded. The proof of the converse is similar.

    (c) If S is bounded away from 0 then—by definition—S has no points in common with the interval $(-\delta, \delta)$, which is of the desired form. If, conversely, S has nothing in common with (a, b), where $a < 0 < b$, then we can take δ to be the smaller of $-a$ and b.

19.   (a) If $a \notin I$, then either $a < 0$ or $a > 1$. In the former case, we can take $\delta = -a/2$; in the latter, $\delta = (a-1)/2$ works.

    (b) The points 2 and 3 are not in J, but J is not bounded away from either 2 or 3.

    (c) $\mathbb{N}$ does indeed have the property in question: if $a \notin \mathbb{N}$. then we can take δ to be the distance to the nearest integer.

1.9   Numbers 103: Completeness

1.   (a) One possibility is the list $\dfrac{1}{508}, \dfrac{1}{509}, \dfrac{1}{510}, \ldots, \dfrac{1}{507+n}, \ldots$

    (b) One possibility is the list $\dfrac{1}{3} + \dfrac{1}{7}, \dfrac{1}{3} + \dfrac{1}{8}, \dfrac{1}{3} + \dfrac{1}{9}, \ldots, \dfrac{1}{3} + \dfrac{1}{7+n}, \ldots$

(c) One possibility is to let $h = b - a$, and use the list $a + h/2$, $a + h/3$, $a + h/4$, ..., $a + h/n$, ... "Successive averaging" is another possibility.

3. For given $\epsilon > 0$, consider the positive number $1/\epsilon$. Since $\mathbb{N}$ is unbounded, there exists an integer n with $n > 1/\epsilon$. But then $0 < 1/n < \epsilon$, as desired.

5. (a) The statement is false if $a = 0$ and $b = 1$.

   (b) The statement is true. If $0 < a < b$ then the the Archimedean principle applies. If $a < b < 0$ and $n = -1$, then $na > b$. The case $a < 0 < b$ is left to you.

   (c) Clearly $b + 1 > b$, so we can choose $n \in \mathbb{R}$ with $na = b + 1$; i.e., $n = (b + 1)/a$. Nothing Archimedean needed.

7. (a) The italicized statement is true; this is essentially the well-ordering principle.

   (b) The italicized statement is now false. The set S of rationals less than π has supremum π, but $\pi \notin \mathbb{Q}$.

9. (a) $I_n = (3 - 1/n, 3 + 1/n)$ works.

   (b) $I_n = [-3 - 1/n, 42 + 1/n]$ works.

   (c) $I_n = (-3 - 1/n, 42 + 1/n)$ works.

11. The hint explains the problem: If (a, b) is contained in all the I_i, then $[a, b]$ is contained in all these intervals, too.

13. (a) The set S is certainly bounded above; by completeness there's a least upper bound, β.

   (b) Note that $(\beta + h)^2 = \beta^2 + 2\beta h + h^2 < \beta^2 + 2\beta h + h = \beta^2 + (2\beta + 1)h$. Now use property (ii) of h.

   (c) The preceding calculation shows $\beta + h$—a number larger than β—has square less than 2. This contradicts the supremum property of β.

   (d) Note that $(\beta - k)^2 = \beta^2 - 2\beta k + k^2 < \beta^2 - 2\beta k$. Substitute for k and simplify.

15. The point is simply that $\sqrt[4]{a} = \sqrt{\sqrt{a}}$, $\sqrt[8]{a} = \sqrt{\sqrt[4]{a}}$, and so on.

17. We have, for instance, $\sqrt[6]{a} = \sqrt{\sqrt[3]{a}}$; other parts are similar.

2.1   Sequences and Convergence

1. For the three sequences given:

		ϵ	1.000	0.100	0.010	0.001
(a)	N		1	$\sqrt{10} \approx 3.16$	10	$\sqrt{1000} \approx 31.6$
(b)	N		1	100	10^4	10^6
(c)	N		1	$e^9 \approx 8103$	e^{99}	e^{999}

3. Suppose toward contradiction that $L > b$. Set $\epsilon = b - L$ and choose N as in the definition of convergence. Then (explain why!) we must have $x_n > b$ for $n > N$, which is absurd. A similar argument shows that $L \geq a$.

5. (a) Set $\epsilon = 0.1$ and choose a corresponding N as in the definition. All x_n with $n > N$ lie in the desired interval.

    (b) Set $\epsilon = 0.001$ and choose a corresponding N as in the definition. Then all x_n with $n > N$ lie in $(4.999, 5.001)$, and therefore exceed 4.999.

    (c) If we set $\epsilon = 1$ and choose a corresponding N, then all x_n with $n > N$ lie in $(4, 6)$. Thus it is possible that $x_n > 6$ only for members of the finite set $\{x_1, x_2, \ldots x_N\}$.

7. (a) Yes, it's possible. We might have $x_n = 4 - 1/n$, for instance.

    (b) Yes, it's possible. Our series could have the pattern $4.1, 3.9, 4.01, 3.99, 4.001, 3.999, \ldots$.

9. (a) Algebraic manipulation gives

$$|a_n - L| = \left| \frac{2n}{3n + 5} - \frac{2}{3} \right| = \frac{10}{9n + 15}$$

and

$$\frac{10}{9n + 15} < \epsilon \quad \Longleftrightarrow \quad \frac{10/\epsilon - 15}{9} < n.$$

Thus, for given $\epsilon > 0$, the value $N = \frac{10/\epsilon - 15}{9}$ works in the definition. A formal proof resembles that in Example 2.

    (b) The proof is like that for (a), except that now we have

$$|b_n - L| = \frac{899900}{9n + 15} < \epsilon \quad \Longleftrightarrow \quad \frac{899900/\epsilon - 15}{9} < n.$$

Thus $N = \frac{899900/\epsilon - 15}{9}$ works for any given $\epsilon > 0$.

    (c) We'll show $c_n \to 0$. Observe first that

$$|c_n - L| = \frac{2n}{3n^2 + 5} < \frac{2n}{3n^2} = \frac{2}{3n},$$

and

$$\frac{2}{3n} < \epsilon \quad \Longleftrightarrow \quad \frac{2}{3\epsilon} < n.$$

Here's the formal proof: Let $\epsilon > 0$ be given; set $N = \frac{2}{3\epsilon}$. This N works, since if $n > N$ then

$$|c_n - L| = \frac{2n}{3n^2 + 5} < \frac{2}{3n} < \frac{2}{3N} = \epsilon.$$

11. Here's a sketch. Let $\epsilon > 0$ be given, and set $\epsilon' = \epsilon/17$. Since $\epsilon' > 0$ we can choose N so that $n > N$ implies $|x_n - 1| < \epsilon'$. This same N works in the definition of convergence of $\{17a_n\}$ to 17.

13. Suppose $\{a_n\}$ is decreasing. Since $\{a_n\}$ is bounded, this set has an infimum; call it L. To see L is also the limit, let $\epsilon > 0$ be given. Because L is the infimum, $L + \epsilon$ is *not* a lower bound for $\{a_n\}$, and so there is some a_N with $L + \epsilon > a_N$. This N "works": If $n > N$, then $a_n \leq a_N$ (because $\{a_n\}$ is decreasing) and so $L \leq a_N < L + \epsilon$, as desired.

15. Suppose $x_n \to 5$. To show $y_n \to 0$, let $\epsilon > 0$ be given. Since $x_n \to 5$ we can choose N so that $n > N \implies |x_n - 5| < \epsilon$. This is just another way of saying that $|y_n| < \epsilon$, and so $y_n \to 0$ as desired. The converse is almost identical.

17.   (a) For any given $\epsilon > 0$ we have $|1/n - 0| \geq \epsilon$ only when $n \leq 1/\epsilon$. There are only finitely many such n.

     (b) Let $\epsilon > 0$ be given, and consider the set $F = \{n \mid |x_n - L| \geq \epsilon\}$. By hypothesis, F is finite. If F is nonempty, then F has a largest element, say N; this N works in the definition of convergence. If F is empty, then $N = 0$ works.

     (c) A sequence $\{x_n\}$ does not converge to zero if, for some $\epsilon > 0$, we have $|x_n| \geq \epsilon$ for infinitely many n.

     (d) A sequence $\{x_n\}$ does not converge to zero if, for some $\epsilon > 0$, there is no N such that $|x_n| < \epsilon$ whenever $n > N$.

19. Every constant sequence has such a table.

21.   (a) true

     (b) true; 0 is a lower bound

     (c) false; $n = 43$ is a counterexample

     (d) true; $N = 10000$ works

     (e) true; $N = 1$ works, for instance

     (f) false

     (g) true; for a given ϵ we can choose $N = 1/\epsilon$ (or the next integer if $1/\epsilon$ isn't an integer)

23.   (a) true

     (b) true

     (c) false; use technology to find a counterexample, like $n = 45$

     (d) true; $N = 10000$ works

     (e) true; $N = 20000$ works ... use the triangle inequality

     (f) false; see (d)

     (g) true; $N = 1/\epsilon$ (or the next integer) works

25. The limits L are $0, 0, 0, 3, 42$, respectively. Possible tables:

		1.000	0.100	0.010	0.001	10^{-10}
	ϵ					
(a)	N	1	100	10000	10^6	10^{20}
(b)	N	1	$\sqrt{10}$	$\sqrt{100}$	$\sqrt{1000}$	10^5
(c)	N	1	$\sqrt{10}$	$\sqrt{100}$	$\sqrt{1000}$	10^5
(d)	N	6	60	600	6000	$6 \cdot 10^{10}$
(e)	N	41	41	41	41	41

27.   (a) We can check explicitly that $1.5^{11} \approx 86.5 < 89 = f_{11}$, and $1.5^{12} \approx 129.7 < 144 = f_{12}$. For the inductive step, note that
$$f_{k+1} > 1.5^{k-1} + 1.5^k = 1.5^{k-1} \cdot 2.5 > 1.5^{k-1} \cdot 1.5^2 = 1.5^{k+1},$$
as desired. (Note that the proof worked because $1 + 1.5 > 1.5^2$. A similar result can be shown to hold if 1.5 is replaced by any number b with $1 + b > b^2$.)

   (b) Using technology one can check that $n \geq 72$ works.

29. Implications are as follows: .

   (a) The statement says that $a_n \to \pi$.

   (b) The statement says that for some n, $a_n = \pi$ for $n > N$.

   (c) The statement says that $a_n = \pi$ for $n > 1$.

In particular, (c) $\implies$ (b) $\implies$ (a).

2.2   Working with Sequences

1. Here is a sketch. For any constant $c \neq 0$, we have
$$|ca_n - ca| = |c| \, |a_n - a| < \epsilon \iff |a_n - a| < \frac{\epsilon}{|c|}.$$
Now for any given $\epsilon > 0$, we can choose N so that $|a_n - a| < \epsilon/|c|$ for $n > N$. (Why can we do this?) This N "works" for the given ϵ and the original sequence $\{ca_n\}$. (Assemble the pieces into an efficient proof.)

3. Both equivalences follows directly from the $\epsilon - \mathbb{N}$ definition. The second also follows from (a) of Proposition 2.7, page 99.

5. Imitate the (partial) proof given for Lemma 2.9.

7.   (a) We can define $\{b_n\}$ by $b_n = 1 - \frac{1}{n}$.

   (b) We can define $\{b_n\}$ by $b_n = 1$ for all n.

(c) Because $\beta - \frac{1}{n}$ is not an upper bound for C, there must exist b_n as desired. (Note the strict inequality.)

(d) Apply the squeeze principle to $\beta - \frac{1}{n} < b_n \leq \beta$.

9. The $n = 1$ (base) case is easy: $s_2 = 1.5 < 2 = s_1$. The inductive step is to show that if $s_n - s_{n-1} < 0$, then $s_{n+1} - s_n < 0$, too. Doing so involves careful but straightforward algebra.

11. It is clear that $\{h_n\}$ is increasing, and the inequality $h_{2n} \geq h_n + \frac{1}{2}$ implies that $\{h_n\}$ is unbounded. Since $h_1 = 1$, the inequality means that $h_2 \geq 1.5$, $h_4 \geq 2.0$, $h_8 \geq 2.5$, $h_{2^{1000}} > 501$, etc.

13. (a) Let $M > 0$ be given. Since $x_n \to \infty$, we can choose N so $x_n > M$ whenever $n > N$. But then also $-x_n < -M$ when $n > N$. Thus $-x_n \to -\infty$.

(b) Suppose $x_n \to \infty$. Let $\epsilon > 0$ be given; we need to find N so $n > N$ implies $\left| \frac{1}{x_n} \right| = \frac{1}{x_n} < \epsilon$. Well, if we set $M = 1/\epsilon$ then (since $x_n \to \infty$) we can choose N so $n > N$ implies $x_n > M = 1/\epsilon$. But this implies $1/x_n < \epsilon$, as desired.

(c) Suppose $x_n < 0$ for all n. Then $x_n \to -\infty$ if and only if $1/x_n \to 0$. This follows from parts (a) and (b): $x_n \to -\infty \iff -x_n \to \infty \iff -\frac{1}{x_n} \to 0 \iff \frac{1}{x_n} \to 0$.

15. (a) The limit is $2/5$.

(b) The sequence diverges to ∞.

(c) The sequence converges to $1/2$.

(d) The sequence diverges to ∞. One strategy is to observe that $\frac{n^2 + \arctan n}{n+2} > \frac{n^2 - 4}{n+2} = n - 2$.

17. (a) One possibility is to let $a_n = (-1)^n$ and $b_n = (-1)^{n+1}$ for all n.

(b) We'll show that $c_n \to \max\{a, b\}$.

One approach is to use the curious fact (which appears in a later section) that for any numbers a_n and b_n,

$$c_n = \max\{a_n, b_n\} = \frac{a_n + b_n + |a_n - b_n|}{2}.$$

Invoking Theorem 2.5 (on algebra with convergent sequences) and Proposition 2.7.b (on how absolute values play nicely with limits) we see that

$$c_n = \frac{a_n + b_n + |a_n - b_n|}{2} \text{ converges to } c = \frac{a + b + |a - b|}{2} = \max\{a, b\},$$

as desired.

Here is another approach. We assume WLOG that $a \leq b$ and show that $c_n \to b$.

Suppose first that $a = b$. Then for given $\epsilon > 0$ we can choose N_a and N_b such that $n > N_a$ implies $|a_n - b| < \epsilon$, and $n > N_b$ implies $|b_n - b| < \epsilon$.

Let $N = \max\{N_a, N_b\}$. If $n > N$, then we have *both* $|a_n - b| < \epsilon$ and $|b_n - b| < \epsilon$. Now recall that, for all n, either $c_n = a_n$ or $c_n = b_n$. Thus $|c_n - b| < \epsilon$ for all $n > N$, and we're done.

The other possibility is that $a < b$. Set $\epsilon = \frac{b-a}{2}$ and choose N (as above) so that both $|a_n - a| < \epsilon$ and $|b_n - b| < \epsilon$ hold whenever $n > N$. This means (draw a picture if this isn't clear) that $a_n < b_n$ for all $n > N$. In other words $c_n = b_n$ for all $n > N$, and so surely $\{c_n\}$ and $\{b_n\}$ converge to the same limit, b.

19. (a) We have $x_1 = 1$, $x_2 = 2$, $x_3 = 3/2$, $x_4 = 5/3$, $x_5 = 8/5$, $x_6 = 13/8$, $x_7 = 21/13$, $x_8 = 34/21$. Note the connection to the Fibonacci sequence $1, 1, 2, 3, 5, 8, 13, \ldots$. The sequence is not monotone, but has the pattern $x_1 < x_2 > x_3 < x_4 > x_5 < \ldots$.

(b) We know $x_{n+1} = 1 + \dfrac{1}{x_n}$. Since $\{x_n\}$ converges to L we can take limits on both sides to get $L = 1 + 1/L$. Solving for L gives $L^2 - L - 1 = 0$. The quadratic formula says $L = (1 \pm \sqrt{5})/2$; from context it's clear we want the positive root $L \approx 1.61803$.

21. (a) Check first that since $x_1 = 3$, we have $1 < x_n < 4$ for all n. Also, $x_{n+1} = (x_n^2 + 4)/5 < x_n \iff x_n^2 - 5x_n + 4 < 0$, and it's easy to see that this holds when $1 < x_n < 4$.

(b) The sequence $\{x_n\}$ converges because it's monotone and bounded.

(c) Because the sequence converges, we must have $L = (L^2 + 4)/5$, so $L = 1$ or $L = 4$. In context it's clear that $L = 1$.

23. We'll show $|x_n/y_n| \to 0$. (Why is this enough?) First choose $M > 0$ such that $|x_n| < M$ for all n. (Why can we do this?) Now choose $\epsilon > 0$. Since $y_n \to \infty$ we can find N so that $n > N$ implies $|y_n| > \frac{M}{\epsilon}$. (Why can we do this?) This N works for the sequence $\{|x_n/y_n|\}$: If $n > N$, then $\left|\dfrac{x_n}{y_n}\right| < \dfrac{M}{y_n} < \dfrac{M}{M/\epsilon} = \epsilon$.

2.3   Subsequences

1.   (a) The sequence $1, 2, 3, 1, 2, 3, \ldots$ works.

(b) The sequence $1, -1, 1/2, -1/2, 1/3, -1/3, \ldots$ works.

(c) The sequence $0, 1, 0, 2, 0, 3, \ldots$ works.

(d) The sequence $1, 1, 1/2, 2, 1/3, 3, \ldots$ works.

(e) The sequence $1, 1, 2, 1, 2, 3, 1, 2, 3, 4, \ldots$ works.

3.   (a) Given a sequence that lists the rationals, we can just form the subsequence of nonnegative rationals.

(b) Look at any two successive terms, say p_1 and p_2. Between these two rational numbers like other rationals. If the sequence were monotone, these other rationals would lie between p_1 and p_2 in the sequence.

(c) We can choose n_1 so that $p_{n_1} = 1$. Then we choose n_2 with $n_2 > n_1$ and $p_{n_2} \geq 2$. Continuing this process completes the proof; details are left to the reader.

5. In general it's possible for a subsequence to behave differently from its parent. The point here is that this particular kind of subsequence—the "tail" of a given sequence—behaves in a more special way.

Here's the idea: Let $\epsilon > 0$ be given, and suppose that some N_1, say $N_1 = 1000$, works for the *subsequence*. Note that the 1001th term of the subsequence is x_{5242}, and we know that from then on, all terms of the subsequence are within ϵ of L. But this is just another way of saying that all terms past x_{5242} in the *original* sequence are within ϵ of L; that is $N = 5242$ works for the given ϵ in the parent sequence. In general, if N_1 works for the subsequence, then $N = 4242 + N_1$ works in the parent sequence. Clean this up a little to get an impeccable proof.

7. Statements (a), (c), (e), and (f) are all equivalent to each other. Statements (b) and (d) are also equivalent.

9. (a) The sequence $0, 3, 0, 3, 0, 3, \ldots$ is one example.

    (b) The statement $x_n \to 3$ means that, for every $\epsilon > 0$, all but finitely many x_n are within ϵ of 3. Negating this condition means that, for some $\epsilon > 0$, infinitely many x_n are at least ϵ away from 3. These $\{x_n\}$ can be taken in order to give the desired subsequence.

11. (a) For $M > 0$, choose N such that $x_n > M$ whenever $n > N$. The same N works for the subsequence $\{x_{n_k}\}$, since if $k > N$, then $n_k \geq k > N$, and so $x_{n_k} > M$, as desired.

    (b) The contrapositive of Theorem 2.12(a) says that if some subsequence $\{x_{n_k}\}$ does *not* converge to L, then the original sequence $\{x_n\}$ doesn't converge to L either. This implies Theorem 2.12(c), because if L is *any* number, then at least one subsequence fails to converge to L, and so L can't be the limit.

13. The sequence $\{x_{11}, x_{101}, x_{1001}, \ldots\}$ does the job. For this sequence we have $n_k = 10^k + 1$.

15. Notice first that $z_n = b_{n/2}$ for even n and $z_n = a_{(n+1)/2}$ for odd n. In particular, both sequences $\{a_n\}$ and $\{b_n\}$ are subsequences of $\{z_n\}$, so if $z_n \to L$ we must have $a_n \to L$ and $b_n \to L$, too.

For the converse, we suppose that both $a_n \to L$ and $b_n \to L$. Let $\epsilon > 0$ be given. We can choose N_1 and N_2 such that $n > N_1 \implies |a_n - L| < \epsilon$, and $n > N_2 \implies |b_n - L| < \epsilon$. Now let $N = 2\max\{N_1, N_2\}$. This N "works" for $\{z_n\}$, because if $n > N$ and n is even, then $n/2 > N \geq N_2$, so $|z_n - L| = |b_{n/2} - L| < \epsilon$. Similarly, if $n > N$ and n is odd, then $(n+1)/2 > N \geq N_1$, so $|z_n - L| = |a_{(n+1)/2} - L| < \epsilon$.

2.4   Cauchy Sequences

1. (a) For given $\epsilon > 0$ we can choose $N = 2/\epsilon$. This N works because if $m, n > N$ then $|x_n - x_m| \leq |x_n| + |x_m| = \dfrac{1}{n} + \dfrac{1}{m} < \dfrac{\epsilon}{2} + \dfrac{\epsilon}{2} = \epsilon$.

(b) The sequence is not Cauchy. If we choose, say $\epsilon = 0.001$, then no N works, since if N is any positive number, no matter how big, then with $m = N + 1$ and $n = N + 2$ we have $|y_n - y_m| = 2/1234 > 0.001$.

(c) The sequence is Cauchy. Note that if $n > m$, we have $|z_n - z_m| = \left| \dfrac{n}{n+1} \dfrac{m}{m+1} \right| = \dfrac{n-m}{(n+1)(m+1)} < \dfrac{n-m}{nm} < \dfrac{n}{nm} = \dfrac{1}{m}$. It follows that for $\epsilon > 0$ we can choose $N = 1/\epsilon$.

(d) The sequence is Cauchy. If $n > m$, we have $|w_n - w_m| = \left| \dfrac{\sin n}{n^2 + 1} - \dfrac{\sin m}{m^2 + 1} \right| \leq \left| \dfrac{\sin n}{n^2 + 1} \right| + \left| \dfrac{\sin m}{m^2 + 1} \right| \leq \dfrac{2}{m^2 + 1} < \dfrac{2}{m^2}$, and $\dfrac{2}{m^2} < \epsilon$ if $m > \sqrt{2/\epsilon}$. Thus, $N = \sqrt{2/\epsilon}$ works in the definition.

3. For $\epsilon > 0$ choose N that works for ϵ in the sense of Definition 2.17. This same N works in the definition of convergence to zero. To see why, let $m > N$ be given. Then choose any n of the form $n = 10^k$, with $n > m$ ($n = 10^m$ is one possibility). Then we have $n > m > N$, and so $|x_m - x_n| = |x_m - 0| < \epsilon$, as desired.

5. (a) Any sequence of rationals tending to an irrational will do.

(b) Yes. The only Cauchy sequences of integers are eventually constant.

(c) $[0, 1]$ is complete because if $\{x_n\}$ is any Cauchy sequence in $[0, 1]$, then $\{x_n\}$ tends to some limit L. Because $0 \leq x_n \leq 1$ for all n, the limit L is in $[0, 1]$, too.

$(0, 1]$ is not complete because $\{1/n\}$ is a Cauchy sequence in $(0, 1]$, but its limit lies outside $(0, 1]$.

7. Let $\epsilon > 0$ be given and set $\epsilon' = \epsilon/42$. Because $\{a_n\}$ is Cauchy sequence there is some N such that $|a_n - a_m| < \epsilon'$ whenever $n > N$. This N works for the sequence $\{x_n\}$; details left to you.

9. Use the fact that two numbers with the same first n decimal places can differ by no more than 10^{-n}.

11. The key point is that $\dfrac{1}{2^{m+1}} + \dfrac{1}{2^{m+2}} + \cdots + \dfrac{1}{2^n} < \dfrac{1}{2^m}$, regardless of n. This fact can be parlayed into a proof.

2.5   Series 101: Basic Ideas

1. (a) $S_n = 0$ for all n, and so the series converges to 0.

(b) $S_n = 42n$ for all n; this diverges (to ∞).

(c) The partial sum sequence $\{S_n\}$ has the form $-1, 0, -1, 0, \ldots$; this diverges.

(d) $S_n = \dfrac{n(n+1)}{2}$ for all n; this diverges (to ∞).

(e) The series is geometric, with $r = 0.99$, so $S_n = \dfrac{1 - r^n}{1 - r} = 100 - 100 \cdot 0.99^n$, and $S_n \to 100$.

(f) One shows by induction that $S_n = \dfrac{n}{n+1}$. Thus $S_n \to 1$.

3. (a) Since $H_{2^n} > \dfrac{n}{2}$ for all n and the right-had sequence diverges to infinity, we must have $H_{2^n} \to \infty$, too. Since $\{H_n\}$ has a divergent subsequence, $\{H_n\}$ must diverge, too.

(b) Suppose toward contradiction that $H_n \to H$ for some finite number H. Since $\{H_{2n}\}$ is a subsequence, we'd have $H_{2n} \to H$, too. But we also have $H_{2n} > \dfrac{1}{2} + H_n$; taking limits of both sides gives $H \geq \dfrac{1}{2} + H$, which is absurd.

5. (a) The series converges absolutely by comparison to $\displaystyle\sum_{k=1}^{\infty} \dfrac{1}{k^2}$.

(b) The series converges absolutely by comparison to $\displaystyle\sum_{k=1}^{\infty} \dfrac{1}{k^2}$.

(c) The series converges absolutely by comparison to the geometric series $\displaystyle\sum_{k=1}^{\infty} \dfrac{1}{3^k}$.

(d) The series diverges by the nth term test—all terms exceed $1/3$.

7. Let $\sum a_k$ and $\sum b_k$ be series, and let $\sum c_k$ be the "sum series," defined by $c_k = a_k + b_k$ for all k. Let A_n, B_n, and C_n denote the partial sums for these series. The key point is that—thanks to the commutative law for addition of *finitely* many numbers—$C_n = A_n + B_n$ for all n. It follows that if $A_n \to A$ and $B_n \to B$, then we must also have $C_n \to A + B$, which is what we wanted to prove.

9. (a) It's enough to show (i) if $n > 1000$ then $S_n > S_{1000}$; (ii) if $n > 1000$ then $S_n \leq S_{1001}$. Claim (i) amounts to observing that *every* string of the form
$$\dfrac{1}{1001} - \dfrac{1}{1002} + \dfrac{1}{1003} - \cdots \pm \dfrac{1}{1000 + n}$$
adds up to a *positive* result; group summands in *pairs* to see why. Claim (ii) holds because every string of the form $-\dfrac{1}{1002} + \dfrac{1}{1003} - \dfrac{1}{1004} + \cdots \pm \dfrac{1}{1000 + n}$ has *negative* sum; again, group summands in pairs to see why.

(b) For given $\epsilon > 0$ we can take any integer $N \geq 1/\epsilon$. Then $|S_N - S_{N+1}| = \dfrac{1}{N+1} < \epsilon$.

(c) For given $\epsilon > 0$ choose N as in the preceding part. Then for $n > m > N$ we have both S_n and S_m between S_N and S_{N+1}, and thus within ϵ of each other.

11. (a) Let A_n and B_n be the partial sums for $\sum a_k$ and $\sum b_k$. The hypothesis boils down to the fact that for $n > 42$ we have $B_n = A_n - A_{42} + B_{42}$. Thus, the sequences $\{A_n\}$ and $\{B_n\}$ differ (for large n) by an additive constant, and so both converge or both diverge.

(b) We know (see the preceding part) that $B_n = A_n - A_{42} + B_{42}$. Because $b_k - a_k = k^2$, we must have $B_{42} - A_{42} = \sum_{k=1}^{4} 2k^2 = 25585$. Thus, $B_n = A_n + 25585$, and since $A_n \to 100$ we must have $B_n \to 25685$.

13. A similar problem appears in Section 1.5.

2.6 Series 102: Testing for Convergence and Estimating Limits

1. (a) $\sum_{k=1}^{\infty} \dfrac{1}{2^k - 1}$ converges by comparison limit comparison to $\sum_{k=1}^{\infty} \dfrac{1}{2^k}$.

   (b) $\sum_{k=1}^{\infty} \dfrac{k}{2k^2 - 1}$ diverges; one approach is limit comparison to $\sum_{k=1}^{\infty} \dfrac{1}{k}$.

   (c) $\sum_{k=1}^{\infty} \dfrac{k}{3^k}$ converges by the ratio test.

   (d) $\sum_{k=1}^{\infty} \dfrac{k^2 + 2}{3k^2 - 2}$ diverges; use the nth term test.

3. (a) The series diverges by the ratio test.

   (b) The series converges by the ratio test.

   (c) The series converges by the ratio test; a little algebra is needed.

5. (a) The series converges absolutely for $-1 < x < 1$; it's geometric.

   (b) The series converges absolutely for $-1 < x < 1$; use the ratio test when $x \neq 0$.

   (c) The series converges absolutely for $-1 < x < 1$; use the ratio test when $x \neq 0$. converges conditionally when $x = -1$.)

   (d) The series converges absolutely for all x; use the ratio test when $x \neq 0$.

7. (a) If $\sum a_k = \sum \dfrac{1}{\sqrt{k}}$ then $\sum a_k^2$ diverges. If $\sum a_k = \sum \dfrac{1}{k}$ then $\sum a_k^2$ converges.

   (b) Use the limit comparison test: Here $\dfrac{a_k^2}{a_k} = a_k$, and this tends to zero by the kth term test. Hence $\sum a_k^2$ converges.

9. If $\sum b_k$ converges to B, then $\sum a_k$ converges to $B - \sum_{k=1}^{K} b_k + \sum_{k=1}^{K} a_k$. This is similar to a problem in the preceding section, where $K = 42$.

11. Notice that the sum $S_n = \dfrac{1}{\sqrt{1}} + \dfrac{1}{\sqrt{2}} + \cdots + \dfrac{1}{\sqrt{n}} > \dfrac{1}{\sqrt{n}} + \dfrac{1}{\sqrt{n}} + \cdots + \dfrac{1}{\sqrt{n}} = \dfrac{n}{\sqrt{n}} = \sqrt{n}$. Since $\sqrt{n} \to \infty$, we must have $S_n \to \infty$, too.

13. (a) The series converges by comparison to $\displaystyle\sum_{k=1}^{\infty} \dfrac{1}{2^{k-1}}$, which converges to 2.

   (b) We have $R_{10} = \displaystyle\sum_{k=11}^{\infty} \dfrac{1}{2^{k-1}+1} < \sum_{k=11}^{\infty} \dfrac{1}{2^{k-1}} = \dfrac{1}{2^9}$.

   (c) We have $S = S_{10} + R_{10} < 1.26255 + \dfrac{1}{2^9} \approx 1.26450$. This means that S lies somewhere in the interval $[1.26255, 1.26450]$.

15. (a) See your favorite calculus text.

   (b) We need to choose n so $R_n < 0.001$. By the given inequality, this holds if n is large enough so that $\displaystyle\int_n^{\infty} \dfrac{dx}{x^3} < 0.001$. A calculus calculation shows that $\displaystyle\int_n^{\infty} \dfrac{dx}{x^3} = \dfrac{1}{2n^2}$, and the last quantity is less than 0.001 if $n \geq 23$. This means that $S_{23} \approx 1.2012$ is within 0.001 of the true answer.

3.1   Limits of Functions

1. Both parts are easy exercises with the definition. In (a) we can choose *any* δ for given ϵ. In (b) we can choose $\delta = \epsilon$.

3. (a) $\lim_{x \to 1}(2x + 3) = 5$. Note that $|f(x) - 5| = 2,|x - 1|$. It follows that for given $\epsilon > 0$ we can set $\delta = \epsilon/2$.

   (b) Factoring the numerator gives $\lim_{x \to -1} \dfrac{x^2 - 1}{x + 1} = \lim_{x \to -1}(x - 1) = -2$. (For given $\epsilon > 0$ we can set $\delta = \epsilon$.)

5. (a) For $f(x) = x^2$, we have $\lim_{x \to 42} f(x) = \lim_{x \to 42} x^2 = 42^2 = f(42)$, so the condition does hold at $a = 42$.

   (b) For $f(x) = x^2$ and *any* input a, we have $\lim_{x \to a} f(x) = \lim_{x \to a} x^2 = a^2 = f(a)$, so the condition holds at every a.

   (c) Since the function $f(x) = \frac{x^2-4}{x-2}$ is not defined at $a = 2$, the condition can't possibly hold.

   (d) We have $\lim_{x \to 3} f(x) = \lim_{x \to 3} \dfrac{x^2 - 4}{x - 2} = \dfrac{3^2 - 4}{3 - 1} = f(3)$, as desired.

7. Let $\{x_n\}$ be any sequence with $x_n \to a$ and $x_n \neq a$ for all n. Then $\{f(x_n)\}$, $\{g(x_n)\}$, and $\{h(x_n)\}$ are all *sequences*, and the hypotheses imply that (i) $f(x_n) \leq g(x_n) \leq h(x_n)$ for all n; (ii) $f(x_n) \to L$ and $h(x_n) \to L$. Now the sequence version of the squeezing theorem implies that $g(x_n) \to L$, too. Since this applies

to *any* sequence $\{x_n\}$ of the given type, Lemma 3.2 implies that $\lim\limits_{x \to a} g(x) = L$, as desired.

9. (a) We say $\lim\limits_{x \to -\infty} f(x) = L$ if for all $\epsilon > 0$ there exists $M < 0$ such that $|f(x) - L| < \epsilon$ whenever $x < M$.

   (b) We say $\lim\limits_{x \to 0^-} g(x) = L$ if for all $\epsilon > 0$ there exists $\delta > 0$ such that $|g(x) - L| < \epsilon$ whenever $-\delta < x < 0$.

   (c) Here's the *idea* of a proof that $\lim\limits_{x \to -\infty} f(x) = L$ implies that $\lim\limits_{x \to 0^-} f(1/x) = L$. The converse is proved similarly.

   Let $\epsilon > 0$ be given. We want to show that $|f(1/x) - L| < \epsilon$ for x in some interval $(\delta, 0)$. By hypothesis, there is some $M < 0$ so that $|f(t) - L| < \epsilon$ when $t < M$. Writing $t = 1/x$, this means that $|f(1/x) - L| < \epsilon$ when $1/x < M < 0$, or, equivalently, when $0 > x > 1/M$. (The inequality algebra is a bit tricky because both M and x are negative.) This implies that we can take $-\delta = 1/M$, or $\delta = -1/M$.

11. (a) We can use $\delta \approx 0.0024$ (or less). Note $\delta \approx \epsilon/4$.

    (b) We can use $\delta \approx 0.00005$ (or less). Note $\delta \approx \epsilon/200$.

    (c) With $\epsilon = 0.01$, the value $\delta = 0.001$ does not quite work at $a = 10$. (The value $\delta = 0.0005$ does work.)

13. First suppose $a \in \mathbb{Z}0$. Let $\epsilon > 0$ be given. Set $\delta = 1$. This δ works, since if $0 < |x - 0| < \delta$, then $x \notin \mathbb{Z}$, and so $|f(x) - 0| = 0 < \epsilon$, as desired.

    If $a \notin \mathbb{Z}$, then we can take δ to be the (positive!) distance from a to the nearest integer.

15. The key idea is that, because S is a finite set, there is a smallest distance, say d, between any two points in S. Therefore, to show that $\lim_{x \to a} j(x) = 0$ when $a \in S$, we can use $\delta = d$ works for *any* $\epsilon > 0$.

    To show that $\lim_{x \to a} j(x) = 0$ when $a \notin S$, we can choose any $\epsilon > 0$ and let δ be the smallest distance from a to any point of S. Such a δ exists because S is finite.

17. (a) The point is that all three quantities in the claimed inequality are *even* functions of θ—i.e., they have the same value for θ and for $-\theta$.

    (b) Draw the right triangles indicated, and note that θ is the area of the pie-shaped wedge with angle θ at the origin.

3.2   Continuous Functions

1. For given $\epsilon > 0$, $\delta = \epsilon$ works.

3. Give details as needed.

   (a) $\delta = 0.001/345$ works.

   (b) $\delta = \epsilon/345$ works.

(c) $\delta = \epsilon/345$ works regardless of the value of a.

5. (a) The idea: If $f + g$ is continuous, then $(f + g) - f = g$ is continuous, too. Give details.

    (b) There are many possiblities.

7. Suppose first that $A \neq 0$. Let $\epsilon > 0$ be given. Set $\delta = \dfrac{\epsilon}{|A|}$; note that $\delta > 0$. If $|x - c| < \delta$, then $|f(x) - f(c)| = |Ax + B - (Ac + B)| = |A||x - c| < |A|\delta = \epsilon$, as desired.

    If $A = 0$ then f is a constant function. In this case *any* number $\delta > 0$ works for a given $\epsilon > 0$.

9. (a) Let $\epsilon > 0$. Since f is continuous at 42 we can choose $\delta > 0$ so $|x - 42| < \delta \implies |f(x) - f(42)| < \epsilon$. The same δ works for g, because $|g(x) - g(42)| < |f(x) - f(42)|$.

    (b) For given $\epsilon > 0$ the same δ works for f and for h.

    (c) For given $\epsilon > 0$ the same δ works for f and for k.

11. The composite function $\cos(\sin(x))$ is continuous; see the relevant fact on page 163.

13. By the continuity hypothesis, $f(0) = \lim_{n \to \infty} f(1/n)$. Since all terms of the sequence $f(1/n)$ are positive, the limit can't be negative. (One can argue this more formally using general properties of sequences. Or see Problem 3, page 91.)

15. Note that $\lim_{x \to 3} f(x) = f(3)$. Let $\epsilon = f(3) - 5$. If we choose any $\delta > 0$ that works for this ϵ, we're done.

17. (a) The discontinuity of $f(x) = 1/x$ at $a = 0$ is not removable; $\lim_{x \to 0} f(x)$ does not exist.

    (b) The discontinuity is removed if we set $f(-2) = -4 = \lim_{x \to -2} f(x)$.

    (c) The discontinuity is removed if we set $f(0) = 0 = \lim_{x \to 0} f(x)$.

19. We need to show that $f(c) = 0$ if c is any irrational number. Recall (why?) that there is a sequence $\{r_n\}$ of *rationals* such that $r_n \to c$; note that $f(r_n) = 0$ for all n. Continuity of f at c requires that $f(r_n) \to f(c)$. In other words, $f(c) = \lim_{n \to \infty} f(r_n) = \lim_{n \to \infty} 0 = 0$.

3.3   Why Continuity Matters: Value Theorems

1. (a) There are many possibilities; one has the tent-shaped graph through $(0, 0)$, $(1, 1)$ and $(2, 0)$.

    (b) If $f(0) = f(1)$, $f(2) = f(1)$, or $f(0) = f(2)$ we're done. We might as well assume, then, that either $f(0) < f(2) < f(1)$ or $f(2) < f(0) < f(1)$. In the former case, by the IVT, there exists c with $0 < c < 1$ and $f(c) = f(2)$, so indeed f is not one-to-one. The remaining case is similar.

3. (a) First write

$$p(x) = x^n + a_{n-1}x^{n-1} + \cdots + a_1 x + a_0$$
$$= x^n \left(1 + \frac{a_{n-1}}{x} + \frac{a_{n-2}}{x^2} + \cdots + \frac{a_1}{x^{n-1}} + \frac{a_0}{x^n}\right).$$

Now as $x \to \infty$ the quantity in parentheses tends to 1 and so the product tends to ∞.

(b) Let $M = 0$. Then there exists N so that $f(x) > M = 0$ whenever $x > N$. Thus if b is any number with $b > N$, we have $f(b) > 0$.

(c) Numerical experiments reveal that $q(-1) < 0$, $q(0) = 1$, $q(1) = 0$, $q(1.5) < 0$, and $q(2) > 0$. This means, by the IVT, that there must be a root between -1 and 0 and another between 1.5 and 2. Factoring gives $q(x) = (x - 1)(x^2 + x + 1)(x^2 - x - 1)$, from which we can find the real roots exactly.

5. (a) The minimum value of p on $[0, 3]$ is $p(0) = 0$. The EVT guarantees that p must assume such a minimum value.

(b) The minimum value of q on $(0, 3]$ is $q(3) = 0$. The EVT doesn't apply here, since $(0, 3]$ is not a closed interval.

(c) Factoring gives $p(a) - 4 = a(3 - a)^2 - 4 = (a - 4)(a - 1)^2$. This shows that $p(1) = 4$ and $p(a) - 4 \le 0$ when $0 \le a \le 3$. In other words, $p(1)$ is the maximum value.

7. (a) $f(x) = |2x - 1|$ is one possibility

(b) $f(x) = \dfrac{\sin(2\pi x) + 1}{2}$ is one possibility.

(c) No such function exists. A continuous function on $[0, 1]$ must *achieve* a minimum value.

9 (a) $f(x) = -(x - 0.5)^2$ is one possibility.

(b) No such function g can exist. Note that we have $f(a) = e < 3 < \pi = f(b)$ for some a and b in $[0, 1]$. Thus 3 is an "intermediate value" and so must be attained for some $c \in [0, 1]$.

(c) $p(x) = -3 + x^2$ is one possibility.

(d) No such function q can exist. If q has odd degree then q is unbounded above and below. If q has even degree then q may be bounded below, but in this case q achieves a minimum value rather than just approaching a minimum.

(e) $f(x) = e^x - 3$ is one possibility.

11. Suppose f attains a maximum value at some $c \in (a, b)$. If $f(c) = f(a)$ or $f(c) = f(b)$ then we're done already, so assume that $f(c) > f(a)$ and $f(c) > f(b)$.

Let v be any number less than $f(c)$ but greater than both $f(a)$ and $f(b)$. (There are infinitely many possibilities.) By the IVT, $f(d) = v$ for some $d \in (a, c)$, and $f(e) = v$ for some $e \in (c, b)$. Thus $f(d) = f(e)$, so f is not one-to-one.

13. Suppose toward contradiction $f(2) \leq f(1)$. If $f(2) = f(1)$ or $f(2) = f(0)$, then f is not one-to-one, a contradiction. Thus we have either $f(0) < f(2) < f(1)$ or $f(2) < f(0) < f(1)$. In the former case, by the intermediate value theorem there exists c with $0 < c < 1$ and $f(c) = 2$, which again contradicts the one-to-one property. The remaining case, $f(2) < f(0) < f(1)$ is similar.

Alternatively, just note that Problem 12 says that f is strictly monotone, and therefore strictly increasing.

15. If f is not constant then the range of f is an interval. But every interval contains both rational and irrational numbers.

3.4 Uniform Continuity

1 (a) For given $\epsilon > 0$ we can use *any* positive δ, such as $\delta = 1$.

(b) For any s and t, we have $|g(s) - g(t)| = |2s + 7 - 2t - 7| = 2|s - t|$. This implies that for any given $\epsilon > 0$ the value $\delta = \epsilon/2$ works.

3. (a) Let $\epsilon > 0$. The value $\delta = \epsilon$ "works" because if $|x-y| < \delta$ then $|f(x) - f(y)| \leq |x - y| < \delta = \epsilon$.

(b) We have $|f(x) - f(y)| = \left| \dfrac{1}{x^2 + 1} - \dfrac{1}{y^2 + 1} \right|$. Work with a common denominator inside the absolute value to complete the proof.

5. Show first that for x and y in $[1/4, 10]$, $|f(x) - f(y)| = 5\dfrac{|x - y|}{xy} \leq 90|x - y|$. Then, for given $\epsilon > 0$ we can set $\delta = \epsilon/90$.

7. In part (c), values of δ that work for f at $c = 1$, $c = 10$, $c = 100$, and $c = 1000$ are $\delta \approx 0.414$, $\delta \approx 0.05$, $\delta \approx 0.005$, and $\delta \approx 0.0005$, respectively. The fact that these δ's decrease to zero means they can't be chosen "uniformly" in $\mathbb{R}$. On the other hand, $\delta = 0.0005$ works for $\epsilon = 1$ on the entire interval $[1, 1000]$.

The difference for g is that $\delta = 0.1$ works for all of $c = 1$, $c = 10$, $c = 100$, and $c = 1000$, and would work for all c. This is what uniform continuity on $\mathbb{R}$ means.

9. (a) Start by squaring both sides of the desired inequality.

(b) For given $\epsilon > 0$ it works to set $\delta = \epsilon^2$. If $0 < x-y < \delta$ then $|f(x) - f(y)| = \sqrt{x} - \sqrt{y} \leq \sqrt{x - y} < \sqrt{\epsilon^2} = \epsilon$.

11. (a) Note that if a and b are in $[-3, 42)$, then $|a^2 + ab + b^2| \leq 3 \cdot 42^2 = 5292$. The hint implies, therefore, that $|f(s) - f(t)| = |s^3 - t^3| = 5292|s - t|$. This implies that for any $\epsilon > 0$ the value $\delta = \epsilon/5292$ works in the definition of uniform convergence on $[-3, 42)$.

(b) The function $f(x) = x^3$ is continuous on the closed and bounded interval $[-3, 42]$, and is therefore also uniformly continuous on the same interval, by Theorem 3.22. Since $[-3, 42) \subset [-3, 42]$, f is also uniformly continuous on the smaller interval.

13. (a) For given $\epsilon > 0$. Choose δ_1 and δ_2 such that, for $x, y \in I$,

$$|x - y| < \delta_1 \implies |f(x) - f(y)| < \frac{\epsilon}{2};$$

$$|x - y| < \delta_2 \implies |g(x) - g(y)| < \frac{\epsilon}{2}.$$

Now show that $\delta = \min(\delta_1, \delta_2)$ works for $f + g$.

(b) One of many possibilities is $f(x) = x = g(x)$ on $I = \mathbb{R}$.

15. (a) For given $\epsilon > 0$ the number $\delta = \epsilon/K$ works in the definition.

(b) If $A = 0$ then L is constant, and $K = 0$ works. Otherwise, for given $\epsilon > 0$ the number $\delta = \epsilon/|A|$ works.

(c) The ratio in question is bounded by K.

(d) For $g(x) = \sqrt{x}$, the ratio $\left| \dfrac{g(x) - g(0)}{x - 0} \right| = \dfrac{\sqrt{x}}{x} = \dfrac{1}{\sqrt{x}}$ blows up as $x \to 0^+$. By the preceding part, g is not Lipschitz continuous on $[0, 1]$.

17. (a) The function f must be constant (and hence uniformly continuous) on $\mathbb{R}$.

(b) The function f need only be bounded (and not necessarily uniformly continuous) on $\mathbb{R}$.

(c) The function f must be constant (and hence uniformly continuous) on $\mathbb{R}$.

(d) For some δ, values of f can't rise or fall value by more than 1 on intervals of length δ. Such a function could have jumps, and so need not even be continuous.

3.5   Topology of the Reals

1. By De Morgan, the intersection of two closed sets is the complement of the union of two open sets.

3. (a) Let $U_i = (-1 - 1/i, 1 + 1/i)$.

(b) Let $A_i = [-1 + 1/i, 1 - 1/i]$.

5. (a) The value $\epsilon = 0.0042$ works.

(b) The value $\epsilon = p$ works.

(c) Suppose U is open and $p \in U$. Since U is a union of open intervals, we must have $p \in (a, b) \subseteq U$ for some open interval (a, b). Since $a < p < b$ we can take ϵ to be the smaller of $p - a$ and $b - p$.

7. Let $\epsilon > 0$ be given, and consider the open interval $U = (a - \epsilon, a + \epsilon)$. Our assumption means that there is a largest N for which $a_N \notin U$. This N works in the usual definition of sequence convergence.

9. (a) All points in $(0, 1)$ are interior to $[0, 1]$.

(b) Every point p in U has $N_\epsilon(p) \subset U$ for some $\epsilon > 0$.

(c) A finite set contains no open intervals.

(d) No points of $\mathbb{Q}$ are interior, because $\mathbb{Q}$ contains no intervals.

11. (a) $x \in f^{-1}(\mathbb{R} \setminus S) \iff f(x) \notin S \iff x \in \mathbb{R} \setminus S$

(b) Suppose $f : \mathbb{R} to \mathbb{R}$ is continuous, and $A \subseteq \mathbb{R}$ is closed. Then $U = \mathbb{R} \setminus A$ is open, so $f^{-1}(U)$ is open. By the previous part, $f^{-1}(A) = \mathbb{R} \setminus f^{-1}(U)$, which is closed. The converse is similar.

13. (a) As we know, every interval (a, b) contains both rationals and irrationals.

(b) The set S is not dense in $\mathbb{R}$. The interval $(0, .0005)$, for instance, has empty intersection with S.

15. Proofs are routine. Show that every element of the right-hand set is an element of the left-hand set, and vice versa.

17. (a) $(0, 1) = B_{0.5}(0.5)$

(b) $\mathbb{R} = B_1(0) \cup B_2(0) \cup B_3(0) \cup \ldots$

(c) $\mathbb{R} \setminus \mathbb{Z} = B_{0.5}(0.5) \cup B_{0.5}(-0.5) \cup B_{0.5}(1.5) \cup B_{0.5}(-1.5) \ldots$

(d) $(0, \infty) = B_1(1) \cup B_2(2) \cup B_3(3) \cup \ldots$

19. (a) All parts of Definition 3.30 are easily checked, including the triangle inequality.

(b) $B_1(p) = (p - 1, p + 1)$; $B_2(p) = \mathbb{R}$.

(c) This topology has the same open sets as with the usual metric.

3.6   Compactness

1. (a) Each x_i is in some U_i, so it takes at most n such U_i to cover F.

(b) We can cover $\mathbb{R}$ with open sets $U_i = (-i, i)$ for $i \in \mathbb{N}$. A finite subcollection covers only a bounded subset of $\mathbb{R}$.

(c) It takes *zero* open sets to cover $\emptyset$; that's certainly finite!

3. (a) Cover each term $1/k$ with a small open interval that contains no other term of the sequence.

(b) The set $\{a_n\}$ is compact in this case, because it contains its own limit.

(c) Many sequences are possible. One is $0, 1, 0, 1, \ldots$.

5. Find a linear (therefore certainly continuous) function $f : \mathbb{R} \to \mathbb{R}$ that maps $[0, 1]$ onto $[a, b]$.

7. The U_i are open because they are inverse images of the open intervals $(-i, i)$. The U_i cover all of $\mathbb{R}$, and so all of K. Compactness of K means that we need only finitely many U_i to cover K. This amounts to boundedness of f on K.

9. Proposition 3.41 says that the image of a compact set is compact.

4.1   Defining the Derivative

1. The difference quotients for $g'(a)$ and for $f'(a)$ are identical, so their limits are equal.

3. (a) Draw your own picture.

   (b) The difference quotient calculations are straightforward.

   (c) The graphs of f and g are reflections of each other in the line $y = x$. The same is true of the respective tangent lines. In particular, the two tangent lines have reciprocal slopes.

5. (a) Calculation with the difference quotient gives $f'(0) = 0$.

   (b) Calculation with the difference quotient gives $f'(c) = \dfrac{2c}{(c^2 + 5)^2}$.

7. (a) The difference quotient $\dfrac{f(x) - f(0)}{x - 0}$ is strictly positive for all x, and so $f'(0)$ must be nonnegative.

   (b) We have $-1 \le \dfrac{f(x) - f(0)}{x - 0} \le 1$ for all x, and so $-1 \le f'(0) \le 1$.

9. Definition 4.1 gives $(fg)'(0) = \lim\limits_{x \to 0} \dfrac{f(x)g(x)}{x} = \lim\limits_{x \to 0} g(x) \lim\limits_{x \to 0} g(x) \dfrac{f(x)}{x} = g(0)f'(0) = 0$. Both limits exist by hypothesis.

11. The limit-of-difference-quotient calculation for $g'(2)$ is like that for f, except for a factor of 4. Note that the g-graph is formed by vertical stretching of the f-graph.

13. (a) It's easy to show that both $\lim\limits_{x \to 0^-} f(x) = 0$ and $\lim\limits_{x \to 0^+} f(x) = 0$; thus $\lim\limits_{x \to 0} f(x) = 0 = f(0)$, which means that f is continuous at 0.

   (b) Yes, f *is* differentiable at $x = 0$. Again, we have $\lim\limits_{x \to 0^-} \dfrac{f(x) - f(0)}{x - 0} = \lim\limits_{x \to 0^+} \dfrac{f(x) - f(0)}{x - 0} = 0$.

15. Note first that $f(0) = 0$. Thus, $f'(0) = \lim\limits_{x \to 0} \dfrac{f(x)}{x}$. This limit exists, and is zero, by squeezing: $|f(x)| \le x^2$ implies $-|x| \le \dfrac{f(x)}{x} \le |x|$. Since the right and left quantities tend to zero, so must the middle.

17. We need to show that $\lim\limits_{x \to 0} x \sin(1/x) = 0$. Exactly this limit is discussed in Example 6, page 155.

19. (a) This is essentially the Principle of Persistent Inequalities. Assume toward contradiction that $f(0) < 42$. Then, says the PoPI, for some $\delta > 0$ we have $f(x) < 42$ for $x \in (-\delta, \delta)$, contradicting our assumption.

   (b) Clearly $h(0) = 0$. Now $h'(0) = \lim\limits_{x \to 0} \dfrac{h(x)}{x} = \lim\limits_{x \to 0} g(x) = 42$. (The last equality holds because g is continuous.)

21. (a) Claim (iv) of Lemma 2 says that since $f'(0) = 100 > 0$, we must have $f(x) < 0$ on $(-\delta, 0)$ and $f(x) > 0$ on $(0, \delta)$ for suitable small δ.

   (b) $\delta = \pi/100 \approx 0.03$ works in this case.

4.2 Calculating Derivatives

1. One can show that $x^n - a^n = (x-a)\left(x^{n-1} + ax^{n-2} + \cdots + a^{n-1}\right)$; there are n summands. Taking the limit as $x \to a$ gives $na^n - 1$ as desired.

3. The derivative $h'(a)$ is defined by $\lim\limits_{x \to a} \dfrac{h(x) - h(a)}{x - a}$. We can evaluate this using the definition of h from f and g and algebraic properties of the limit:

$$\lim_{x \to a} \frac{h(x) - h(a)}{x - a} = \lim_{x \to a} \frac{3f(x) - 42g(x) - 3f(a) + 42g(a)}{x - a}$$
$$= 3 \lim_{x \to a} \frac{f(x) - f(a)}{x - a} - 42 \lim_{x \to a} \frac{g(x) - g(a)}{x - a} = 3f'(a) - 42g'(a).$$

5. (a) The chain rule says that if $h(x) = 1/g(x)$, then $h'(a) = -g'(a)/g(a)^2$.

    (b) Applying the product rule to $f(x)/g(x) = f(x) \cdot h(x)$ gives the derivative $f'(a) \cdot h(a) + f(a)h'(a) = f'(a)/g(a) - f(a)g'(a)/g(a)^2$, which is (after a bit of algebra) seen to be the desired expression.

7. All parts are straightforward—but perhaps messy—calculations.

9. Clearly, $g(a) = f(a)^2 = 0$. Also the chain rule gives $g'(a) = 2f(a)f'(a) = 0$.

11. (a) Notice first that f is continuous at $x = 17$, by Theorem 4.3, page 213. We also know, or can easily prove, that $h(x) = |x|$ is continuous everywhere. Since $g(x)$ is the composition $h(f(x))$, we know that $g(x)$ must also be continuous.

    (b) If $f(x) = x - 17$, then $g(x) = |x - 17|$, which is clearly not differentiable at $x = 17$.

    (c) Note that $g(x)^2$ is really the same thing as $f(x)^2$, which is the product of two differentiable functions, and therefore differentiable.

4.3 The Mean Value Theorem

1. If a and b are any two roots of f, then Rolle's theorem implies that $f'(c) = 0$ for some c between a and b. Our claim follows.

   If $f(x) = 2 + \sin x$, then f has *no* real roots, but $f'(x) = \cos x$ has infinitely many roots.

3. (a) Every odd-degree polynomial has at least one root; we've shown this using the intermediate value theorem.

    (b) There are many possibilities. The functions $f_1(x) = x^3$ (with $a = b = 0$), $f_2(x) = x^3 - 3x + 2$ (with $a = -3$ and $b = 2$), and $f_3(x) = x^3 - 3x + 1$ (with $a = -3$ and $b = 1$) have one, two, and three roots, respectively.

    (c) Rolle's theorem implies that between any two distinct roots of f lies a root of f'. If f had at least 4 roots, f' (a quadratic polynomial) would have at least 3 roots, which is impossible.

(d) We know f has at least one root. If $a > 0$ then $f'(x) = 3x^2 + a$, so $f'(x) > 0$ for all x. Thus f' has *no* roots, and so f can have *at most* one root.

5. (a) Rolle's theorem gives x_1 between a and c and x_2 between c and b.

(b) Rolle's theorem applied to f' on the interval $[x_1, x_2]$ gives the desired x_0.

(c) If $f : [a, c] \to \mathbb{R}$ has continuous derivatives f', f'', ..., $f^{(42)}$ on $[a, c]$, and there are inputs $x_1, x_2, \ldots, x_{43}$ with $f(x_1) = f(x_2) = \cdots = f(x_{43})$, then the desired x_0 exists.

7. Rolle's theorem implies that f' has a root in each interval $(0, a)$, $(a, 2a)$, $(2a, 3a)$, ... Since f' is also periodic with period a, $(f')' = f''$ also has roots in each of these intervals, and similarly for higher-order derivatives.

9. (a) Every number c in $(1, 7)$ works.

(b) The only solution is $c = 4$.

(c) In this case the MVT equation $f'(c) = \dfrac{f(b) - f(a)}{b - a}$ reduces to $100c^{99} = 1$, which gives $c = \dfrac{1}{100^{1/99}} \approx 0.955$.

(d) In this case the equation $f'(c) = \dfrac{f(b) - f(a)}{b - a}$ reduces to $6c + 5 = 3a + 3b + 5$, which gives $c = (a + b)/2$, the midpoint of (a, b).

(e) If $q(x) = Ax^2 + Bx + C$ and $[a, b]$ is any interval, then the mean value equation holds if and only if $c = (a + b)/2$. To see why, note that in this case the equation $f'(c) = \dfrac{f(b) - f(a)}{b - a}$ reduces to $B + 2Ac = B + A(a + b)$.

11. If a car has the same velocity at times a and b, then there must be some intermediate time c at which the acceleration is zero.

13. (a) Consider $h(x) = f(x) - g(x)$. By hypothesis, $h'(x) = 0$ for all $x \in (a, b)$. By Proposition 4.11, h is constant, as desired.

(b) Let $h(x) = f(x) - g(x) - 5x$. Then $h'(x) = 0$ for all $x \in (a, b)$, and so $h(x) = C$ for some constant C. It follows that $f(x) = g(x) + 5x + C$.

15. If we write $m = \dfrac{f(b) - f(a)}{b - a}$, then $L(x) = m(x - a) + f(a)$, and the rest follows.

17. (a) This should be straightforward.

(b) The proof of (i) can be essentially rerun with strict inequalities.

(c) If f were not monotone, then we'd have $f(a) = f(b)$ for some a and b in I. Rolle's theorem now says there is $c \in I$ with $f'(c) = 0$, which contradicts the hypothesis.

19. Suppose $f(b) \geq b$ for some $b > 0$. Then the MVT implies that $f'(c) = \dfrac{f(b) - f(0)}{b - 0} = \dfrac{f(b)}{b} \geq 1$ for some c between 0 and b; this contradicts our assumption.

21. (a) The function f is continuous, and so must attain a minimum on the interval $[s, t]$ by the extreme value theorem. The minimum can't be at either endpoint because $f'(s) < 0$ and $f'(t) > 0$.

(b) The function $g(x) = f(x) - vx$ satisifies the conditions in (a), so there is some c between s and t with $0 = g'(c) = f'(c) - v$. Thus $f'(c) = v$ as desired.

23. (a) All parts follow by plugging $t = 0$ into f and p_2.

(b) That $g(0) = 0 = g(1)$ is an easy calculation. The t_1 in question exists by Rolle's theorem applied to g.

(c) That $g'(0) = 0 = g'(t_1)$ is an easy calculation. The t_2 in question exists by Rolle's theorem applied to g'.

(d) As in the previous parts, the claimed t_3 exists by Rolle's theorem applied to g''.

4.4   Sequences of Functions

1. (a) For any real number x and $\epsilon > 0$, let $N = 1/\epsilon$. If $n > N$ then $|f_n(x) - f(x)| = f_n(x) = \dfrac{1}{n} < \epsilon$. Thus $f_n \to f$ pointwise on $\mathbb{R}$. Because the same inequality holds for *all* x, the convergence is also uniform.

(b) For any fixed real number x we have $f_n(x) = \frac{x}{n} \to 0$ as $n \to \infty$, so $f_n \to f$ pointwise on $\mathbb{R}$. The convergence is not uniform on $\mathbb{R}$. For $\epsilon = 1$, for instance, no N works: for every n we have $|f_n(x) - f(x)| = |x/n| > 1$ whenever $|x| > n$.

(c) For any fixed real number x we have $f_n(x) = \frac{\sin x}{n} \to 0$ as $n \to \infty$. Thus $f_n \to f$ pointwise on $\mathbb{R}$. The convergence is also uniform; the proof is similar to that in (a).

3. (a) Fix $a \in \mathbb{R}$ and let $a_n = a + 1/n$. Note that $a_n \to a$ and that $f(a_n) = f_n(a)$ for all n. Since f is continuous at a, $f_n(a) = f(a_n) \to f(a)$, as desired.

(c) Let $\epsilon > 0$ be given. Choose $\delta > 0$ that works for ϵ in the definition of uniform continuity, and choose an integer N with $1/N < \delta$. This N works for ϵ in the definition of uniform convergence: if $n > N$ then—for all real x—we have

$$|f_n(x) - f(x)| = |f(x + 1/n) - f(x)| < \epsilon.$$

(The last inequality holds because $|(x + 1/n) - x| < \delta$.)

5. The result is just a restatement of the analogous fact for Cauchy sequences of numbers.

7. (a) The statement is true. A convergent sequence of integers must eventually be constant.

(b) The statement is false. The sequence $\{\sqrt{2}/n\}$ is irrational, but converges to 0.

(c) The statement is true by a "squeezing" property of limits.

9. (a) The claim follows from analogous properties of number sequences.

(b) Here is the key identity: $\left|(x + 1/n)^2 - x^2\right| = \dfrac{2x}{n} + \dfrac{1}{n^2}$. The right-hand side is large for large $|x|$.

11. Let $\epsilon > 0$ be given. Since h is uniformly continuous on $\mathbb{R}$ we can choose $\delta > 0$ such that $|h(s) - h(t)| < \epsilon$ whenever $|s - t| < \delta$. Because $f_n \to f$ uniformly on I we can choose N such that $|f_n(x) - f(x)| < \delta$ holds for all $x \in I$ whenever $n > N$. By our choice of δ, this guarantees that $|h(f_n(x)) - h(f(x))| < \epsilon$ for all $x \in I$, as desired.

13. (a) The ratio calculation is slightly messy, but it works.

    (b) We get $f'' = -f$. This combined with $f(0) = 0$ and $f'(0) = 1$ says that $f(x) = \sin(x)$.

    (c) Differentiate f term by term to find a power series for f'. What famous function is this? How do you know?

    (d) The graphs should resemble the sine and cosine functions, respectively.

5.1   The Riemann Integral: Definition and Examples

1. (a) Calculation gives $\mathrm{RS}(f, \mathcal{P}, \mathcal{S}) = 0 \cdot 0.8 + 1 \cdot 0.2 + 2 \cdot 0.7 + 3 \cdot 0.3 + 4 \cdot 0.9 + 9 \cdot 0.1 = 7.0$; $\|\mathcal{P}\|$ is 0.9, the largest width.

    (b) To make $\mathrm{RS}(f, \mathcal{P}, \mathcal{S})$ as *small* as possible, choose left endpoint samples: $\mathcal{S} = \{0, 0.8, 1.0, 1.7, 2, 2.9\}$. This gives $\mathrm{RS}(f, \mathcal{P}, \mathcal{S}) = 0 \cdot 0.8 + 0.8^2 \cdot 0.2 + 1 \cdot 0.7 + 1.7^2 \cdot 0.3 + 4 \cdot 0.9 + 2.9^2 \cdot 0.1 = 6.136$.

    (c) To make $\mathrm{RS}(f, \mathcal{P}, \mathcal{S})$ as *bad* as possible, choose right endpoint samples: $\mathcal{S} = \{0.8, 1.0, 1.7, 2, 2.9, 3\}$. This gives $\mathrm{RS}(f, \mathcal{P}, \mathcal{S}) = 0.8^2 \cdot 0.8 + 1^2 \cdot 0.2 + 1.7^2 \cdot 0.7 + 2^2 \cdot 0.3 + 2.9^2 \cdot 0.9 + 3^2 \cdot 0.1 = 12.404$, an error of 3.04.

    (d) The sample set $\mathcal{S} = \{\sqrt{3}\}$ gives $\mathrm{RS}(f, \mathcal{P}, \mathcal{S}) = 3 \cdot 3 = 9$.

3. The fact that $f(x) \geq 3$ for all $x \in [0, 2]$ implies that $\mathrm{RS}(f, \mathcal{P}, \mathcal{S}) \geq 6$ for every partition $\mathcal{P}$ and sample points $\mathcal{S}$. This implies, as in the proof of Proposition 5.3, that the integral itself cannot be less than 6.

7. Notice first that the inequality $0 \leq f(x) \leq g(x)$ implies that $0 \leq \mathrm{RS}(f, \mathcal{P}, \mathcal{S}) \leq \mathrm{RS}(g, \mathcal{P}, \mathcal{S})$ holds for all suitable partitions $\mathcal{P}$ and sample sets $\mathcal{S}$. To show f is integrable on $[a, b]$, let $\epsilon > 0$ be given. Since g is integrable, there is some $\delta > 0$ that works for g (and the value $I = 0$) in Definition 5.1. The preceding inequality shows that the same $\delta > 0$ works for f.

9. Let $\epsilon > 0$ be given. Since I_1 satisfies the definition, can choose $\delta_1 > 0$ that works for I_1 and ϵ. Since I_2 satisfies the definition, we can choose $\delta_2 > 0$ that works for I_2 and ϵ. Let δ be the smaller of δ_1 and δ_2, and calculate any Riemann sum $\mathrm{RS}(f, \mathcal{P}, \mathcal{S})$ for $\|\mathcal{P}\| < \delta$. Then both I_1 and I_2 are within ϵ of $\mathrm{RS}(f, \mathcal{P}, \mathcal{S})$, and so within 2ϵ of each other. Since this holds for all positive ϵ, we must have $I_1 = I_2$.

11. (a) The condition holds by *uniform continuity* of f on $[a, b]$.

(b) Note that

$$|\mathrm{RS}(f, \mathcal{P}, \mathcal{S}) - \mathrm{RS}(f, \mathcal{P}, \mathcal{T})| = \left| \sum_{i=1}^{n} (f(s_i) - f(t_i)) \Delta x_i \right|$$

$$\leq \sum_{i=1}^{n} |f(s_i) - f(t_i)| \Delta x_i < \sum_{i=1}^{n} \epsilon \Delta x_i = \epsilon(b - a).$$

13. All the calculations work the same as in Problem 12, except that we need $\delta = \epsilon/(b-a)$.

5.2   Properties of the Integral

1. (a) The function h is integrable because it is the difference of two integrable functions.

   (b) In the situation of Corollary 5.6, we can set $h(x) = f(x) - g(x)$. Theorem 5.5 says that h is integrable and that $\int_a^b f - \int_a^b g = \int_a^b h$. By (a) the last quantity is nonnegative, which is what we wanted to show.

3. The result of Problem 2 implies that the integral lies between -6 and 12.

5. (a) By the extreme value theorem, f assumes a minimum value, say $f(c) = m$, for some $c \in [a, b]$. By hypothesis, $m > 0$. Now $f(x) > m$ for all $x \in [a, b]$, and so Corollary 5.6 gives

$$\int_a^b f \geq \int_a^b m = m(b - a) > 0,$$

as claimed.

   (b) Part (a) implies that $f(c) \leq 0$ must hold for some $c \in [a, b]$. For similar reasons, $f(d) \geq 0$ must hold for some $d \in [a, b]$. If either $f(c) = 0$ or $f(d) = 0$, we're done; otherwise, the intermediate value theorem says that $f(c) = 0$ for some e between c and d.

   (c) Let $f(x) = 1$ for $x \geq 0$ and $f(x) = 1$ for $x < 0$. Then $\int_{-1}^{1} f = 0$, but $f(c) \neq 0$ for all c.

7. The claim amounts to saying that $-\int_a^b |f| \leq \int_a^b f \leq \int_a^b |f|$. This follows from Corollary 5.6, and from the fact that $-|f(x)| \leq f(x) \leq |f(x)|$ for all $x \in [a, b]$.

9. (a) The function f is integrable on every interval $[a, b]$ because it is built up from integrable functions as allowed by Theorem 5.8, page 274.

   (b) A look at the graph shows $\int_{-1}^{1} f = -1$.

   (c) A look at the graph shows $\int_{-n}^{n} f = -n$.

11. All parts are basic elementary calculus calculations.

13. Suppose $f(x) \leq g(x)$ holds except at some finite list $x_1, x_2, \ldots x_n$ of points in $[a, b]$. If we redefine f at these points (we could set $f(x_i) = g(x_i)$, for instance, without changing the integral) then $f(x) \leq g(x)$ holds throughout $[a, b]$, and so Corollary 5.6 gives the desired result.

5.3   Integrability

1. All of the concocted functions are continuous.

3. Here is one possibility: Set $h = \epsilon/100$ and use $\mathcal{P} = \{0,\, h,\, 1-h,\, 1+h,\, \ldots,\, 10-h,\, 10\}$.

7. Given any box sum for $\int_0^2 f$ with value less than ϵ, add (if necessary) one more point at $x = 1$ to get another box sum with value less than ϵ. This same box sum works for $\int_0^1 f$; just ignore boxes to the right of $x = 1$.

11. Let m_i and M_i be the inf and sup of f on $[x_{i-1}, x_i]$. For each i, it is clear that $m_i \le f(s_i) \le M_i$, and hence that $m_i \Delta x_i \le f(s_i)\Delta x_i \le M_i \Delta x_i$. Summing over i gives the desired result.

13. The function f_{big} is constant on each partition subinterval, and is therefore integrable on $[a, b]$ by Theorem 5.8, page 274. By construction, moreover, the integral $\int_a^b f_{\text{big}}$ has the same value as the upper sum $\text{US}(f, \mathcal{P})$.

    It is also clear that $f_{\text{big}}(x) \ge f(x)$ for all $x \in [a, b]$, and so (by Corollary 5.6, page 270) we have

    $$\text{US}(f, \mathcal{P}) = \int_a^b f_{\text{big}} \ge \int_a^b f = I,$$

    The proof that $\text{LS}(f, \mathcal{P})$ underestimates I is almost identical.

5.4   Some Fundamental Theorems

1. The first claim follows immediately from the mean value theorem for integrals. Finding examples for the second part is easy.

3. The new average value is $5 \cdot 3 + 7 = 22$; the average value behaves "linearly."

5.   (a) $b = 9/4$.

     (b) $b = 2$.

     (c) $b = \sqrt{3} \approx 1.732$.

     (d) $b \approx 1.094$.

7. Let M be the maximum value of f. Then $h(x) = M - f(x) \ge 0$ for all $x \in [a, b]$. The hypotheses imply that $\int_a^b h = 0$; hence $h(x) = 0$ for all $x \in [a, b]$.

    The result need not hold if f is not continuous. Consider the function f given by $f(0) = 1$ and $f(x) = 0$ on $[0, 1]$.

9. The hypothesis means that $F(b) = \int_0^b f = b^2/2$ for all x. Thus $f(b) = F'(b) = b$ all b.

11. The interval $[0, 8/3]$ is one possibility.

13.   (a) Note that $h_n(0) = 0$ for all n, so clearly $h_n(0) \to h(0) = 0$. If $x > 0$, then $h_n(x) = 0$ for all x with $x < 1/n$, so again $h_n(x) \to h(x) = 0$. Show that the sequence $\{h_n\}$ converges pointwise to h on $[0, 1]$. By contrast, $\int_0^1 h_n = 1$ for all n, while $\int_0^1 h = 0$.

(b) The sequence $\{h_n\}$ converges uniformly to h on $[0.1, 1]$, because $h_n(x) = 0$ for all $x \in [0.1, 1]$ whenever $n > 10$. Similarly, $\int_{0.1}^{1} h_n = 0$ for all $n > 10$, so the integrals converge as well.

15. (a) The uniform limit function f is continuous on $[0, 1]$ and hence is also integrable there.

(b) For given $\epsilon > 0$, choose N such that $|f_n(x) - f(x)| < \epsilon$ for all $n > N$. Now

$$\left| \int_0^1 f_n - \int_0^1 f \right| = \left| \int_0^1 (f_n - f) \right| \leq \int_0^1 |f_n - f| < \int_0^1 \epsilon = \epsilon,$$

which completes the proof.

Index